Lubrication of Gearing

Lubrication of Gearing

Lubricants and their properties
Design of gears
Practical gear lubrication
Failure analysis

Wilfried J. Bartz

Translated from the German by Pam Chatterley, BA, MITI
English translation edited by Dr A. J. Moore

Mechanical Engineering Publications Limited
London

First published (in German) 1988
© Expert Verlag, 7044 Ehningen bei Böblingen

English language Edition 1993
© Mechanical Engineering Publications Limited

ISBN 0 85298 831 1

A CIP catalogue record for this book is available from the British Library

Typeset by Santype International Limited,
Netherhampton Road, Salisbury, Wiltshire
Printed by Page Bros. (Norwich) Ltd.

CONTENTS

v

Contents

Contents

AUTHOR'S PREFACE

The gearing designer must emphasize long-term reliability during the concept, design, and production stages. This ensures that the user of the gearing enjoys operation which is as free of failure and interruption as possible during the design life – assuming satisfactory care and maintenance and the use of suitable lubricants.

In this volume, the lubricant is considered not just as a functional element, but also as a design element. It is shown how the lubricant can be taken into account in gear design on the basis of its viscosity and surfactant characteristics, and how it can be limited in certain applications.

The gear lubricants are discussed and practical rules for the design, commissioning, maintenance, and operation of gear mechanisms are presented. Gearing and lubrication experts, as well as operations engineers, can thus become familiar with the tribological factors involved in gear lubrication. Particular attention is paid to the analysis and explanation of, and prevention of damage to, gear mechanisms.

The book contains the following main sections:

- principles of the lubrication of toothed gear sets;
- load-bearing capacity and its analysis;
- lubricants for gear mechanisms;
- selection of lubricants to DIN 51 509;
- gear lubrication in special conditions;
- solid lubricants for gears;
- lubricant supply;
- failure of gear wheels and mechanisms.

The concept, content, and level of the book make it suitable not only as a reference book for the practising engineer but also as a text book for students and teachers. This volume will also be of use to designers of gears and plant, as well as operators of gear mechanisms and users of gear lubricants. The book will also be a useful aid to employees of the technical services and the development laboratories of gear lubricant manufacturers.

W. J. Bartz
Ostfildern, Germany

TRANSLATION EDITOR'S PREFACE

Despite the enormous advances that have been made in the theoretical treatment of gear lubrication over the last thirty years or so, successful operation of gear sets remains highly dependent on empirical knowledge. This book has encapsulated, into as small a volume as might reasonably be expected, a body of such knowledge drawn from research studies and practical experience. Although the essential elements of the subject are all outlined here, there is also sufficient depth to satisfy the many different types of technologist who may find themselves either dipping into, or totally immersed in, this demanding area of tribology. Students, research workers, design and development engineers, and lubricant technologists should all find something of interest and value to them.

Lubricants have to service a wide variety of gear sets with very different lubrication requirements; the most severe applications will test even high performance products to the limits of their load-carrying and endurance capabilities. Whilst some readers will wish to understand the origin and nature of these limitations in some detail, others will be more concerned with the way in which lubricant selection procedures are designed to cope with them. Both levels of interest are catered for here. For the more detailed level of analysis, some familiarity is needed with the basic tribological concepts of elasto-hydrodynamic lubrication, contact stress analysis, and the effects of frictional heating; these are introduced in the first three chapters. Tools are then available with which empirical relationships between load-carrying performance and the physical and chemical properties of lubricants can be rationalized.

The heart of the book is the chapter on gear lubricants (Chapter 4). Classifications of gear oils are introduced here and the more important forms of lubricant discussed in detail. Areas covered include the refining of mineral oils and the way this influences base oil properties, the structure, manufacture, and properties of synthetic base stocks, and composition and manufacture of greases, and the function and chemical composition of key additive components. Standard laboratory tests used to define many of the physical and chemical properties of lubricants are then described, followed by lubricant standards and specifications. Selection considerations are further elaborated in Chapters 5–7.

The last two chapters deal with lubricant supply guidelines and the analysis of gear failure. Thirty-nine different examples of gear failure in the concluding chapter provides fair warning that the lubrication task in question should not be underestimated.

A. J. Moore
July, 1993

xi

1 Introduction

Mechanical or fluid transmissions are necessary to transfer force (or moments) between two shafts. Mechanical gears transmit forces (or moments) between two unaligned shafts which generally rotate at different speeds. In the case of mechanical gears, one must differentiate between uniform and non-uniform motion gears with stepped or continuous transmission. With uniform-motion gears, the force or moment transmission may be positive or non-positive (Fig. 1.1).

Toothed gear sets are positive uniform-motion gears with stepped transmission; further considerations will be restricted to this type. Toothed gear sets made of light alloy and plastic, and dry, unlubricated or self-lubricating gears will not be covered.

As in most areas of engineering, the trend in gearing is towards increasingly difficult operating conditions. This applies to industrial gear mechanisms as well as to vehicle gears. In order to produce ever greater outputs while, at the same time, reducing installed size and weight, it is necessary to explore new avenues. This includes, among other things, using different tooth shapes, more expensive materials, improving surface finishes, using new gear-cutting methods, introducing better hardening methods, and using high-performance lubricants.

The purpose of all these measures is to ensure long-term reliability of gears and to prevent gear damage. The main causes of gear damage are: design errors, unsuitable materials, incorrect material pairings, production and hardening errors, assembly errors, excessively severe operating conditions, foreign bodies, the use of incorrect lubricants, unsuitable lubrication methods, and lack of lubricant.

Clearly, the lubricant is only one of many factors, but it is very important in maintaining gear reliability. Appropriate and prudent selection of the correct lubricant for a gear mechanism is, therefore, of great significance.

In order to establish or specify the requirements which gear lubricants or gear lubrication must fulfil, it is essential to analyse all factors which affect the lubrication process and thus the lubricants. This becomes clear if one considers the numerous types and sizes of gear mechanisms, and the noticeably different contact, meshing, load, and speed conditions. The most important groups of factors include the design and configuration as well as the material of the gear teeth and the casing, the manu-

1

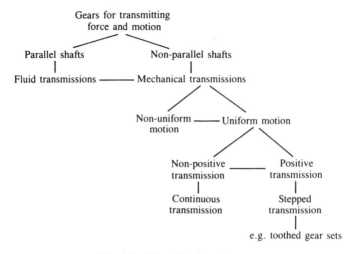

Fig. 1.1   Classification of gears

facture of the gears and the operating conditions. Furthermore, the gear lubricant not only has to supply the gears, but also has to carry out other tasks.

Table 1.1 summarizes the factors affecting gear lubrication. Further details of these factors appear in Table 1.2 (1)–(4).

Table 1.1   Factors affecting gear lubrication

Influential factors:
  Design and configuration of gears and casing
  Materials pairs
  Manufacture and surface quality
  Operating conditions
  Other friction partners and materials

Material groups:
  Heavy metals: steel, cast iron, non-ferrous heavy metals
  Light alloys
  Plastics

Type of lubrication:
  Oil or grease lubrication
  Dry or self-lubrication
  No lubrication

Table 1.2   Important factors affecting the reliability of gears

- Design and construction: power transmitted, rotational speed, ratio of sliding speed to circumferential speed, transmission ratio, pinion offset, material, homogeneity of material, etc.

- Manufacture: production precision of gear teeth, roughness of tooth flanks, hardness of tooth flanks (surface hardness, depth of case hardening), heat treatment, etc.

- Gear casing: rigidity under load, thermal distortion due to temperature increase, change in pinion pre-load due to deformations of casing, etc.

- Operating conditions: load, stem stress, load change, rotational speed, temperature, external heating, etc.

2 Principles of Gear Lubrication

2.1 TOOTH ENGAGEMENT AND CONTACT FILM FORMATION

The lubrication of two meshing gearwheels is fundamentally different from that of a plain bearing. When the direction of rotation is continuous and the load direction is unchanged, there is always a separating and load-bearing oil film in the gap of a correctly designed and constructed plain bearing. This can, therefore, be considered as *continuous* lubrication. Even in plain bearings subject to transient loading, the contact film is generally never broken. In contrast, with pairs of gearwheels, the contact film must be reformed with each tooth engagement. Figure 2.1 gives a schematic illustration of this. This process may be regarded as *discontinuous* lubrication. In very high speed gear sets, often only a short time exists for a contact film to form and separate the flanks, and the pressure build-up is not aided by the geometric conditions. Nevertheless, an oil film which is under pressure can form, and transmits at least part of the load.

Figure 2.2 shows the hydrodynamic, or rather elasto-hydrodynamic, pressure build-up (5)(6), which, for an experimental gear mechanism, increases with rotational speed and falls with load. Figure 2.3 demonstrates the elasto-hydrodynamic effects between the meshing tooth flanks (7). The load, falling with increasing speed, which can be supported without damage, i.e., without scuffing and increased wear, indicates the presence of considerable mixed friction components (left-hand sections of the curves in Fig. 2.3). A load-bearing lubricant film only starts to form at a certain minimum speed. As the speed increases further, the pressure in the lubricant film also increases and so does the load which can be supported (right-hand sections of the curves in Fig. 2.3). As the viscosity increases, the load–velocity curves move towards higher loads. Hydrodynamic or elasto-hydrodynamic effects can also be recognized.

The film thickness between the tooth flanks cannot be calculated using classical hydrodynamic lubrication theory. It is necessary to take account of the pressure viscosity of the oil and the elasticity of the tooth flanks in order to estimate the contact film thickness (elasto-hydrodynamic lubrication).

5

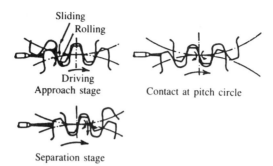

Fig. 2.1   Tooth engagement and contact film formation, schematic

In addition to sufficient viscosity, and narrowing of the gap in the direction of motion – the latter is not optimal where the teeth engage – the formation of a contact film requires a relative velocity. There are two fundamentally different types of motion, namely rolling and sliding. Plain bearings are designed with good conformity (Fig. 2.4(a)), and the surface pressures are relatively small. It is true that the coefficient of sliding friction is quite high, especially at low speeds – a disadvantage which can be overcome to a certain extent by selecting an optimal material pairing with good sliding characteristics. In contrast, in rolling bearings, conformity between the rollers and the race is poor (Fig. 2.4(b))

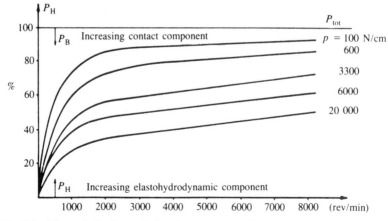

Fig. 2.2   Elasto-hydrodynamic bearing component between two tooth flanks (5)(6)

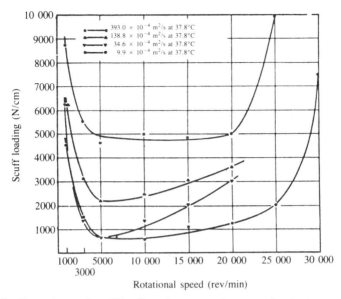

Fig. 2.3 **Dependence of scuff loading of a gear pair on speed and viscosity (from Borsoff (7))**

and hence the surface pressure is high. Yet the coefficient of rolling friction is generally lower than that of sliding friction. In toothed gears, the two types of motion are present and there is a combination of sliding and rolling. As a rule, surface pressures are high and poor conformity does not aid the formation of a contact film. For reasons of strength, however, in many cases it is not possible to use materials combinations

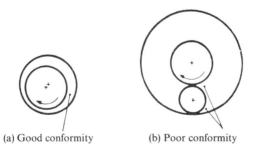

(a) Good conformity   (b) Poor conformity

Fig. 2.4 **Schematic representation of narrowing of gap with good and poor lubrication**

which aid the sliding process. In these circumstances, it may be concluded that the selection of a suitable lubricant is an important factor in gear lubrication.

2.2 GEAR TYPES

2.2.1 Fundamentals

The stress to which the gear oil is subjected depends on the motion and load conditions at the tooth sides and the temperatures of the gears. The combination of these individual stresses depends to a certain extent on the type of gear or mechanism.

Gears may be classified according to the toothing (profile shape), the shape of the tooth flank (pitch line), or the tooth shape, etc. Classification according to the position of the shaft axes is particularly suitable for the purposes of lubrication. Spur gears have parallel axes, while normal bevel gears have intersecting axes. Both types of gear can be considered as rolling gears. If the axes cross, as with the offset bevel gears, worm gears, and crossed-axis helical gears, they are called rolling crossed-axis gears. Figure 2.5 schematically shows these basic gear types which are associated with certain velocity characteristics.

2.2.2 Description

The individual gear and mechanism types are described briefly below. The gear illustrations are taken from the BP brochure *Lubrication of industrial gears* (8).

2.2.2.1 General concepts

The general concepts and terms for gears, gear pairs, and gear mechanisms are defined in DIN 868. The most important basic concepts are explained in this Standard.

A gear is a machine part which rotates about an axis and which consists of a gearwheel, contact surfaces, and teeth. Depending on the position of the teeth on the gearwheel, the gearing may be internal or external (Fig. 2.6). A pair of gears consists of two gear wheels, the smaller pinion, and the larger wheel (Fig. 2.7), which are separated by the centre distance (Fig. 2.8).

The ratio is the relationship between the rotational speed of the first driving gear and the speed of the last driven gear.

The gear set is a unit consisting of a pair of gears (single-stage) or of several pairs of gears (multi-stage) (Fig. 2.9).

Rolling gears

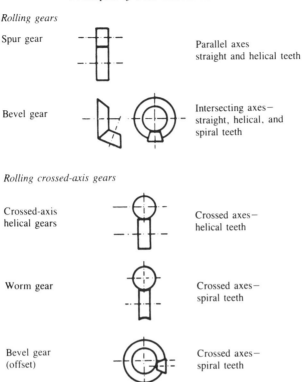

Spur gear — Parallel axes straight and helical teeth

Bevel gear — Intersecting axes— straight, helical, and spiral teeth

Rolling crossed-axis gears

Crossed-axis helical gears — Crossed axes— helical teeth

Worm gear — Crossed axes— spiral teeth

Bevel gear (offset) — Crossed axes— spiral teeth

Fig. 2.5   Basic gear types (schematic)

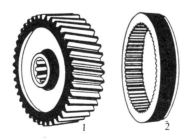

Fig. 2.6   External teeth (1) and internal teeth (2)

Lubrication of Gearing

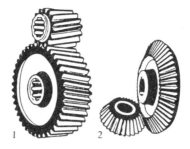

Fig. 2.7   Pinion and gear pairs: (1) spur gears; (2) bevel gears

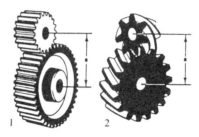

Fig. 2.8   Definition of centre distance for parallel (1) and crossed (2) axes

Fig. 2.9   Two-stage gear set

2.2.2.2 Rolling gears

In rolling gears, the motion at the contact surfaces is primarily rolling, with a relatively small sliding component. At the pitch point the sliding component is zero.

Spur gears

These are characterized by parallel axes, line contact and a low sliding component (Fig. 2.10). Spur gears are used in straight, helical, and herringbone forms. The most common type is an external, straight spur gear pair. There are no axial forces, and the bearings can be simple in design.

To produce quieter operation with increasing peripheral speed, helical gears are used. However, the resultant axial forces require more complex bearings, (Fig. 2.11).

With herringbone gears, the axial forces are eliminated, and the overall tooth width is twice as great as with helical gears. The considerably more complex manufacture is a disadvantage (Fig. 2.12).

Rack and pinion drive

This is characterized by parallel axes, line contact, and a low sliding content (Fig. 2.13). The rack is a special form of an infinitely large, external spur gear.

Bevel gears

These gears have intersecting axes, line contact, and a low sliding content (Fig. 2.14). A special form is the bevel face gear (Fig. 2.15).

Fig. 2.10   Straight-spur gear

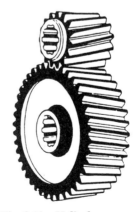

Fig. 2.11   Helical spur gear

Fig. 2.12   Herringbone spur gear

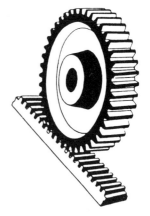

Fig. 2.13   Rack and pinion

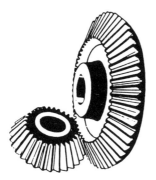

Fig. 2.14   Bevel gear

Fig. 2.15   Face bevel gear

2.2.2.3   Rolling crossed-axis gears

Due to the special shape of gear and tooth flanks of rolling crossed-axis gears, there is a mixture of rolling and sliding at the contact surfaces. The sliding component may be very large and hence lubrication of this type of gear requires particular attention.

Cylindrical worm gears

These gears have crossed axes, line contact and a very large sliding component (Fig. 2.16). They permit the transmission of high ratios in one stage with quiet, low-shock, low-vibration operation.

Helical crossed-axis gears

These have crossed axes, point contact and a large sliding component (Fig. 2.17). Helical crossed-axis gears are only suitable for small transmission ratios and low power transmission owing to the point contact.

Offset bevel gears (*crossed-axis bevel gears*)

Offset bevel gears have crossed axes, elliptical contact, and a large sliding component (Fig. 2.18), and are also known as hypoid gears. These gears allow the centre distance to be reduced. The sliding component is larger than the rolling component; it increases with increasing axis offset. Hypoid gears are used mainly in vehicle drive shafts.

Fig. 2.16 Cylindrical worm gear   Fig. 2.17 Helical crossed-axis gear

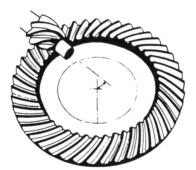

Fig. 2.18 Offset bevel gear (crossed-axis bevel gear, 'hypoid' gear)

Table 2.1   Performance comparison of various types of gears

Position of axes	Gear type	Maximum transmission ratio	Maximum peripheral speed	Maximum output (kN)	Maximum torque (kNm)
Parallel	Spur			2250	
	Straight	10		7500	
	Helical	10	5	22500	9000
		10	25	22500	2300
		10	200	22500	600
Intersecting	Bevel				
	Straight	7	2.5	375	90
	Skew	7	2.5	3750	90
		7	60	3750	45
Crossing at 90°	Worm	50	60	750	300
	Crossed helical	50		75	170
	Hypoid	50		750	
Crossing at 80–100° (not 90°)	Worm	50	50		115
	Crossed helical	50			170

2.2.3   Performance of gear sets

Table 2.1 compares transmission ratios, peripheral speeds, power and torque for various gear sets.

2.3   STRESS IN GEAR SETS

Stress on the gear lubricant is dependent on the motion and load conditions on the tooth flanks as well as on the temperatures of the gear wheels and the oil. The combined effect of these individual stresses depends to a certain extent on the type of gear.

2.3.1   Velocities

If a sphere rolls on a plate (Fig. 2.19(a)), the point of contact between the sphere and the plate is continuously changing. This motion is called rolling motion (see rolling bearings). If, however, the sphere rolls on the spot (Fig. 2.19(b)), the same area of the plate is always stressed; this is called sliding motion (see sliding bearings). It can easily be seen that the specific stress of the sphere/plate system is greater with sliding motion.

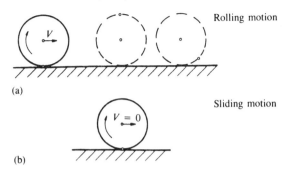

(a)

(b)

Fig. 2.19 Schematic representation of sliding and rolling motion

When two gearwheels mesh, there is a combination of sliding and rolling motion (Fig. 2.20).

The ratio of sliding velocity to rolling velocity is important for the lubricant stress. Figure 2.21 illustrates the different velocity ratios for rolling gears and rolling crossed-axis gears. It can be seen that with rolling gears (curve b) there is only a sliding velocity component in the vertical direction. At the rolling point it is equal to zero and increases in the directions of the base and the tip. In the case of crossed-axis rolling gears (curve a), on the other hand, there is also a sliding velocity component in the flank direction, so there is some sliding velocity at the rolling point. For this reason, the ratio of sliding velocity to rolling velocity is relatively large in the case of worm gears and offset bevel gears (hypoid gears). The direction of the sliding velocity is not constant,

Same marks: points on surface
which coincide at contact point

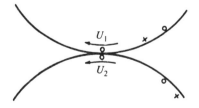

o $U_1 = U_2$, i.e., pure rolling
× $U_1 \neq U_2$, i.e., rolling and sliding

Fig. 2.20 Schematic representation of combined sliding and rolling motion

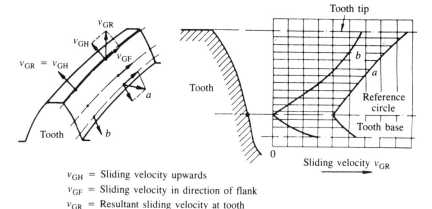

v_{GH} = Sliding velocity upwards
v_{GF} = Sliding velocity in direction of flank
v_{GR} = Resultant sliding velocity at tooth

Fig. 2.21   Sliding velocities on the tooth flanks of rolling and crossed-axis rolling gears

but changes in the vertical direction of the tooth flank. The 'wiping' motion is a considerable hindrance to the formation of an oil film under pressure.

Particular attention must, therefore, be paid to the selection of a lubricant for worm gears and offset bevel gears. Furthermore, if the gears are subjected to excessive load with respect to their size, as is the case with worm and hypoid gears (offset axis bevel gears for vehicle axles), the high stresses necessitate the development of special gear oils. With regard to average guide values, it can be assumed that the ratio of sliding velocity to rolling velocity in rolling gears is less than 0.7 as a rule, while with crossed-axis rolling gears it is generally more than one, and values of six or more can be achieved (**9**)–(**11**).

The extent of the influence of the axis offset on the maximum sliding velocity is shown by Fig. 2.22. While, in the case of a non-offset bevel gear, a given gear ratio gives a maximum sliding speed of about 5 m/s for a pinion speed of 5000 min^{-1}, this rises to about 17 m/s with an axis offset of 40 mm – a normal value in practice. Figure 2.23 shows that the values calculated for Fig. 2.22 can actually occur in practice (**10**)(**11**).

2.3.2   Loads

Even with the best possible production, it is impossible to maintain such excellent tooth flank accuracy that the teeth are uniformly loaded over the whole width, especially in large plants. Due to errors in axis parallel-

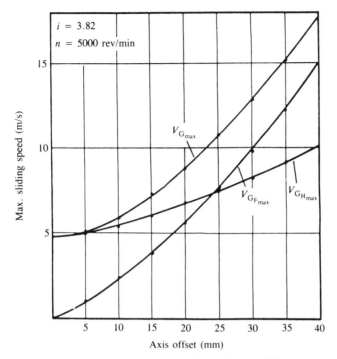

Fig. 2.22 Effect of axis offset on the maximum sliding speed

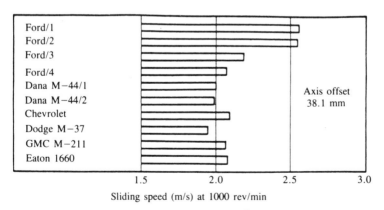

Fig. 2.23 Maximum sliding speeds at 1000 min⁻¹ in various vehicle rear axles

ism, helix errors, deformations under load, etc., the effective tooth width must be smaller than the actual width of the tooth. Furthermore, the total force acting on a tooth is generally higher than the value calculated from the power transmitted. Meshing errors, deformations, and wear of the flanks cause load changes and shocks, which increase with the peripheral speed of the gear. There are also other dynamic forces which occur in the gear set due to out-of-balance effects and torque fluctuations at the input and output side.

When designing a gear set and evaluating the lubricant, the operating performance must, therefore, be considered to be the product of rated output and a shock factor (6). For these purposes it is advisable to use the torque acting at the driving gear. In the case of vehicle gears, there is a special factor here which is responsible for extremely high loading of the gearwheels and the lubricant. High torque peaks occur at the transition from idling (overrun) to load (traction). These are transient processes, and the maximum torques can reach values which are 50–150 percent higher than the tractive torques (10)(11). It is very likely that a lot of flank damage is due to these torque peaks. The example in Fig. 2.24 illustrates this situation.

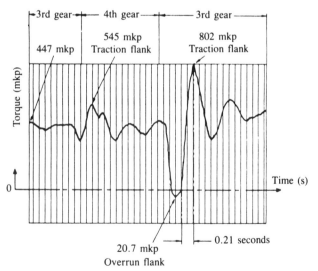

Fig. 2.24   Torque peaks during gear changes (Powell and Hoyl (11))

1 kp (kilopond) = 1 kgF = 9.81 N

2.3.3 Flank stress

Figure 2.25 gives a schematic representation of the load on the tooth flanks of two meshing gears caused by stresses and sliding (Dudley and Winter (6)). This load applies to the region of the contact point, which becomes flattened elastically under the influence of the forces being transmitted. The surface pressure is greatest in the centre of the flattened area (Fig. 2.26). Below this point – at a distance of 0.78 a (a = semi-width of the flattened area) – is where the greatest shear stress also occurs, enclosing an angle of 45 degrees with the surface. During the course of a rolling process, this increases from zero to 0.3 p_0 (p_0 = maximum surface pressure) and then falls to zero again; it is, therefore, a pulsating stress. The sliding friction components cause further surface stresses, compressive stresses forming just before the contact zone, and tensile stresses forming just after it. The resultant shear stresses run parallel to the surface and are called orthogonal shear stresses; during a rolling process these reach maximum values of $\pm 0.25 \, p_0$ at a depth of 0.5 a. In this area, therefore, the tooth flank is subjected to an alternating stress on each contact, which can contribute to surface cracks and plastic deformation. On the other hand, the internal shear stresses cause cracks *under* the surface.

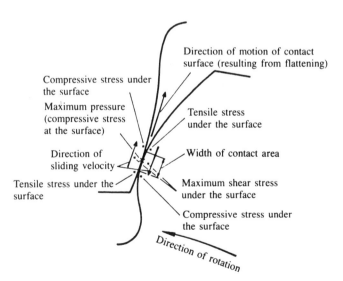

Fig. 2.25   Flank load of two meshing gears – schematic (Dudley and Winter (6))

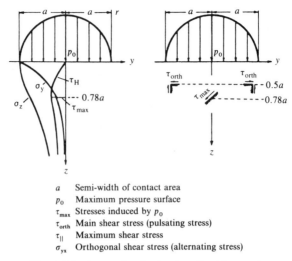

a   Semi-width of contact area
p_0 Maximum pressure surface
τ_{max} Stresses induced by p_0
τ_{orth} Main shear stress (pulsating stress)
$\tau_{\parallel}$ Maximum shear stress
σ_{yx} Orthogonal shear stress (alternating stress)

Fig. 2.26   Stress distribution in roller/plate pair

Even if no contact film can be formed due to unfavourable dynamic conditions, the lubricant still has to reduce the stresses produced by sliding motion by aiding the dynamic process.

2.3.4   Temperatures

All velocity and load factors influence the tooth flank temperatures. Lechner has shown (12) that the so-called continuous tooth flank temperature (which can be regarded as being equal to the average gear body temperature) is responsible for scuffing of the tooth flanks. Figure 2.27 shows these characteristic temperatures of a gearwheel, while Fig. 2.28 contains some measured temperatures.

In addition to the tooth centre temperature and the tooth centre temperature at idle, there is also the tooth flank temperature, which varies in meshing. The continuous tooth flank temperature is also of interest; this is the temperature to which the tooth flanks are always subjected, even when they are not meshing. It is approximately the same as the tooth centre temperature. The empirical equations for the connection between temperature and velocity and load given in Table 2.2 have been developed experimentally and in theoretical studies (12)(13). When comparing the equations for maximum tooth flank temperatures in Niemann and Lechner and Blok, it is necessary to remember that these authors did not

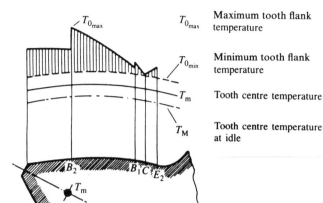

$T_{0_{max}}$	$T_{0_{max}}$	Maximum tooth flank temperature
	$T_{0_{min}}$	Minimum tooth flank temperature
	T_m	Tooth centre temperature
	T_M	Tooth centre temperature at idle

Fig. 2.27   Characteristic temperatures of a gear (Niemann and Lechner (12))

Table 2.2   Empirical equations for determining gear temperatures from Niemann and Lechner (12) and Theyse (13)

$$\vartheta_M = 75 + 4.5u^{1/3}$$
$$\vartheta_m = \vartheta_M + Au^{1/3}p^{0.9}$$
$$\vartheta_{0_{max}} = \vartheta_M + A^*v^{1/3}p^{1/2}$$

from Niemann and Lechner

$$\vartheta_{flash} = \vartheta_M + Cv^{1/2}p^{3/4} \qquad \text{from Blok}$$

u = peripheral velocity (m/s)
v = sliding velocity, relative (m/s)
ϑ_M = tooth centre temperature at idle (°C)
ϑ_m = tooth centre temperature (°C)
$\vartheta_{0_{max}}, \vartheta_{flash}$ = maximum tooth flank temperature (°C)
A, A^* = dimensional temperature factor
  $= f$(geometry, roughness, friction coefficient, lubricant)
$C = 0.63\mu \sqrt[4]{\{E'/(\delta r)\}}$
μ = friction coefficient
E' = reduced elastic modulus (kP/cm²)
r' = reduced radius (cm)
$\delta = \lambda \zeta c$
  = thermal conductivity × room specific heat
p = normal tooth force (kP/cm)

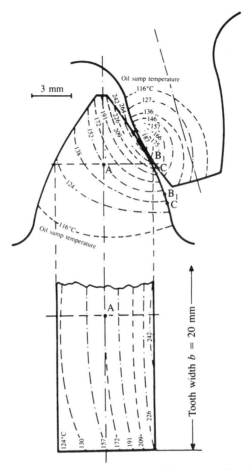

Fig. 2.28   Measured gear temperatures (Niemann and Lechner (12))

use the sliding velocity which is important for the temperature rise. Instead they used the peripheral velocity, and allowed for the influence of sliding by means of the temperature factors A and A^*. The disadvantage of Blok's equation is that the factor C contains the unknown friction coefficient.

Figure 2.29 shows the basic temperature variation calculated with these equations as a function of load – the influence of velocity and load can be seen. The actual temperatures were calculated for the conditions

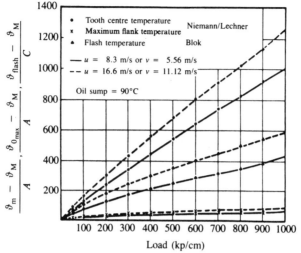

Fig. 2.29 **Characteristic temperature curves calculated with equations in Table 2.2**

of a test gear pair, with a friction coefficient of 0.075. Figure 2.30 gives the results. The high gear temperatures which can occur with higher loads and velocities can be seen clearly.

The continuous tooth flank temperature, i.e., the average surface temperature existing continuously, even when the teeth are not meshing, can be calculated as follows (12)

$$\vartheta_m = \vartheta_M \, A u^{1/3} w^{0.9}$$

where

ϑ_m = continuous tooth flank temperature
    = tooth centre temperature
ϑ_M = tooth centre temperature at idle
u = peripheral velocity
w = normal force per unit tooth width
A = temperature factor

The temperature factor A depends on the tooth geometry, flank roughness, the materials pair, and the lubricant (Fig. 2.31).

Although tests have shown that the continuous tooth flank temperature does not have to be absolutely identical with the tooth centre

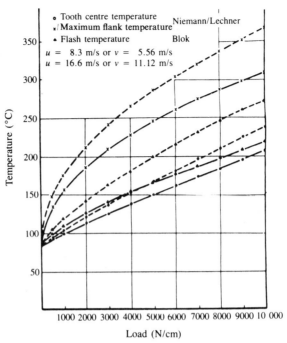

Fig. 2.30 Calculated gear temperatures for a test gear

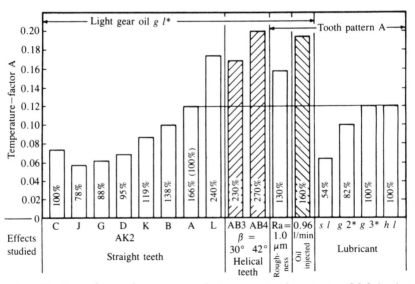

Fig. 2.31 Dependence of temperature factor A on tooth pattern and lubrication (from Niemann and Lechner (12))

24

temperature (**14**), the above equation has proved acceptable for this purpose because it contains no values which are unknown at the time of the analysis of a gear, such as friction coefficient or power loss. The assessments and tests discussed here were based on the same gear pairs as used by Lechner, so that adequate information would be available on the temperature factor in order to be able to estimate the desired temperatures.

The good correlation between the measured temperatures and those calculated in these tests can be seen in Fig. 2.32 (**12**). It should be emphasized, however, that this usable correlation was only achieved by using the same toothing and similar types of oil. For the FZG test tooth patterns A and C we thus get the continuous tooth flank temperatures

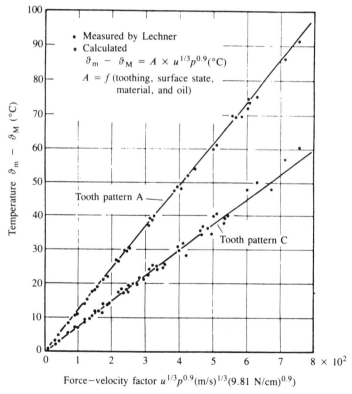

Fig. 2.32   **Comparison of measured (12) and calculated continuous tooth flank temperatures for FZG A and C toothing**

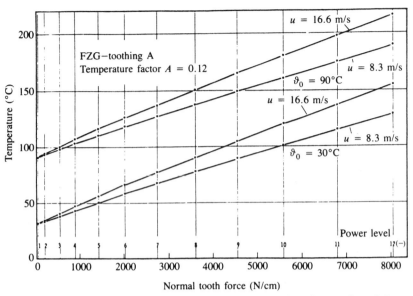

Fig. 2.33 **Continuous tooth flank temperature as a function of normal tooth force in FZG A toothing for two different peripheral velocities and oil sump temperatures**

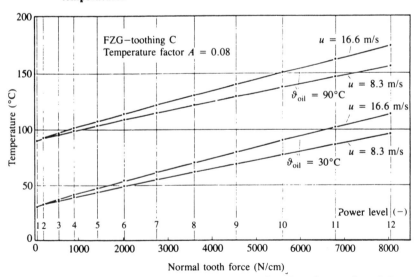

Fig. 2.34 **Continuous tooth flank temperature as a function of normal tooth force on FZG C toothing for two different peripheral velocities and oil sump temperatures**

26

shown in Figs 2.33 and 2.34 as a function of the normal tooth force. These were calculated for two peripheral velocities and two oil sump temperatures. The lower temperature level of tooth pattern *C* which was expected in view of the lower sliding content can be clearly seen.

As the tooth flank temperatures are of decisive importance for maintaining a reliable vibration state, one aims to keep gear temperatures low or to dissipate the heat produced by means of an adequate volume of oil. In the case of rolling gears, and also crossed-axis rolling gears used in industry which are not subjected to excessive stress relative to their size, the operating temperature of the gear can generally be controlled well. Once again, the crossed-axis rolling gears for vehicles occupy a special position. As the sliding velocity is greater in hypoid and worm gears than in spur gears and normal bevel gears at the same rotational speed, greater friction must be expected. The gear temperatures rise accordingly. As vehicle gears are made smaller to save weight and space, it is not possible to provide a large volume of oil to dissipate heat. Oil temperatures often reach 130°C, and in exceptional cases 150°C or more. However, at this temperature, the oxidative and thermal resistance of the gear oils and their additives is not always reliable.

2.4 FRICTION AND LUBRICATION CONDITIONS ON TOOTH FLANKS

As the gear temperatures and strains are largely characterized by the friction conditions on the tooth flanks, which in turn are dependent on the lubrication conditions, the relevant factors will be briefly described.

2.4.1 Friction and lubrication conditions

Friction of bare surfaces (Fig. 2.35(a))

This friction state is characterized by direct contact of surfaces which have no surface coatings. It can be regarded as dry friction. As the friction process is not favoured in any way, the values for friction and wear are high. This friction state can only be achieved genuinely in a vacuum. Thus the associated characteristics can play a part in the lubrication of units in space. This friction state can also be experienced briefly when machining metal, and in extreme wear and scuffing processes, when bare metal surfaces are exposed intentionally or unintentionally.

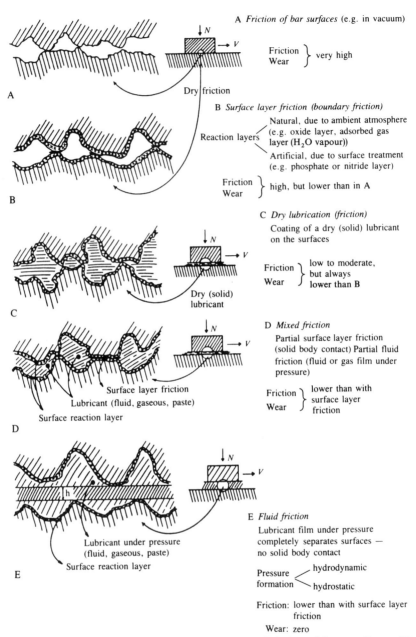

A *Friction of bar surfaces* (e.g. in vacuum)

Friction
Wear } very high

A Dry friction

B *Surface layer friction (boundary friction)*
Reaction layers
Natural, due to ambient atmosphere (e.g. oxide layer, adsorbed gas layer (H_2O vapour))
Artificial, due to surface treatment (e.g. phosphate or nitride layer)

Friction
Wear } high, but lower than in A

B

C *Dry lubrication (friction)*
Coating of a dry (solid) lubricant on the surfaces

Friction
Wear } low to moderate, but always lower than B

Dry (solid) lubricant

C

D *Mixed friction*
Partial surface layer friction (solid body contact) Partial fluid friction (fluid or gas film under pressure)

Surface layer friction
Lubricant (fluid, gaseous, paste)
Surface reaction layer

Friction
Wear } lower than with surface layer friction

D

E *Fluid friction*
Lubricant film under pressure completely separates surfaces — no solid body contact

Lubricant under pressure (fluid, gaseous, paste)
Surface reaction layer

Pressure formation < hydrodynamic / hydrostatic

Friction: lower than with surface layer friction
Wear: zero

E

Fig. 2.35 Friction and lubrication conditions: (a) Friction of bare surfaces; (b) Surface layer friction; (c) Dry lubrication/friction; (d) Mixed friction; (e) Fluid friction

Surface coating friction (Fig. 2.35(b))

In normal atmospheric conditions, all engineering surfaces – metallic surfaces are primarily of interest here – have coatings of adsorbed gas and/or fluid and/or chemical reaction coatings. They favour the friction process and reduce friction and wear; as their shear strength is lower than that of the base material, they are continually rubbed off and reformed. They thus protect the surface of the base material from excessive wear and subsequent destruction. This favourable behaviour is utilized by intentionally creating such protective surface coatings. One example is phosphate layers applied by phosphatizing, which can considerably facilitate certain forming processes. It is an open question whether such measures can be regarded as 'lubrication' in the widest sense.

Applying a layer of solid lubricant, e.g., graphite or molybdenum disulphide, can undoubtedly be defined as a lubrication process, i.e., dry lubrication (Fig. 2.35(c)). As, in this case, the surfaces are not coated with solid layers, but rather the contact pattern of the surfaces is altered by 'filling' the surface roughness, this process must be differentiated clearly from the surface layer friction state characterized by adsorption and reaction layers.

Mixed friction (Fig. 2.35(d))

The friction process is aided by the presence of small quantities of lubricant, which may be fluid or paste, at the point of friction. It is true that direct contact of the friction partners cannot be excluded, but these contact points alternate with areas in which a lubricant film separates the surfaces. In some areas there can even be a build-up of hydrodynamic pressure. By hydrodynamics we mean the creation of pressure in the lubricant film at the friction point. As the friction process is characterized by solid-body friction as well as by fluid friction, it is called mixed friction. In this area, friction and wear are influenced both by the ability of the lubricant to create protective films on friction surfaces with chemical and physical reactions, and by its viscous characteristics.

Fluid and gas friction (Fig. 2.35(e))

If the pressures produced in the lubricant layer between the friction pair are sufficient to balance the external force, the surfaces are completely separated and there is no direct contact at all. The friction process is displaced from the contact plane between two solid bodies to a fluid layer. The welcome result is wear-free operation in this area. This state is called hydrodynamic lubrication as long as the contact film is produced within the friction point. The only significant lubricant characteristic is the viscosity or another value which characterizes the flow process.

2.4.2 Friction on tooth flanks

2.4.2.1 Efficiency

The total efficiency of a gear set η_G is found from the input power P_1 and the output power $P_2 = P_1 - P_v$ with the power loss P_v (15). This gives

$$\eta_G = \frac{P_2}{P_1} = \frac{P_1 - P_v}{P_1} = 1 - \frac{P_v}{P_1} \leqslant 1$$

Table 2.3 contains some examples of the efficiency of gears (15). The high values especially for spur gears are clear.

2.4.2.2 Power loss

2.4.2.2.1 Basic factors

The total power loss is found from the individual losses for the toothing, the bearings, the gaskets, and other causes of friction (15)(16). Then

$$P_V = P_{VZ} + P_{VL} + P_{VD} + P_{VX}$$

where

$P_{VZ} = $ toothing power loss
$P_{VL} = $ bearing power loss
$P_{VD} = $ gasket power loss
$P_{VX} = $ other power loss components

Table 2.3 Guide values for efficiencies of gear sets (15)

Type of gear teeth	Efficiency (percent)
Crossed helical	84
Worm	30–97
Bevel:	
axes not offset	97
axes offset	94–96
Spur:	
turbine gears	97–98
vehicle gears	97
planetary gears	98–99

The toothing and bearing power losses, in particular, must be divided into idle losses and load losses, so

$$P_{VZ} = P_{VZO} + P_{VZP}$$

and

$$P_{VL} = P_{VLO} + P_{VLP}$$

where

ZO, LO = indices for idle losses
ZP, LP = indices for load losses

The total idle power loss can be defined as follows:

$$P_{VO} + P_{VOZ} + P_{VOPI} + P_{VOL}$$

where

P_{VOZ} = idle power loss of gears
P_{VOPI} = splash power loss of gears with splash lubrication
P_{VOL} = idle power loss of bearings

Toothing load losses P_{VZP} certainly account for most of the losses. The following calculation for tooth pattern load losses is determined over the contact path

$$P_{VZP} = F_{R(x)} v_g(x)$$

where

F_R = friction force
v_g = sliding velocity
x = coordinate in the direction of the contact path

Furthermore

$$F_{R(x)} = F_{N(x)} \mu(x)$$

where

F_N = normal force, gear loading
μ = coefficient of friction

This gives

$$P_{VZP} = F_{N(x)} \mu(x) v_g(x)$$

and after transformation

$$P_{\mathrm{VZP}} = F_{\mathrm{N_{max}}} \frac{F_{\mathrm{N}(x)}}{F_{\mathrm{N_{max}}}} \, \mu_{\mathrm{m}} \, v \, \frac{v_{\mathrm{g}}(x)}{v}$$

where

μ_{m} = mean coefficient of friction.

Furthermore

$$F_{\mathrm{N_{max}}} = \frac{F_{\mathrm{max}}}{\cos \alpha_{\mathrm{ZW}} \, \zeta_{\mathrm{ex}}}$$

This gives

$$P_{\mathrm{VZP}} = F_{\mathrm{Z_{max}}} \frac{1}{\cos \alpha_{\mathrm{ZW}}} \frac{1}{\zeta_{\mathrm{ex}}} \, \mu_{\mathrm{m}} \, v \int_{\mathrm{A}}^{\mathrm{E}} \frac{F_{\mathrm{N}(x)}}{F_{\mathrm{N_{max}}}} \frac{v_{\mathrm{g}}(x)}{v} \, \mathrm{d}x$$

with

$$P_{\mathrm{a}} = F_{\mathrm{Z_{max}}} \, v$$

$$H_{\mathrm{V}} = \int_{\mathrm{A}}^{\mathrm{E}} \cdots \, \mathrm{d}x = \text{tooth loss factor}$$

we get

$$P_{\mathrm{VZP}} = P_{\mathrm{a}} \, \mu_{\mathrm{m}} \, H_{\mathrm{V}}$$

Figure 2.36 shows the relationships (16).

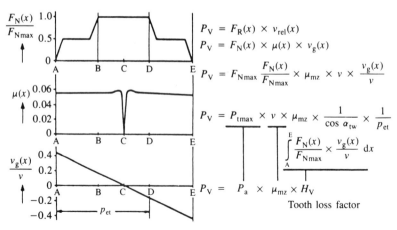

Fig. 2.36   Determination of load losses (from Michaelis (16))

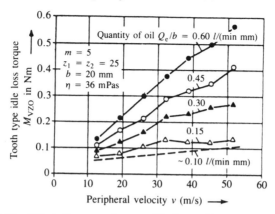

Fig. 2.37   Effect of spray volume on idle losses (from Ohlendorf (17))

2.4.2.2.2   Factors influencing idle losses

Idle losses are not greatly influenced by the tooth pattern, but increase considerably with peripheral velocity. If the velocity is tripled, the losses are roughly doubled (**16**).

In the case of spray lubrication, the quantity of oil has a considerable effect on idle losses. Figure 2.37 shows some test results (**17**). On the other hand, in the case of splash lubrication, a greater immersion depth can be expected to increase idle losses considerably, especially at higher peripheral speeds, as Fig. 2.38 shows (**16**).

On the lubricant side, the viscosity is important, while the chemical structure of the oils is not, Fig. 2.39 (**16**).

2.4.2.2.3   Factors influencing load losses

The tooth form, characterized by meshing angle, contact ratio, module, and transmission, has a clear effect on load losses (Fig. 2.40), which decrease with increasing peripheral velocity, as Fig. 2.41 shows (**17**).

Different gear oils also produce different load losses, as can be seen in Fig. 2.42 (**16**). It is noticeable that the losses are considerably lower with a partially synthetic gear oil than with mineral oils.

2.4.3   Friction coefficient

Friction coefficients, which naturally vary over the contact path, are significant for friction losses. An empirical equation for estimating the mean

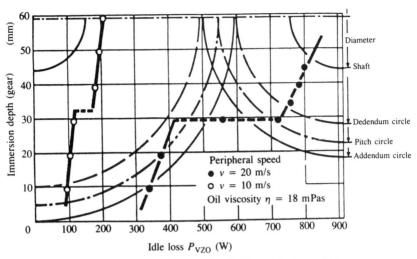

Fig. 2.38   Effect of immersion depth on idle losses (16)

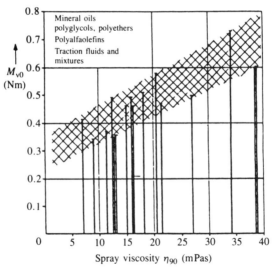

Fig. 2.39   Total idle loss torque as a function of spray viscosity for $v = 10$ **m/s**
(16)

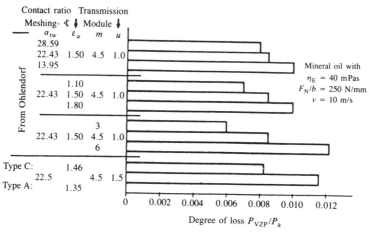

Fig. 2.40 Effect of tooth form on degree of loss (16)

friction coefficient has been derived for mineral oils on the basis of gear tests (**16**).

It is as follows

$$\mu_{\text{m}} = 0.045 \frac{(K_{\text{A}} F_{\text{bt}}/b)^{0.2 - 0.05}}{v_{\Sigma\text{c}} \zeta_{\text{c}}} \eta_{\text{oil}} \times X_{\text{R}}$$

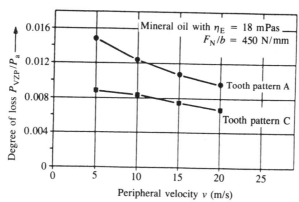

Fig. 2.41 Effect of peripheral velocity and tooth form on degree of loss (16)

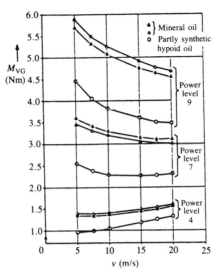

Fig. 2.42   Measured loss moment M_{VG} of a gear with various hypoid gear oils (16)

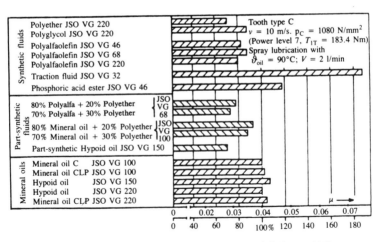

Fig. 2.43   Friction coefficients of various lubricants (16)

where

F_{bt}/b = peripheral force related to tooth width
K_A = load factor to DIN 3990, Part 1
v_Σ = total velocity
ς = radius of curvature
η_{oil} = viscosity at oil temperature
X_R = roughness factor

Comparison with measured friction coefficients shows good correlation for mineral oils. However, lower friction coefficients were found for synthetic fluids. This can also be seen in Fig. 2.43, which contains friction coefficients measured for various lubricating oils. The lower friction coefficients of the polyglycols, polyethers and polyalphaolefins and of their mixtures with one another and with mineral oils, compared with those of mineral oils, are obvious. The extremely high friction coefficients for special fluids used in friction gear sets can also be seen. Additives which reduce wear and protect against scuffing do *not* reduce friction. This can be seen from the comparison of mineral oils C, CLP, and hypoid oils.

3 Load-Bearing Capacity and its Calculation

3.1 GENERAL

Once the toothing data have been provisionally established during the process of designing a gear set, it is necessary to calculate the load-bearing capacity of the gear. Here it is desirable to take account of the effect of the lubricant, if it has any influence on the load-bearing capacity. This calculation produces a nominal load-bearing capacity which must be greater than the actual load on the gear and the lubricant. This includes the calculated resistance to tooth fracture (tooth root strength), to pitting (tooth flank strength), to scuffing (scuffing resistance), and to excessive heat (6)(15).

Figure 3.1 shows, in a simplified form, the effect of the material, the design, and the operating conditions as well as the lubricant on the load-bearing capacity of a gear. It can be seen that the fracture resistance cannot be influenced by the lubricant. On the other hand, pitting can depend – at least indirectly – on the gear oil, insofar as it is possible to aid the sliding process with the lubricant. Scuffing of the tooth side can be prevented to a certain extent by selecting a suitable gear oil.

The load-bearing capacity of a tooth system can, therefore, be characterized by resistance to tooth fracture (tooth root strength), to pitting (tooth flank strength), and to scuffing (scuffing resistance); scuffing resistance includes resistance to excessive wear. The following components of the load-bearing capacity of a tooth system can be influenced by the lubricant:

	Contact film under pressure	Physico-chemical reaction film
Scuffing/wear	Direct	Direct
Pitting	Direct	Limited

Viscosity is the lubricant factor which is decisive in the formation of a contact film under pressure. The formation of physico-chemical reaction films depends on the surface-active behaviour of the lubricant, i.e., on its additive package. The effects associated with these characteristics are dealt with in the following sections.

Influenced by — Resistance to	Material	Design operating conditions	Lubricant
Tooth fracture (tooth root strength)	x	x	–
Pitting (tooth side strength)	x	x	(x)
Scuffing (scuff resistance)	(x)	x	x
Excessive heating	–	x	x
x may be influenced, (x) may be influenced to a limited degree, — not influenced			

Fig. 3.1 Effect of material, design, and operating conditions on load-bearing capacity of gears

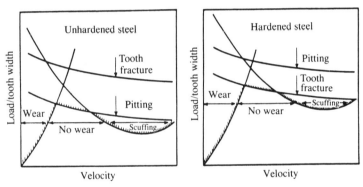

Fig. 3.2 Types of damage and operating conditions

The type of damage to be expected in certain load/velocity conditions can be determined via the schematic representation shown in Fig. 3.2.

3.2 PITTING RESISTANCE

3.2.1 Principles

The causes of pitting are normal and shear stresses due to Hertzian stress with maximum values under the surface, as well as compression and tensile stresses due to friction forces on the surface.

Pitting is a fatigue failure, whereby after a minimum number of rolling actions on the tooth flank, particles of material separate from the surface.

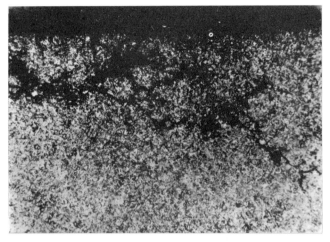

Fig. 3.3   Fatigue damage to the material structure – precursor of pitting

This separation is preceded by considerable damage to the material structure under the surface caused by crack formation (Fig. 3.3).

In Fig. 3.4 one can clearly see the outline of the material which is about to separate. The bond to the underlying material is not yet broken completely, but separation is inevitable due to cracks which are more or less parallel to the surface (Fig. 3.5).

One can differentiate between initial pitting and progressive pitting on the basis of the significance of the appearance of the damage and the necessity of taking countermeasures. When assessing pitting with regard to possible damage, it is not easy to predict whether its condition will stay as it is or not. The pitting is slight if the pits are so small and their number is so few that the gear's operation is not impaired. Severe pitting very quickly leads to complete destruction of the tooth flank. As a result of the excessive specific load on the teeth, tooth fractures can very quickly terminate the serviceability of the gearwheel (**18**)(**19**).

3.2.2   Factors influencing pitting

Taking account of the mechanisms of formation discussed above, the following factors influence pitting on tooth flanks: material, surface treatment, surface quality, tooth type and shape, operating conditions, and lubricant.

The factors influencing tooth flank load-bearing capacity and thus pitting are summarized in Fig. 3.6 (**5**). It is true that pitting is influenced

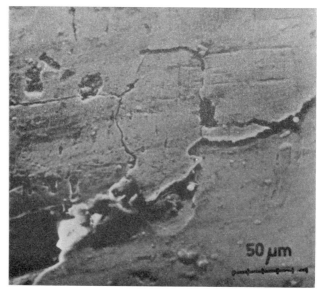

Fig. 3.4 Outline of a surface particle which will shortly separate due to material fatigue

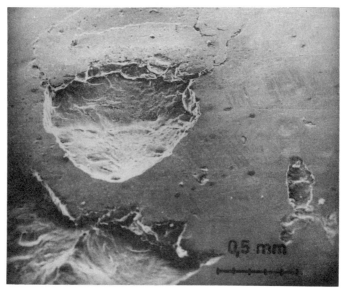

Fig. 3.5 Conchoidal pit due to material fatigue

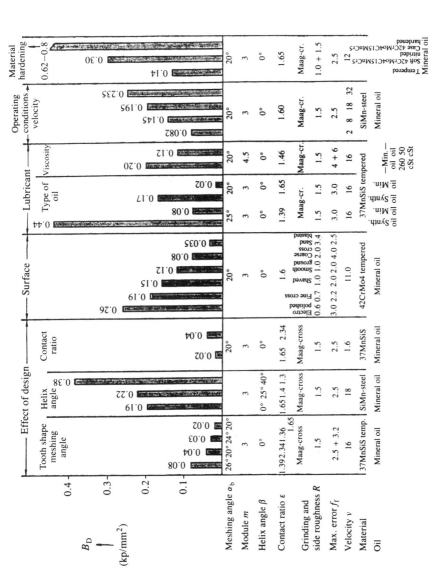

Fig. 3.6   Factors influencing load-bearing capacity of flanks (pitting) of gear teeth
(5)

by lubricant viscosity and the type of lubricant (mineral oils, synthetic oils), but the results of tests on the effect of lubricant additives on pitting are very contradictory.

The load-bearing capacity of the tooth flank indicates the resistance to premature failure due to pitting resulting from material fatigue. The cause of pitting is primarily that the maximum Hertzian stress for the material is exceeded.

Table 3.1 summarizes the possible influence of the lubricant on pitting, classified on the basis of whether crack formation is influenced from the structure or from the interior of the material.

Surface fatigue due to pulsating and alternating stress can occur through two mechanisms. High shear stresses resulting from the rolling contact cause cracks *under* the surface, while tangential stresses resulting from the sliding effect cause cracks which start *on* the surface.

It appears that pitting caused by internal cracking can only be influenced by a reduction in the Hertzian stress, i.e., by a reduction in the pulsating and alternating stresses under the surface. With a given load this can be achieved by increasing the elasto-hydrodynamic contact component, i.e., with a higher effective viscosity – this can be achieved by means of a higher nominal viscosity, a greater pressure viscosity coefficient, and a lower viscosity temperature dependence. The positive effect of higher viscosity on reducing the danger of pitting is not in question.

Table 3.1　Possible effects of the lubricant on pitting

Reduction of crack formation from inside

Reduction of the Hertzian stress resulting from an increase in the hydrodynamic or elasto-hydrodynamic contact area by means of
- a higher nominal velocity
- increased pressure coefficient of the viscosity and
- reduced temperature dependence of the viscosity

Reduction of crack formation starting on the surface

Reduction of the dry friction coefficient by means of
- a suitable operating viscosity
- a suitable lubricant type
- suitable lubricant additives

Reduction of the explosive effect at cracks by means of
- a higher operating viscosity
- suitable lubricant additives

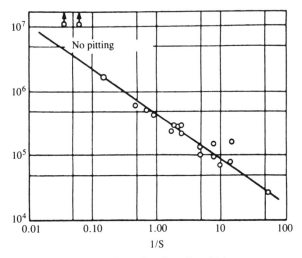

Fig. 3.7 Pitting related to film thickness

The effect of the oil film thickness in relation to surface roughness and thus the significance of oil viscosity for pitting has been demonstrated (**20**). The results in Fig. 3.7 indicate the number of revolutions of a rolling gear pair necessary for the occurrence of pitting as a function of the ratio

$$S = \frac{\text{sum of surface roughness}}{\text{oil film thickness}}$$

In these tests, the oil film thickness was changed by the oil viscosity. Table 3.2 indicates the roughness values occurring in practice.

Figure 3.8 shows the effect of base oil viscosity on pitting using S/N curves (**21**).

One way of delaying the development of cracks which start on the surface is to reduce the sliding friction coefficient at the contact point; this would have the effect of reducing the tangential stresses on the surface. This effect could also be achieved with a higher effective viscosity. Lubricants with particularly good sliding friction properties would also have a positive effect. Tests with case-hardened and with quenched and tempered gears have shown that polyglycols give better results with regard to pitting resistance than ester fluids and mineral oils.

The behaviour of extreme pressure and anti-wear additives is confusing; in view of their good sliding friction characteristics, one would

Table 3.2 Relationship between roughness, machining method, and gear size (from Hoesch Work Sheet 50, 'Toothed Gear Lubrication')

Surfaces	Greatest average peak-to-valley height, R_a* (μm)		Greatest surface roughness, R_t† (μm)	
m	1–8	>8	1–8	>8
Milled tooth-flanks of spur gears**	1.6	3.2	5–12.5	10–25
Stamped tooth-flanks of spur gears	3.2	3.2	10–25	10–25
Shaved tooth-flanks of spur gears	0.8	1.6	3.2–8	5–12.5
Ground tooth-flanks of spur gears	0.4	0.8	1.6–4	3.2–8
Lapped tooth-flanks of bevel gears	0.8	1.6	4–8	5–12.5
Ground tooth-flanks of worm gears	0.4	0.8	1.6–4	3.2–8
Milled tooth-flanks of worm gears of worm gears	1.6	3.2	5–12.5	10–25

* Roughness values given can only be achieved with material tensile strength up to 850 N/mm²
† Converted in accordance with DIN 4767

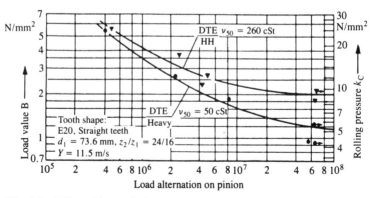

Fig. 3.8 Effect of base oil viscosity on pitting, based on S–N curves (21)

expect them to reduce the danger of pitting. Tests have shown that both neutral and negative behaviour can be expected. Only in certain cases does the presence of such additives in a base oil result in a definite delay in the formation of pits. Positive effects have been achieved with, for example, some additives containing sulphur and phosphorus, and also with high concentrations of MoS_2 in tests with both case-hardened and quenched and tempered gears. It must be noted, however, that the effects of viscosity and the state of the lubricant mask those of the additives.

Figures 3.9–3.15 give test results regarding the effects of additives and lubricant type on pitting (**22**). These results do not allow any definite conclusions, so it is not possible to include lubricant characteristics in the calculation of tooth flank load-bearing capacity.

3.2.3 Calculation of resistance to pitting

3.2.3.1 General
Calculation of tooth flank load-bearing capacity to DIN 3990, Part 2, (**23**), is based on the Hertzian stress which must remain under a certain maximum value to ensure freedom from damage. Other influences are taken into account by means of specific factors for the allowable stress and the selection of material characteristics. The effect of the lubricant, especially its viscosity, is included in the calculation of the allowable flank stress by means of the lubricant factor Z_L.

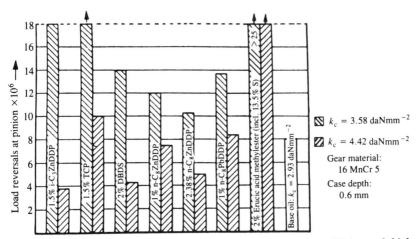

Fig. 3.9   Tests with case-hardened experimental gears and additives of high chemical purity

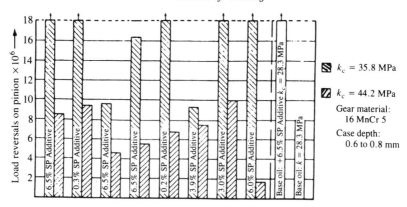

Fig. 3.10   Tests on case-hardened experimental gears and commercial gear additive packages

3.2.3.2   Calculation of flank stress (Hertzian stress)

Calculation of the tooth flank load-bearing capacity is based on the Hertzian stress at the pitch circle.

The following equation is used to determine the flank stress at the pitch circle

$$\sigma_H = \sigma_{HO} \times \sqrt{(K_A K_V K_{H\beta} K_{H\alpha})}$$

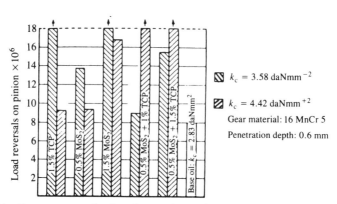

Fig. 3.11   Tests on case-hardened experimental gears to determine the effectiveness of molybdenum disulphide

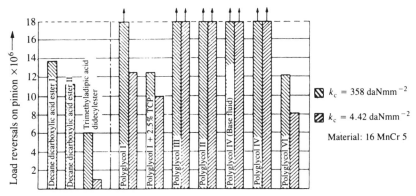

Fig. 3.12   Tests on synthetic lubricants and case-hardened experimental gears

where

σ_{HO} = nominal value of flank stress
K_A = application factor
K_V = dynamic factor
$K_{H\beta}$ = width factor for flank stress
$K_{H\alpha}$ = face factor for flank stress

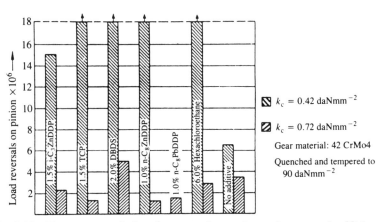

Fig. 3.13   Tests on quenched and tempered experimental gears and additives of high chemical purity

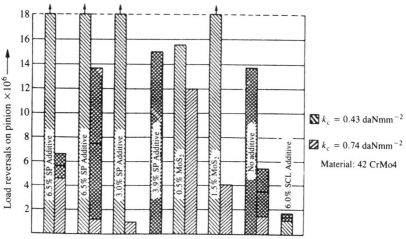

Fig. 3.14   Tests on quenched and tempered experimental gears and commercial gear oil additive packages

The factors K_A, K_V, $K_{H\beta}$, and $K_{H\alpha}$ are calculated in accordance with DIN 3990, Part 1. The following equation is used to calculate σ_{HO}

$$\sigma_{HO} = Z_H Z_E Z_\varepsilon Z_\beta \sqrt{\left\{ \left(\frac{F_Z}{d_1 b} \right) \left(\frac{u+1}{u} \right) \right\}}$$

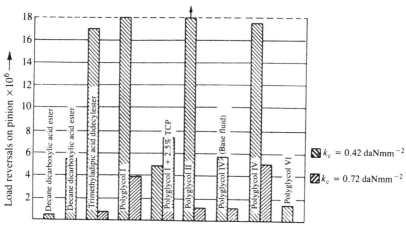

Fig. 3.15   Tests on quenched and tempered experimental gears and synthetic lubricants

where

Z_H = zone factor
Z_E = elasticity factor
Z_ε = contact factor
Z_β = helix factor
F_Z = nominal peripheral force
b = tooth width
d_1 = pitch diameter of pinion
u = gear ratio Z_2/Z_1

The factors Z_H, Z_E, Z_ε, and Z_β are determined in accordance with DIN 3990, Part 2, and F_t is determined in accordance with Part 1 of the same standard.

The allowable flank stress should be calculated for both pinion and gear. The following equation is used

$$\sigma_{HP} = \frac{\sigma_{H\,lim} Z_N}{S_{H\,min}} Z_L Z_R Z_V Z_W Z_X$$

where

$S_{H\,min}$ = specified minimum safety factor for flank stress
$\sigma_{H\,lim}$ = durability factor for flank stress
Z_N = life factor for flank stress
Z_X = size factor for flank stress
Z_L = lubricant factor
Z_R = roughness factor
Z_V = velocity factor
Z_W = material pair factor

DIN 3990, Part 2 indicates how to determine $\sigma_{H\,min}$, Z_N, Z_W, and Z_X.

The factors Z_L, Z_R, and Z_V jointly cover the effect of the oil film on tooth side load-bearing capacity. They are calculated in Section 3.2.3.4.

3.2.3.3 Resistance to pitting (tooth flank load-bearing capacity)
The equations derived above give the calculated resistance factor for flank stress (to pitting)

$$S_H = \frac{\sigma_{HP}}{\sigma_H}$$

Hence

$$H = \frac{\sigma_{H\,\lim} Z_N}{\sigma_{HO}} \frac{Z_L Z_R Z_V Z_W Z_X}{\sqrt{(K_A K_V K_{H\beta} K_{H\alpha})}}$$

For gears in which pitting could cause serious or catastrophic damage, or for gears which must achieve very long lives (10^{10} to 10^{11} load reversals), correspondingly high safety factors must be specified so that there is only a very low probability of damage.

Conversely, relatively small safety factors, down to about one, can be permitted for low-speed gears.

3.2.3.4 *The effect of the lubricant on pitting resistance*
The lubricant film between the tooth flanks influences their load-bearing capacity. The formation of the lubricant film and evaluation of a minimum lubricant film thickness depend primarily on the following values (see Section 3.3.4.2):
- viscosity of the lubricant in the gap:
- the sum of the instantaneous velocities of the two tooth flanks;
- load;
- equivalent radius of curvature;
- roughness of flank surface.

In addition to the load (F_t or M_t) and the design values for radii of curvature, the following influencing factors must also be determined:

Z_L = lubricant factor to take account of lubricant viscosity;
Z_V = velocity factor to take account of the peripheral velocity;
Z_R = roughness factor to take account of the peak-to-valley height.

(a) *Lubricant factor Z_L*
The relationship between the lubricant factor Z_L and the oil viscosity is illustrated in Fig. 3.16 (**23**).

The lubricant factor can be calculated as follows

$$Z_L = C_{ZL} + \frac{4(1.0 - C_{ZL})}{\left(1.2 + \dfrac{134}{v_{40}}\right)^2}$$

For the range of $\sigma_{H\,\lim} = 850 \text{ N/mm}^2 - 1200 \text{ N/mm}^2$

$$C_{ZL} = \frac{\sigma_{H\,\lim} - 850}{350} 0.08 + 0.83$$

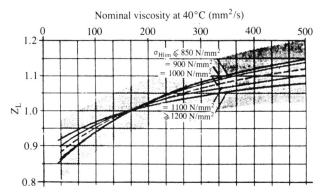

Fig. 3.16 Lubricant factor Z_L against oil viscosity as per DIN 3990, Part 2

Furthermore

v_{40} = kinematic viscosity at 40°C in mm^2/s

These values apply to mineral oils with and without EP additives. For certain synthetic oils which create lower friction, Z_L can be multiplied by a factor 1.1 for case-hardened gears and by a factor 1.4 for quenched and tempered gears.

(*b*) *Velocity factor Z_V*

The relationship between the velocity factor Z_V and the peripheral velocity is shown in Fig. 3.17 (**23**).

The velocity factor can be calculated as follows

$$Z_V = C_{ZV} + \frac{2(1.0 - C_{ZV})}{\sqrt{\left(0.8 + \dfrac{32}{v}\right)}}$$

For the range of $\sigma_{H\,lim}$ 850 N/mm^2 to 1200 N/mm^2

$$C_{ZV} = \frac{\sigma_{H\,lim} - 850}{350} 0.08 + 0.85$$

Furthermore:

v = nominal kinematic viscosity mm^2/s

(*c*) *Roughness factor Z_R*

The relationship between the roughness factor Z_R and the relative peak-to-valley height is shown in Fig. 3.18 (**23**).

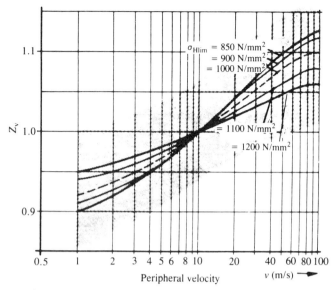

Fig. 3.17 **Velocity factor Z_V against peripheral velocity as per DIN 3990, Part 2**

The averaged peak-to-valley height R_Z is given by

$$R_Z = \frac{R_{Z1} + R_{Z2}}{2}$$

where R_{Z1} is the peak-to-valley height of the pinion and R_{Z2} that of the gear.

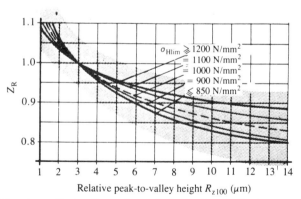

Fig. 3.18 **Roughness factor Z_R against relative peak-to-valley height as per DIN 3990, Part 2**

The averaged relative peak-to-valley height R_{Z100}, relative to an axial spacing of 100 mm, is calculated as

$$R_{Z100} = R_Z \times \sqrt[3]{\left(\frac{100}{a}\right)}$$

with axial spacing a.
The roughness factor can be calculated as follows

$$Z_R = \left(\frac{3}{R_{Z100}}\right)^{C_{ZR}}$$

For the range of $\sigma_{H\,lim}$ 850 N/mm^2 to 1200 N/mm^2

$$C_{ZR} = \frac{1000 - \sigma_{H\,lim}}{5000} + 0.12$$

3.3 RESISTANCE TO SCORING

3.3.1 Principles

Resistance to scoring is provided on the one hand by surface reaction layers and on the other by fluid films under pressure. Of course, the reaction layers on the tooth flank can only withstand certain load and velocity stresses. If certain limits are exceeded, increased wear and scoring will result, despite the presence of such protective layers. Lechner established (12)(24) that lubrication fails if the continuous tooth flank temperature reaches a certain level. The load limit at which scoring of the tooth flanks occurs with a given oil thus depends on the velocity. Figure 3.19(a) shows a schematic representation of a scoring load/ velocity curve (9). When the oil has a greater additive content and when velocities are lower, this curve is displaced towards higher loads. In the discussion of this situation, the descending branch of the curve and the rising branch of the curve will be considered separately. In the operating range relating to the descending branch of the curve, the lubrication state of the tooth flanks is one of mixed friction and dry layer friction. When the peripheral velocity is constant, the continuous tooth flank temperature rises with increasing load (Fig. 3.19(b)), while, at higher temperatures, the allowable load limits of the reaction film which has formed on the tooth flanks are smaller. The point of intersection of the two curves determines the scoring load limit for these conditions. When the peripheral velocity is greater, the continuous tooth flanks temperature

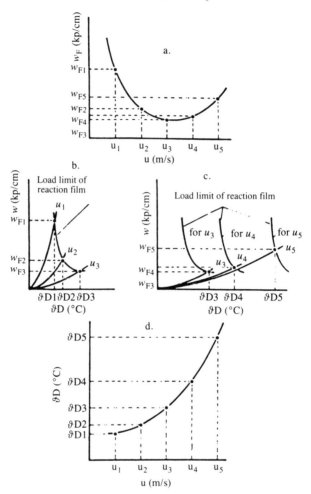

Fig. 3.19 Relationship between load, velocity and temperature with regard to gear lubrication (from Lechner (24))

increases at a greater rate with load, so the point of intersection of the two curves is lower. The scoring load limit is, therefore, now lower as well.

The lubrication state on the rising branch of the curve after a minimum velocity (u_3) has been exceeded is characterized by fluid friction and mixed friction. With increasing peripheral velocity, the force acting directly on the reaction film falls, as an ever greater proportion of the load is borne by an elasto-hydrodynamic contact film. Even if the allowable loading of the reaction film remains constant, the total load to be transmitted rises. As Fig. 3.19(c) schematically shows, the allowable load on the chemical protective film apparently rises as the load limit curve shifts towards higher temperatures with increasing peripheral velocity. So only at higher temperatures is the load limit curve intersected by the curves which indicate the increase in continuous tooth flank temperature with load. Despite a sharper temperature rise with load, the scoring load limits are thus higher again. Figure 3.19(d) shows the resultant rise in continuous tooth flank temperature with increasing peripheral velocities.

The most important conclusion to be drawn from these factors is that the characteristic shape and position of the load limit/velocity curve depend on the EP characteristics of the gear oil.

3.3.2 Effects on scoring resistance

It is reasonable to expect that the scoring load capacity should depend on other factors in addition to the gear oil. Apart from the velocity and the tooth shape, such factors may include tooth geometry, the gear material, its hardness, and the surface condition of the tooth flanks.

Figure 3.20 shows how these factors affect the scoring load capacity of a gear pair (25). While the transition from a 20 degree to a 30 degree meshing angle leads to a rise in load capacity, increases in the module from 3 to 4.5 and an increasing helix angle result in a reduction in scoring load capacity. In addition, chromium and molybdenum alloy steels behave better than unalloyed and austenitic, anti-magnetic steels. With regard to the hardening process, it has been found that hard nitriding and soft nitriding of heat-treatable steels can give better results than case-hardening, so far as scoring resistance is concerned. The grinding process can also influence the scoring load capacity of the tooth flanks. As the scoring load capacity can be raised by phosphating, it is not surprising that such treatment produces a reaction or protective film on the flank surfaces, which is otherwise only formed by the additives in the gear oil during service.

3.3.3 Definition of scoring resistance

On the basis of the comments on the relationship between load capacity and resistance, the following resistance to scoring (including wear) can be

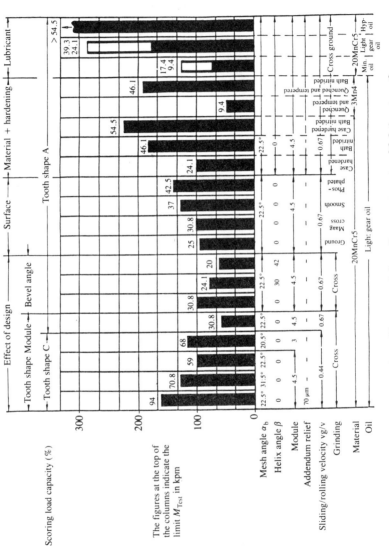

Fig. 3.20 The effect of toothing, material, surface treatment and lubricant on the scoring load capacity (from Lechner (25))

defined

$$\text{Scoring resistance } S_F = \frac{\text{allowable stress}}{\text{existing stress}} > 1$$

In the area of fluid friction, the lubricant film under pressure must transmit the load and ensure resistance. Thus the following scoring/wear resistance can be defined

$$S_{F,\,EHD} = \frac{\text{load limit of EHD contact film}}{\text{gear loading}} = \frac{W_{EHD}}{W}$$

The thickness of the contact film forming between the tooth flanks, which is related to the roughness of the surfaces, is regarded as a measure of the loading limit of the EHD contact film. The hydrodynamic and elasto-hydrodynamic basic equations for calculating this film thickness must be derived.

In the area of mixed friction, the chemical/physical reaction film on the surfaces has to ensure the load capacity and thus the resistance. The following scoring/wear resistance can, therefore, be defined

$$S_{F,\,R} = \frac{\text{load limit of reaction film}}{\text{gear loading}} = \frac{W_R}{W} = \frac{P_{all}}{P}$$

$S_{F,\,R}$ should be between 3 and 5 for satisfactory gear life. The gear oil test in the FZG test machine which forms the basis of this method of calculation has been standardized as DIN 51 354. It should be borne in mind that despite suitable correction, it appears that the test results can be applied only to spur gears.

Figure 3.21 shows tooth sides of test gears after a test of two gear oils. In both cases testing was carried out up to the twelfth load level. While only slight wear occurred with one oil, the other oil could not prevent scoring of the tooth flanks under this stress.

3.3.4 Elasto-hydrodynamic scoring resistance

3.3.4.1 Principles of hydrodynamics

The creation of a hydrodynamic pressure requires a suitably shaped gap, a viscous medium in the gap, and relative motion.

The gap between the two solid body surfaces can be variable (narrowing/widening) or constant (parallel). The direction of the relative

**Fig. 3.21   Tooth flanks after tests with two gear oils on the FZG test machine.
Top: slight wear. Bottom: destruction by scoring**

velocity must be matched to the gap, i.e., it must promote narrowing of
the gap.

The creation of a hydrodynamic pressure requires that the medium in
the gap should present a resistance to displacement. This behaviour is
expressed by the viscosity characteristic. All fluids, e.g., lubricating oils,

water, liquid metals, paste-like substances, e.g., lubricating greases, and gases, e.g., air, possess this property; liquid helium is an exception.

For pressure formation it is also necessary that the medium in the gap should adhere to the solid body surfaces. This is the case with all liquids, paste-like substances, and gases, with the exception of mercury.

A relative velocity produces displacement of the medium in the gap, which presents a resistance to the medium, so there is a build-up of pressure. With variable gaps this relative velocity must act in the direction of the narrowing of the gap and with parallel gaps it must be directed vertically to the gap to create a pressure.

Figure 3.22 gives a simplified representation of the relationship between gap, viscosity, and velocity.

The following assumptions and simplifications are used for derivation of the theoretical principles:

(a) the force of gravity and the forces of inertia are negligibly small;
(b) the surfaces bounding the gap are rigid and smooth;
(c) the curvature of the gap is negligibly small;
(d) the flow in the gap is laminar;
(e) the pressure is constant over the height of the lubricant gap (i.e., in the *y* direction);
(f) the shear stress and velocity gradients are considered only in the direction of the gap height;
(g) the gap height is small in relation to the other dimensions of the gap;
(h) the lubricant adheres to the gap surfaces without slip;

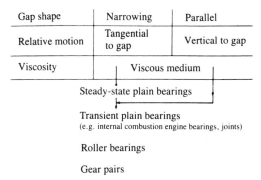

Fig. 3.22 Pre-requisites for the formation of hydrodynamic pressure

(i)   the lubricant possesses Newtonian flow properties;
(j)   the viscosity and density of the lubricant are constant in the gap.

As will be mentioned later, points (b) (rigid and smooth gap surfaces) and (j) (constant viscosity in the gap), and also to a certain extent point (i) (Newtonian lubricant) are very restrictive assumptions which scarcely apply to actual friction points in reality.

By stating the equilibrium of forces in the gap and taking account of the continuity equation, we obtain the following general form of the Reynolds' differential equation of hydrodynamic pressure build-up for incompressible media

$$\frac{\delta}{\delta_x}\left\{\left(\frac{h^3}{\eta}\right)\left(\frac{\delta p}{\delta x}\right)\right\} + \frac{\delta}{\delta z}\left\{\left(\frac{h^3}{\eta}\right)\left(\frac{\delta p}{\delta z}\right)\right\}$$

$$= 6\left(U_2 + U_1\right)\frac{\delta h}{\delta x} + 6h\delta\frac{(U_2 + U_1)}{\delta_x} + 12\frac{\delta h}{\delta t}$$

To simplify the Reynolds' differential equation further, the following assumptions are made:

–   the viscosity in the gap is constant;
–   only one surface moves; the other remains still;
–   it is a case of steady-state motion.

This gives the following form of the equation

$$\frac{\delta}{\delta_x}\left(h^3\frac{\delta p}{\delta x}\right) + \frac{\delta}{\delta z}\left(h^3\frac{\delta p}{\delta z}\right) = 6\eta \times U\frac{\delta h}{\delta x}$$

Approximation solutions exist for this equation, which are significant mainly for plain bearings. This equation can be simplified considerably if it is assumed that the gap is infinitely long. For the pressure distribution in the gap dp/dx we then get the equation

$$\frac{dp}{dx} = 6\eta \times U\frac{h - h^*}{h^3}$$

where

η = viscosity
U = peripheral velocity
h = gap
h^* = gap at the location of the highest pressure

3.3.4.2 Principles of elasto-hydrodynamics

Two of the most significant pre-requisites of classical hydrodynamics, namely a viscosity which is independent of the pressure and rigid surfaces, do indeed permit analysis of plain bearings for a wide range of operating conditions, as will be demonstrated, but they produce unrealistic results when applied to roller bearing lubrication and above all to gear lubrication.

As two surfaces rolling against one another become closer, the gap between them becomes smaller with increasing load, so the pressure in the oil film rises. The pressures created are so large that the dependence of the viscosity on pressure can no longer be disregarded. The increase in viscosity between the surfaces caused by the rise in pressure leads to enlargement of the gap with a given load. As oil pressure rises, any real material would deform elastically and yield laterally. The result is further enlargement of the gap. This situation will be discussed below (**26**)–(**29**).

For dry contact we obtain the situation shown in Fig. 3.23 for pressure development and surface state.

With rigid surfaces there would be point or line contact with infinite pressure (see Fig. 3.23(b)). For real surfaces we get the pressure curve in Fig. 3.23(c) due to elastic deformation.

To explain the situation with lubricated contact, we will assume initially an unloaded state and a film thickness h' (see Fig. 3.24(a)).

After the load N is applied, the gap h' would decrease to the gap h with rigid surfaces (Fig. 3.24(b)). With real surfaces this gap increases due

Dry contact (elastic surfaces)

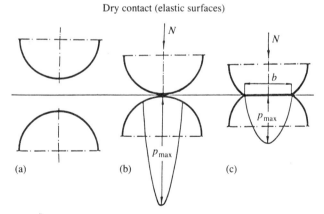

Fig. 3.23   Hertzian surface pressure for dry contact

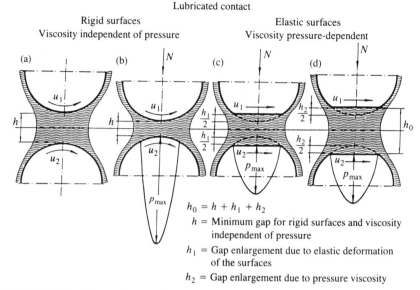

Fig. 3.24   Variation of gap and pressure with lubricated Hertzian contact (greatly simplified)

to elastic deformation by the value twice $h_1/2$ (see Fig. 3.24(c)). As the viscosity increases due to the rise in pressure in the lubricant when the surfaces move together, the gap increases accordingly by the value twice $h_2/2$. The resultant elasto-hydrodynamic gap is then

$$h_0 = h + h_1 + h_2$$

(see Fig. 3.24(d)). The need to take account of pressure viscosity and elasticity will be discussed using the following observations.

Martin solved the hydrodynamic basic equation for the geometrical conditions of a gear pair, using the simplified Reynolds' equation while disregarding the side effect and assuming an infinitely long gap (**30**). This simplified assumption for calculating the pressure development during tooth meshing is permissible because, unlike the situation with a plain bearing, with meshing gears the contact width *tranverse* to the direction of motion is more than an order of magnitude larger than *in* the direction of motion. Hence

$$dp/dx = 12\eta u(h - h^*)/h^3$$

with the effective peripheral velocity $u = u_1 - u_2/2$, the oil pressure p, the viscosity η, the gap h and the gap at the point of greatest pressure h^*; x is the coordinate in the direction of motion (direction of u). For the pre-requisite that the viscosity should not change with pressure, $\eta = \eta_0$, with η_0 as the viscosity at atmospheric pressure and operating temperature.

For the gap width h related to the reduced radius R, Martin reaches the following solution (30)

$$h_0/R = 4.9 \times \eta_0 \times h/w$$

Using this equation, the film thickness at the pitch point was calculated as a function of the specific loading, for a specific gear pair and a given rotational speed. A constant viscosity was assumed, using on the one hand the value for the temperature at the beginning and on the other hand the value for the final temperature of a test performed in parallel with the calculation with the same gear pair.

Figure 3.25 gives the results (28)(29). According to the curve for the initial viscosity, the gap has already decreased to a value in the range of

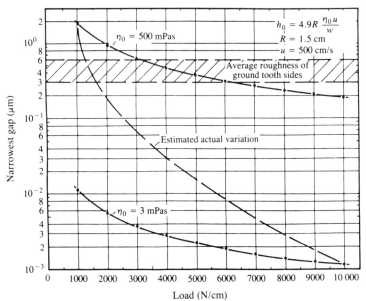

Fig. 3.25 **Gap variation at the pitch point between the tooth sides of a spur gear pair as a function of the load for rigid surfaces and constant viscosity**

the surface roughness at relatively low loads. From the outset, the final viscosity causes gap widths which are orders of magnitude smaller than the peak-to-valley height of the flank surfaces. High wear, if not actual scoring, must have resulted. As the accompanying tests showed, however, this damage did not occur until much higher loads were reached.

Even the fact that the temperature rises and the viscosity falls with increasing load, (resulting in the dashed curve for the dependence of the gap on load), does not alter the discrepancy between the calculation and the practical test. It must be concluded that the assumptions for pressure build-up between the sides of a gear pair, which form the basis for the calculation, cannot be correct, and that the actual gap widths are correspondingly larger.

Because the two surfaces become closer, the gap decreases with increasing load, so the pressure in the oil film increases. The oil pressures can be so high that it is no longer possible to disregard the pressure dependence of the viscosity.

For the purpose of simplification, the following relationship will be assumed for the dependence of the viscosity on pressure

$$\eta_p = \eta_0 \, e^{\alpha p}$$

with the pressure viscosity coefficient for the viscosity α.

The increase in viscosity between the surfaces caused by pressure leads to a widening of the gap with a given load. By insertion and transformation of the equation already given, we obtain

$$(dp/dx) \, e^{-\alpha p} = 12\eta_0 \, u(h - h^*)/h^3$$

By substitution we get

$$(dp/dx)e^{-\alpha p} = dq/dx$$

and

$$dq/dp = e^{-\alpha p}$$

and the solution is finally

$$p = (1/\alpha) \ln(1 - \alpha q)$$

Taking α into account, the actual pressure p can now be calculated for each pressure q. To estimate the resultant gain in load capacity, the following relationship can be used

$$w_{\eta = \eta_p} \approx 2.3 w_{\eta = \eta_0}$$

However, a factor of 2.3 is not sufficient to equalize the discrepancy discussed with the aid of Fig. 3.25.

The pressure p tends towards infinity if the pressure q reaches a limit of $1/\alpha$. However, this would only be conceivable if the material surfaces remained absolutely rigid however close the two surfaces became and whatever the resultant increase in viscosity and pressure. In fact, any real material will deform elastically as the oil film pressure rises. The result is a further increase in the gap between the surfaces.

The elastic deformation of the surfaces can be illustrated by the following equation

$$v = -2/(\pi \times E') \int_{x=s_1}^{x=s_2} p_s \ln(x - s)^2 \, ds + C$$

with the common modulus of elasticity E' for the two surface materials; s indicates a certain distance in the x direction.

The increase in load capacity can be characterized by a factor of about seven by taking account of the elastic deformation, but that is not sufficient to explain the discrepancy in Fig. 3.25.

The above observations indicate that two peripheral conditions apply. Firstly, the gap cannot become zero with increasing load due to the simultaneous increase in viscosity, and secondly the pressure in the oil film cannot become infinite due to the elastic deformation. Thus, an equilibrium is reached between the oil film pressure and the gap width, which should have relatively little dependence on the load.

The following equations are necessary for a mathematical description of this situation

$$dp/dx = 12\eta u(h - h^*)/h^3$$

$$\eta_p = \eta_0 \, e^{\alpha p}$$

$$h = h_0 + x^2/2R + v$$

with the common deformation v' of the two surfaces.

There is no complete solution to this system of equations. There are several approximate solutions, of which those by Dowson and Higginson, solved for the gap related to the reduced radius, agree well with results obtained experimentally. This equation is (**26**)

$$h_0/R = 2.65(\alpha E')^{0.54}(\eta_0 u/E'R)^{0.7}(w/(E'R))^{-0.13}$$

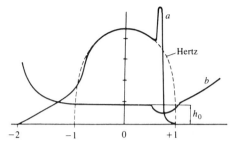

Fig. 3.26 Curves for pressure and gap with isothermal, elasto-hydrodynamic conditions (schematic): (a) pressure; (b) gap

The relatively small influence of the load (exponent 0.13) on the gap width expected from theoretical observations is clear.

This elasto-hydrodynamic film thickness equation leads to the curves for gap and pressure shown in Fig. 3.26, characterized by the contraction at the gap exit with the corresponding pressure peak.

Taking account of the elasticity of the materials involved in the sliding process and of the pressure viscosity of the oil, it is possible to calculate the gap produced by two surfaces rolling against one another, which correlates sufficiently accurately with the experimental results.

Table 3.3 clarifies the significance of this effect. An assumed gap of 2.54 μm could withstand a load of only 1400 N/cm according to classical hydrodynamics, while inclusion of the material's elasticity allows for a load of 9400 N/cm and the inclusion of the pressure viscosity of the oil allows for a load of 5100 N/cm.

According to elasto-hydrodynamic theory, however, it is possible to transmit a load of 90 000 N/cm, which coincides with practical experience.

Table 3.3   Theoretical loading of a contact for an assumed gap width of 2.54 μm

Theory	Load capacity (N/cm contact width)
Classical hydrodynamics without alteration	0.14×10^4
Inclusion of pressure viscosity	0.51×10^4
Inclusion of elasticity of material	0.94×10^4
Elasto-hydrodynamic solution	9.00×10^4

3.3.4.3 Application of the film thickness equation to gear pairs

With the aid of the film thickness equation

$$H_{min} = 2.65G^{0.54}U^{0.70}W^{-0.15}$$

we will calculate the film thickness at the pitch point between two rolling tooth flanks. The calculation will be performed for spur and bevel gears, that is, for rolling gears, with straight and helical teeth.

Table 3.4 contains an explanation of the symbols used. Tables 3.5, 3.6, and 3.7 contain information for calculating the effective radius of curvature, the effective peripheral velocity and the specific normal tooth force.

With this information the film thickness equation can be suitably converted and adapted to the geometrical and operating conditions of tooth systems. Table 3.8 contains the resulting standard formulae for straight and helical spur and bevel gears. In Table 3.9 the formulae are separated for ease of use.

Table 3.4   Symbols used in elasto-hydrodynamic film thickness calculation

$H_{min} = \dfrac{h_{min}}{\rho}$	Gap parameter
$W = \dfrac{w}{E\rho}$	Load parameter
$U = \eta_0 \dfrac{u}{E\rho}$	Velocity parameter
$G = \alpha E$	Material parameter
h_{min}	Minimum film thickness between surfaces
η_0	Operating viscosity at atmospheric pressure
α	Pressure coefficient of viscosity
$u = \frac{1}{2}(u_1 + u_2)$	Effective peripheral velocity of two surfaces
$\rho = \dfrac{\rho_1 \times \rho_2}{\rho_1 + \rho_2}$	Effective radius of curvature of the two surfaces
$E = \dfrac{2E_1 \times E_2}{E_2(1 - \sigma_1^2) + E_1(1 - \sigma_2^2)}$	Effective modulus of elasticity of the two materials
σ_1, σ_2	Poisson's ratio for the two materials
$w = F/b$	Normal tooth force per unit tooth width

Table 3.5   Calculation of the effective radius of curvature

$$\frac{1}{\rho} = \frac{1}{\rho_1} + \frac{1}{\rho_2} = \frac{2}{\sin \beta_n} \times \left(\frac{1}{d_{0,1}} + \frac{1}{d_{0,2}} \right)$$

	Spur gear	*Bevel gear (90 degrees)*
Straight teeth	$\rho = \dfrac{ai \sin \beta_n}{(i \pm 1)^2}$	$\rho = A_m \sin \beta_n \left(\dfrac{i}{i^2 + 1} \right)$
Helical teeth	$\rho_n = \dfrac{ai \sin \beta_n}{\cos^2 Y (i \pm 1)^2}$	$\rho_n = A_m \dfrac{\sin \beta_n}{\cos^2 Y_m} \left(\dfrac{i}{i^2 + 1} \right)$

1 = pinion;   2 = gear;   + = external teeth;   − = internal teeth

ρ = radius of curvature	i = transmission ratio
d_0 = pitch diameter	β = mesh angle
a = centre distance	Y = helix angle
A = cone distance	

3.3.4.4   Graphic estimation of film thickness

For certain values of transmission ratio, load, and viscosity, the film thickness between rolling tooth flanks in relation to gear rotational speed and centre distance can be taken from nomograms in Figs. 3.27 and 3.28.

Table 3.6   Calculation of the effective peripheral velocity

$$u = \tfrac{1}{2}(u_1 + u_2)$$

At pitch point: $u = u_1 = u_2 = \rho_1 \omega_1 = \rho_2 \omega_2$

	Spur gear	*Bevel gear (90 degrees)*
Straight teeth	$u = \dfrac{\pi n_1}{30} \left(\dfrac{a}{i + 1} \right) \sin \beta_n$	$u = \dfrac{\pi n_1}{30i} A_m \sin \beta_n$
Helical teeth	$u = \dfrac{\pi n_1}{30 \cos Y} \left(\dfrac{a}{i + 1} \right) \sin \beta_n$	$u = \dfrac{\pi n_1}{30i \cos Y_m} A_m \sin \beta_n$

1 = pinion;   2 = gear;   + = external teeth;   − = internal teeth

n = rotational speed	i = transmission ratio
A = cone distance	β = mesh angle
a = centre distance	Y = helix angle
u = peripheral velocity	

Table 3.7 Calculation of the specific normal tooth force

$$w = \frac{F}{b}$$

$$F = \frac{F_u}{\cos \beta_b}; \quad \text{straight teeth}$$

$$F = \frac{F_u}{(\cos \beta_b \cos Y_g)}; \quad \text{helical teeth}$$

$$F_u = \frac{2M}{d_b}$$

w = specific normal tooth force $\quad M$ = torque
F = normal tooth force $\quad\quad\quad d_b$ = pitch circle diameter
b = tooth width $\quad\quad\quad\quad\quad\quad \beta$ = mesh angle
F_u = peripheral force $\quad\quad\quad\quad Y$ = helix angle

Table 3.8 Calculation of the elasto-hydrodynamic film thickness – I

$$h_{min} = 2.65 \times \alpha^{0.54}(\eta_0 u)^{0.7}\rho^{0.43}E^{-0.03}w^{-0.13}$$

Spur gear

Straight teeth

$$h_{min} = \frac{2.65 \times \alpha^{0.54}}{E^{0.03}w^{0.13}}\left(\eta_0 \frac{\pi n_1}{30}\right)^{0.7} \times (a \sin \beta_n)^{1.13}\frac{i^{0.43}}{(i \pm 1)^{1.56}}$$

Helical teeth

$$h_{min} = \frac{2.65 \times \alpha^{0.54}}{E^{0.03}w^{0.13}}\left(\eta_0 \frac{\pi n_1}{30}\right)^{0.7} \times \frac{(a \sin \beta_n)^{1.13}}{\cos^{1.56} Y}\frac{i^{0.43}}{(i \pm 1)^{1.56}}$$

Bevel gear

Straight teeth

$$h_{min} = \frac{2.65 \times \alpha^{0.54}}{E^{0.03}w^{0.13}}\left(\eta_0 \frac{\pi n_1}{30}\right)^{0.7}(A_m \sin \beta_n)^{1.13}\frac{i^{0.27}}{(i^2 + 1)^{0.43}}$$

Helical teeth

$$h_{min} = \frac{2.65 \times \alpha^{0.54}}{E^{0.03}w^{0.13}}\left(\eta_0 \frac{\pi n_1}{30}\right)^{0.7}\frac{(A_m \sin \beta_n)^{1.13}}{\cos^{1.56} Y}\frac{i^{0.27}}{(i^2 + 1)^{0.43}}$$

Table 3.9   Calculation of the elasto-hydrodynamic film thickness – II

$$h_{min} = 2.65 \frac{\alpha^{0.54} \eta_0^{0.7}}{E^{0.03} w^{0.13}} XYZ$$

	Spur gear			
	Straight teeth		Helical teeth	
	Internal teeth	External teeth	Internal teeth	External teeth
X	$\{(\pi n_1)/30\}^{0.7}$			
Y	$(a \sin \beta_n)^{1.13}$		$\dfrac{(a \sin \beta_n)^{1.13}}{\cos^{1.56} Y}$	
Z	$\dfrac{i^{0.43}}{(i-1)^{1.56}}$	$\dfrac{i^{0.43}}{(i+1)^{1.56}}$	$\dfrac{i^{0.43}}{(i-1)^{1.56}}$	$\dfrac{i^{0.43}}{(i+1)^{1.56}}$

	Bevel gear	
	Straight teeth	Helical teeth
X	$\{(\pi n_1)/30\}^{0.7}$	
Y	$(A_m \sin \beta_n)^{1.13}$	$\dfrac{(A_m \sin \beta_n)^{1.13}}{\cos^{1.56} Y}$
Z	$\dfrac{i^{0.27}}{(i^2+1)^{0.43}}$	

The factors in Fig. 3.29 can be used to correct this film thickness value for real values of transmission ratio, load, and viscosity.

3.3.4.5   Analytical/graphic determination of film thickness
Another method of determining the film thickness combines analytical and graphical procedures. Firstly the viscosity parameter A and the elasticity parameter B are calculated using the following equations

$$A = \left(\frac{\alpha^2 w^3}{\eta_0 u \rho^2} \right)^{1/2}$$

$$B = \left(\frac{w^2}{\eta_0 u E' \rho} \right)^{1/2}$$

Figure 3.30 shows the ranges for the parameters A and B which are possible in practice for various tribological systems.

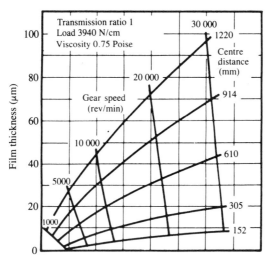

Fig. 3.27 Nomogram for determining film thickness with thick films (from Dowson and Higginson (26))

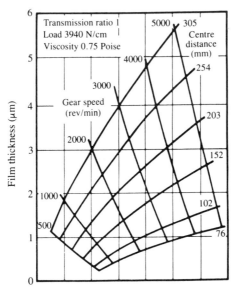

Fig. 3.28 Nomogram for determining film thickness with thin films (from Dowson and Higginson (26))

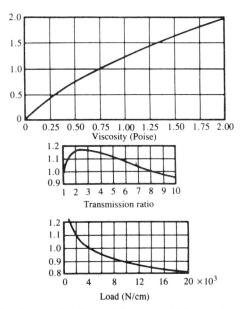

Fig. 3.29 Correction factors for determining film thickness (from Dowson and Higginson (26))

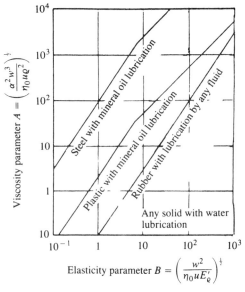

Fig. 3.30 Practical operating ranges for the viscosity and elasticity parameters

74

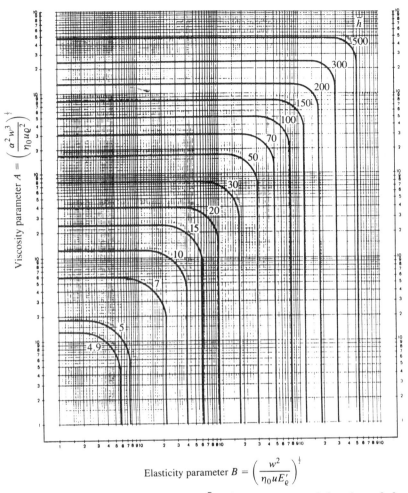

Viscosity parameter $A = \left(\dfrac{a^2 w^3}{\eta_0 \mu Q^2}\right)^{\frac{1}{4}}$

Elasticity parameter $B = \left(\dfrac{w^2}{\eta_0 u E'_Q}\right)^{\frac{1}{4}}$

Fig. 3.31 Dimensionless film thickness $\bar{h}$ against parameters of viscosity and elasticity

Figure 3.31 shows curves of constant dimensionless film thickness $\bar{h}$ as a function of A and B. After calculation of A and B, the associated value of $\bar{h}$ can be found. The actual film thickness is then deduced from the equation

$$h = \frac{\bar{h}\eta_0 u\rho}{w}$$

3.3.4.6   Film thickness and damage criterion

The so-called specific film thickness is defined as the damage criterion, i.e., as the minimum film thickness for prevention of wear and scoring. If it is sufficiently large, scoring wear is unlikely or even impossible. Here it is necessary to make the specific film thickness equal to the elasto-hydrodynamic scoring resistance, so

$$S_{F_{EHD}} = \lambda$$

It is desirable to make λ considerably greater than one, to prevent premature damage.

The specific film thickness λ is defined as follows

$$\lambda = \frac{\text{lubricant film thickness } h}{\text{combined surface roughness } R}$$

The combined surface roughness R is obtained from the individual roughness values R_1 and R_2 of the two flank surfaces. The peak-to-valley heights R_1 and R_2 are RMS values, obtained for ground surfaces from the corresponding CLA values by multiplying them by a factor of 1.3.

$$R = \frac{R_1 + R_2}{2}$$

Table 3.10 gives typical values for the combined peak-to-valley height of gear surfaces before and after running-in.

The following can serve as criteria for possible scoring damage

$$
\begin{aligned}
\lambda \geqslant 3 &\quad - \quad \text{no damage} \\
1.5 < \lambda < 3 &\quad - \quad \text{damage possible} \\
\lambda \leqslant 1.5 &\quad - \quad \text{damage probable}
\end{aligned}
$$

Table 3.10   Peak-to-valley heights before and after running-in for various surface machining processes

Surface finish	*Combined peak-to-valley height*	
	Before running-in	*After running-in*
Milling	2.3–4.6	1.2–2.3
Hobbing	1.2–2.3	0.9–1.7
Grinding	0.7–1.4	0.6–1.2
Lapping	0.6–1.1	0.4–0.9
Honing	0.3–0.6	0.2–0.4

3.3.4.7   Procedure for calculating film thickness

The following procedure should be used when applying elasto-hydrodynamics to gear design

- determine effective values for velocity, load, modulus of elasticity and curvature;
- determine/estimate the effective temperature;
- determine viscosity and the pressure–viscosity coefficient as a function of temperature;
- calculate film thickness;
- compare film thickness with peak-to-valley height and calculate scoring resistance.

Figure 3.32 gives a flow diagram for this procedure.

3.3.4.8   Restrictions on the use of the film thickness equation for estimating scoring resistance

When applying the film thickness equation to gear design, some restrictions should be borne in mind. Firstly, account should be taken of the fact that the film thickness calculation relates only to one point on the tooth flank. In Fig. 3.33, the significant film thickness is shown against contact path for a specific example. It can be seen that the film thickness is greater or smaller in other areas of the tooth flank.

Furthermore, one should remember that the film thickness calculation is performed under steady-state conditions, while, in reality, transient conditions of load, velocity and surface curvature predominate. Whereas the calculation assumes isothermal conditions, the viscous friction in the oil film produces a rise in temperature which causes a reduction in viscosity and thus a reduction in film thickness. This non-isothermal condition can be corrected as follows

$$h_{min_{non\text{-}isoth}} = C_{th}\, h_{min_{isoth}}$$

where

$$C_{th} = \frac{1}{1 + 0.254\, A^{0.62}} \quad \text{and} \quad A = \frac{u^2 \gamma n_0}{\lambda}$$

Here λ $(W/m°C)$ is the thermal conductivity and γ $(°C^{-1})$ is the viscosity–temperature coefficient of the oil. Figure 3.34 shows C_{th} as a function of A.

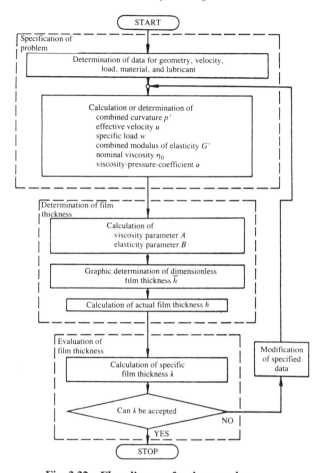

Fig. 3.32   Flow diagram for the procedure

Further differences between the assumptions for deriving the film thickness equation and real conditions are as follows:

- due to the crowning of many machine elements, the contact surface differs from the rectangular shape for line contact;
- the material stress is increased by the viscous friction in EHD contact;
- temperature increases occurring due to friction and relative sliding cause additional stresses in the contact surfaces.

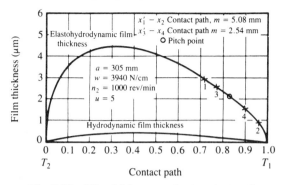

Fig. 3.33   Film thickness against contact path

3.3.4.9   Examples of the application of the film thickness equation

The oils in Table 3.11 were used to study the possibilities and limitations of the application of elasto-hydrodynamics to gear design.

The minimum film thickness with respect to load was calculated for a given gear pair using the viscosity values of these oils. Figure 3.35 gives the results.

The film thickness values at high loads would lead one to expect the failure sequence of oils shown in Fig. 3.36 (top line). Tests carried out with these gear types and these conditions produced the damage

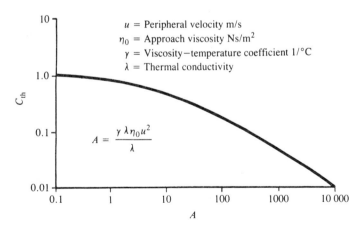

Fig. 3.34   Correction factor C_{th} for taking account of thermal influences on the film thickness

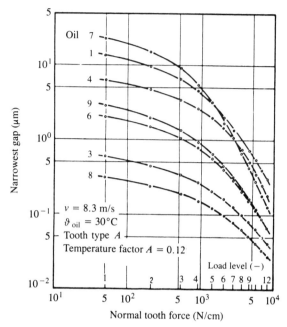

Fig. 3.35   Minimum film thickness as a function of load for various oils

Table 3.11   Test oils for elasto-hydrodynamic studies

No.	Description	Viscosity at 50°C ($m^2/s \times 10^{-4}$)	Density at 20°C ($kg/m^3 \times 10^3$)
1	Mineral oil (mixed)	395.0	0.897
2	Mineral oil (white oil)	38.0	0.888
3 ⎱	Synthetic lubricant	⎰ 12.6	0.946
4 ⎰	(polyalkylglycolether)	⎱ 260.6	0.979
5	Mineral oil (white oil)	11.2	0.842
6 ⎱	Mineral oil	⎰ 39.0	0.881
7 ⎰		⎱ 240.0	1.009
8	Synthetic lubricant Di(2-ethyl-hexyl)sebacate	8.2	0.912
9	Synthetic lubricant mixture of chlorinated aromatic hydrocarbons and aromatic phosphate esters)	32.7	1.368

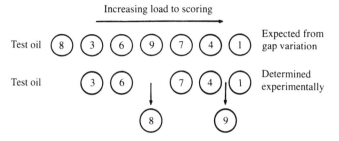

Fig. 3.36 **Damage occurrence sequence of test oils (calculated and measured)**

occurrence sequence shown in the bottom row in Fig. 3.36. The correlation is usable.

Figure 3.37 shows the film thickness as a function of load for oil 1 at various velocities and temperatures. The curve of wear for the tests

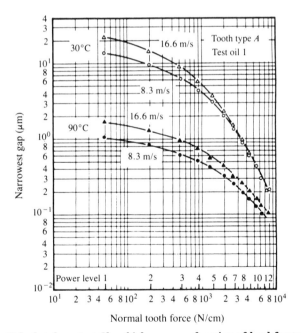

Fig. 3.37 **Calculated contact film thickness as a function of load for two oil sump temperatures and two peripheral velocities**

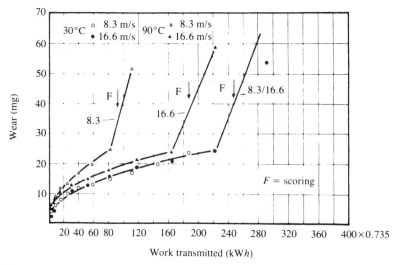

Fig. 3.38   Wear and incipient scoring as a function of work transmitted for two oil sump temperatures and two peripheral velocities

carried out under the same conditions is given in Fig. 3.38. The good correlation between the expected behaviour of the oils and the actual behaviour is clear.

3.3.4.10   Summary and conclusion for the application of elasto-hydrodynamics

There are some restrictions affecting the application of elasto-hydrodynamics, which must be overcome by means of further investigations. The most important areas are as follows.

(a)   The state of the contact film between the flanks before the onset of increased wear or scoring is described by means of the elasto-hydrodynamic gap equation. The latter can, therefore, give information on the application limits of additive-free gear oils, but it cannot indicate the scoring resistance afforded by EP gear oils. This fact represents the decisive difference between methods of calculation for the scoring resistance of gear pairs on the basis of elasto-hydrodynamics and on the basis of characteristic or critical temperatures.

(b)   The relationship between the elasto-hydrodynamic contact film and the onset of flank damage which has been discussed was based on the assumption that increased wear and scoring do not occur so long as a certain minimum gap is maintained. This classical explanation of the onset of scoring by the formation and subsequent breaking of micro-welds, which would require direct contact of the surfaces, must, however, be modified in the light of some test results which have demonstrated scoring with proven separation of the two surfaces.

(c)   The results and observations discussed in this work relate only to the pitch point of rolling gears. However, the tooth flank damage correlated with the gap width at this location generally has its origins at other locations. Calculation of film thickness over the whole tooth flank would have to be performed as a function of temperature distribution, load and velocity. The fact that one average integral temperature could no longer be used as a basis would make the calculation considerably more difficult, and would also make it unreliable.

(d)   The calculation of elasto-hydrodynamic contact film thickness discussed relates to pure rolling without superimposed sliding. Therefore, it can presently only be used for rolling gears, i.e., spur gears and non-offset bevel gears. It is questionable whether it could be applied to rolling crossed-axis gears, i.e., offset bevel gears, worm and crossed-axis helical gears, simply by inserting different effective temperatures, because, in the case of these gear types, the velocity conditions impede the formation of a contact film to a considerable extent. As, however, practical experience largely excludes the use of additive-free gear oils for rolling crossed-axis gears, it can scarcely be necessary to specific calculation methods based on elasto-hydrodynamics.

(e)   An important pre-requisite for a useful method of calculating elasto-hydrodynamic contact film thicknesses is adequate information, at the time of analysis and design of the gears, on the effective temperatures to be expected – a requirement which, of course, applies to all other methods of calculating scoring resistance.

The fact that the viscosity/temperature behaviour and the viscosity/pressure behaviour of the gear oils to be used must be available for a wide temperature range, need only be mentioned to complete the picture.

On the assumption that the restrictions listed can be overcome, elasto-hydrodynamic factors can be included in the analysis of a gear pair as follows, taking a lubricant into account.

(a) For a given oil with a given viscosity/temperature and viscosity/ pressure characteristic, it is possible to calculate the maximum allowable load which can be transmitted by a given gear pair under given operating conditions, with no possibility of tooth flank damage.

(b) For a given gear type and given load and velocity conditions, it is possible to calculate the necessary minimum viscosity of the oil to be used which will ensure damage-free operation. A thermal balance must then be used to determine whether this viscosity can be controlled by the gear's thermal system. If it proves to be too high, a lower viscosity oil must be used. The resultant loss of load capacity must be equalized by extreme pressure additives. This permits the calculation to include the necessary transition from additive-free gear oils (calculation based on elasto-hydrodynamics) to gear oils with additives (calculation based on a critical temperature).

Thus the laws of elasto-hydrodynamic lubrication permit the lubricant to be included in the analysis of gear pairs via its viscosity. In addition to being a functional element, the lubricant is thus assigned the tasks of a real structural element. This 'promotion' at last gives the lubricant a role to which it is entitled, in any case on the basis of its capabilities.

3.3.5 Scoring resistance with a reaction film

3.3.5.1 Basic principles
The scoring resistance provided by the reaction film represents the relationship between the load limit of the lubricant in a given gear set and the applied load. The load-bearing capacity of the reaction layer formed on the tooth flanks by surface-active additives by means of chemical reactions with the material is characteristic of the load limit of the lubricant. Naturally, this load limit depends on the design, material, and operating characteristics of the gear set. For the gear set in question it would, therefore, have to be experimentally determined, which would not be the optimal method for the following reasons:

– large-scale gear sets produced individually or in small numbers cannot be used as experimental gear sets for gear oil testing;
– the load limit of the gear oil is available only for designing the gear set.

Extensive tests on spur and bevel gears, in which the most important parameters affecting scoring wear, such as lubricant, tooth geometry, operating conditions, gear material, manufacturing methods and surface treatment, were varied, together with analysis of scoring damage produced in practical operation, have shown that results obtained with a small experimental gear set can be applied to other rolling gears with larger dimensions and different designs (**31**).

The FZG test machine can be used as an experimental gear set. The test procedure is specified in DIN 51 354 (**32**).

3.3.5.2 Specified load

The actual Hertzian stress P_W for the given operating conditions, taking account of dynamic, operating, and load-distribution factors, is calculated as per DIN 3990, Sheet 1, 'Load capacity calculation for spur and bevel gears' (**23**).

3.3.5.3 Scoring resistance based on the allowable Hertzian stress of the lubricant

3.3.5.3.1 Principles

Scoring resistance based on the allowable Hertzian stress of the lubricant is defined as follows

$$S_{FR} = \frac{\text{allowable Hertzian stress of lubricant}}{\text{actual Hertzian stress in gear set}} = \frac{P_{all}}{P_W}$$

Lechner has devised a method for determining P_{all} (**31**). The principle of the method of calculating scoring resistance can be found in Fig. 3.39; it can be split into the following stages.

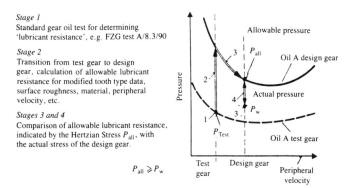

Stage 1
Standard gear oil test for determining 'lubricant resistance', e.g. FZG test A/8.3/90

Stage 2
Transition from test gear to design gear, calculation of allowable lubricant resistance for modified tooth type data, surface roughness, material, peripheral velocity, etc.

Stages 3 and 4
Comparison of allowable lubricant resistance, indicated by the Hertzian Stress P_{all}, with the actual stress of the design gear.

$$P_{all} \geqslant P_W$$

Fig. 3.39   Principle of Lechner's method of analysis (31)

(a) Determination of load capacity of lubricant P_{test} in the FZG test A/8.3/90.
(b) Transition from test gear set to design gear by modifying P_{test} using the coefficients X_x, thus obtaining P_{all}.
(c) Calculation of the actual Hertzian stress P_W at the pitch point of the design gear.
(d) Calculation of the scoring resistance from the quotient of the allowable and the actual Hertzian stress.

3.3.5.3.2   Determination of P_{all}
The following is sufficiently accurate for the allowable Hertzian stress of the gear oil in the relevant operating conditions

$$P_{all} = P_{test}(X_V X_{VZ} X_Z X_\Sigma)^{0.5} 10\left(\frac{N}{mm^2}\right)$$

Here

$P_{test}(N/mm^2)$ = Hertzian stress at scoring load in FZG test A/8.3/90 to DIN 51 354
$X_V, X_{VZ}, X_Z(-)$ = primary coefficients for including effects of velocity and tooth shape
X_Σ = product of secondary coefficients for including:
X_b = tooth width
X_k = tip relief
X_R = roughness
X_0 = surface treatment
X_T = oil temperature
X_Q = volume of oil
X_ψ = oil spray angle
X_D = direction of rotation
X_w = material
X_P = contact

While including the primary coefficients is essential if one wishes to obtain useful results for scoring resistance, the product X_Σ of the secondary coefficients can be made equal to 1 as a first approximation, if no other information is available. This procedure is justified by the observation that this method of determining the scoring resistance is an approximation procedure.

3.3.5.3.3 Determination of primary coefficients

The velocity and tooth shape coefficients serve primarily to facilitate the transition from the test gear set to the gear set being designed.

(a) *Velocity coefficients* X_V *and* X_{VZ}. The effect of the peripheral velocity v on the coefficient X_V and thus on the scoring load capacity can be seen from Fig. 3.40. As the peripheral velocity increases, the scoring load capacity falls at first, and after going through a minimum rises again due to auxiliary elasto-hydrodynamic effects. The following empirical equation has been derived for the coefficient X_V (**31**)

$$X_V = (1 - 1/c)\left(\frac{14.1413}{(5 + v)^{1/3}} + 0.030644v - 6.22302\right) + 1.0$$

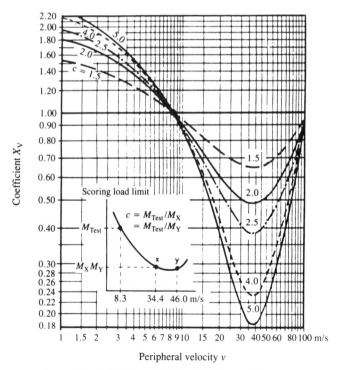

Fig. 3.40 Velocity coefficient X_V from Lechner (31)

Here

v = peripheral velocity (m/s)

c = oil constant; takes account of effect of lubricant quality on the slope of the scoring load/velocity curve. c can also be determined in tests; in this case, FZG tests should also be carried out at $v = 34$ and 46 m/s, i.e., the scoring load/velocity curve has to be determined.

Reference values for the oil constant c are

- viscous ($\geqslant$ SAE 80) and/or high-additive-content oils: $c = 1.5$–2.5
- low-viscosity and/or additive-free oils: $c = 4.0$–5.0

The tooth shape also affects the shape of the scoring load/velocity curve. The higher the load capacity of a tooth type, the steeper is the decline of the scoring load limit with velocity; the converse also applies.

Figure 3.41 shows the effect of peripheral velocity on the tooth shape velocity coefficient X_{ZV}. The following empirical equation has been derived for this coefficient (31)

$$X_{VZ} = \{9.3/(1 + v)\}^{0.5(1 - 0.9\varepsilon_{1,2})}$$

Here ε_1 or ε_2 is the maximum tip contact ratio of the pinion or the gear.

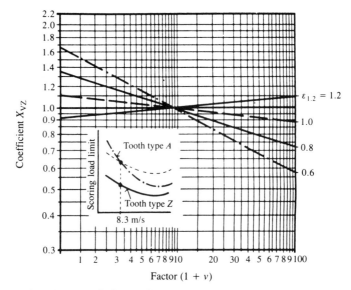

Fig. 3.41   Tooth shape velocity coefficient X_{VZ} from Lechner (31)

(b)   *Tooth shape coefficient* X_Z. The tooth shape coefficient X_Z takes account of the change in the scoring load limit when the tooth profile changes compared with the test tooth type. Allowance is made for the fact that the scoring load limit is inversely proportional to the tooth power loss. The following empirical equations were derived for calculating X_Z; a distinction is made between straight and helical gears.
Straight gears

$$X_Z = \underbrace{\frac{2.572}{a_b \cos \alpha_b} \times \left(\frac{z_1 i}{1+i}\right)^{1.5}}_{A} \underbrace{\left(\frac{1}{1 + \varepsilon_1^2 + \varepsilon_2^2 - \varepsilon_1 - \varepsilon_2}\right)^{1.5}}_{B}$$

$$= \text{calculation factor } A \times B$$

Helical gears

$$X_Z = \underbrace{\frac{2.572}{a_b \cos \alpha_b} \times \left(\frac{z_1 i}{1+i}\right)^{1.5}}_{A} \underbrace{\left(\frac{\varepsilon_1 + \varepsilon_2}{\varepsilon_1^2 + \varepsilon_2^2} \times \cos \beta_g\right)^{1.5}}_{B^*}$$

$$= \text{calculation factor } A \times B^*$$

Here

a_b = operating centre distance (mm)
α_b = operating pressure angle in transverse profile
z_1 = number of teeth on pinion
i = gear ratio z_2/z_1
$\varepsilon_1, \varepsilon_2$ = tip contact ratios of pinion and gear
β_g = base circle helix angle

The factors A, B, and B* for the tooth type data can be found from Figs 3.42, 3.43, and 3.44.

3.3.5.3.4   Determination of secondary coefficients
Table 3.12 contains a summary of the empirical equations derived for estimating the various secondary coefficients. Details of the associated functions and relationships can be found in the original literature (**31**).

3.3.5.3.5   Scoring resistance of bevel gears, including planetary gears
The scoring resistance of bevel gears with straight, helical and spiral teeth can be calculated in a similar manner. A pre-requisite for this is that the tooth type data of the equivalent spur gear pair, related to centre face width and pitch cone, must be introduced.

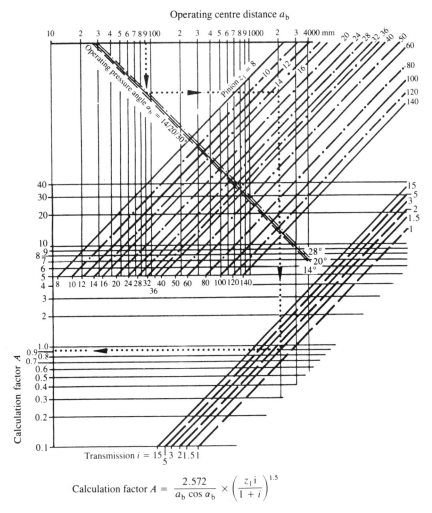

Fig. 3.42   Calculation factor A for the tooth shape coefficients from Lechner (31)

The peripheral velocity used is that value which applied originally to the tooth centre of the bevel pinion.

Unlike with stationary gear sets, in the case of planetary gears, the power to be transmitted is never equal to the input power. Rather this is the sum of the toothing power N_Z and the coupling power N_K.

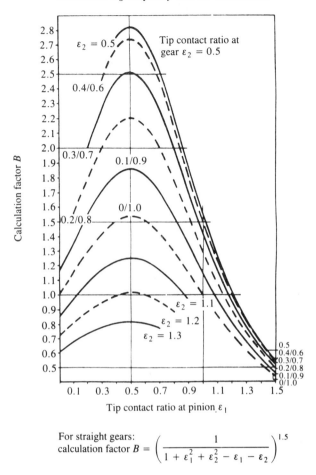

For straight gears:
calculation factor $B = \left(\dfrac{1}{1 + \varepsilon_1^2 + \varepsilon_2^2 - \varepsilon_1 - \varepsilon_2} \right)^{1.5}$

Fig. 3.43   Calculation factor B for straight gears from Lechner (31)

At the same time, it is not the peripheral speed which is decisive for the scoring process, but the rolling velocity. This is the relative velocity between sun wheel and arm, and it occurs at the contact point of the operating pitch circles of a pair of gears consisting of a sun wheel and a planet gear. This pitch velocity is calculated from an equivalent rotation velocity and the pitch circle diameter of planet gear or sun gear, depending on which of the two wheels is the pinion.

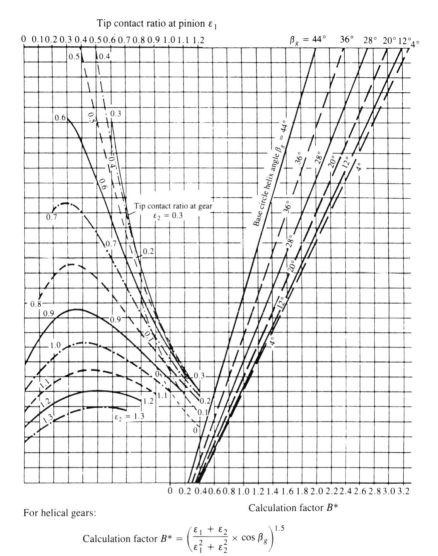

For helical gears:

$$\text{Calculation factor } B^* = \left(\frac{\varepsilon_1 + \varepsilon_2}{\varepsilon_1^2 + \varepsilon_2^2} \times \cos \beta_g \right)^{1.5}$$

Fig. 3.44 Calculation factor B* for helical gears from Lechner (31)

Table 3.12 Formulae for determining the secondary coefficients

Face width	$X_b = 0.7 + 6.0/b(-)$
Tip relief	$X_K = 1.0 + 0.0155\varepsilon_{1,2}^4 K(-)$
Surface roughness	$X_R = (0.0047793d_{b1}/R_a)^{0.25}(-)$
Surface treatment	X_0 from table
Oil temperature	$X_T = 1.0 - (0.0149 - 0.000194M_{test})$ $\times (T_B - 90) \geqslant 0.4(-)$
Oil volume:	
unhardened gears	$X_Q = 12.5Q/b$
hardened gears	$X_Q = 20Q/b \leqslant 1.0$
Oil spray direction	$X_\varphi = 1.0 - 0.0007\varphi(-)$
Direction of rotation	X_D from table
Material hardening:	
austenite content	$X_{W1} = 0.32 + 1.19(1 - 0.01AU)^{2.5}$
heat treatment	X_{W2} from Table
total coefficient	$X_W = X_{W1}X_{W2}(-)$
Power split	$X_P = (1/P)^{0.25}(-)$

3.3.5.3.6   Accuracy of the calculation method

A large random sample was selected and statistically evaluated to determine the accuracy of the method. In this study, not only were the results of bench tests on spur and bevel gears used, but numerous industrial gear sets were also simulated. The standard deviation of scoring resistance found in the analytical process was

$$\sigma \pm 0.2.$$

This means that with a target scoring resistance of 1.0, the actual resistance lies between 0.8 and 1.2, with a 32 percent probability of error (**31**).

3.3.5.4   Scoring resistance based on the integral temperature limit

3.3.5.4.1   Principles

The scoring resistance based on the integral temperature limit is defined as follows as the so-called temperature resistance

$$S_{F_T}V\frac{\delta_{int\,s}}{\delta_{int}} = \frac{\text{temperature limit for scoring}}{\text{temperature on tooth flank}}$$

Michaelis devised a method for determining the scoring resistance (**33**), which also requires test runs on the FZG test machine to DIN 51 354 (**32**). The principle of the process can be divided into the following steps, similar to the method shown above:

(a) determine the damage moment for the lubricant using FZG test run A/8.3/90;
(b) calculate the integral temperature limit, taking account of the material and heat treatment of the design gear and the damage force level of the oil;
(c) calculate the integral temperature taking account of the structural data and the actual load on the design gear;
(d) calculate the scoring resistance from the quotient of the integral temperature limit and the integral temperature.

3.3.5.4.2 Determination of integral temperature

The integral temperature is deemed to be the average, weighted surface temperature of the tooth flank. Figure 3.45 gives information on the

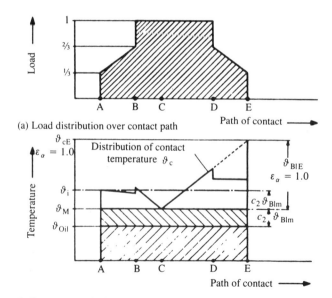

(a) Load distribution over contact path

(b) Temperature distribution over contact path

Fig. 3.45 Determination of integral temperature from Michaelis (34): (a) load distribution over contact; (b) temperature distribution over contact

temperature definitions. The following relationship was derived for the integral temperature of the design gear

$$\delta_{int} = \delta_M + C_2 \delta_{fla\ int}\ (^\circ C)$$

where

δ_M = mass temperature $(^\circ C)$
C_2 = weighting factor = 1.5 for spur gears
$\delta_{fla\ int}$ = average flash temperature over the contact $(^\circ C)$

For the average flash temperature over the contact we get

$$\delta_{fla\ int} = \delta_{fla\ E}\, X_\varepsilon\ (^\circ C)$$

Here

$\delta_{fla\ E}$ = maximum flash temperature at pinion tooth tip with a transverse contact ratio $\varepsilon_\alpha = 1.0$
X_ε = contact ratio factor

For the maximum flash temperature at the pinion tooth tip we get

$$\delta_{fla\ E} = \mu_m X_M X_{BE}\ \frac{w_t^{3/4} v^{1/2}}{a^{1/4}}\ \frac{1}{X_Q X_{Ca}}\ (^\circ C)$$

Here

μ_m = average coefficient of friction
X_M = flash factor
X_{BE} = geometry factor at pinion tooth tip
X_Q = contact factor
X_{Ca} = tip relief factor
w_t = effective peripheral force (N/mm)
v = peripheral velocity (m/s)
a = centre distance (mm)

3.3.5.4.3   Determination of factors influencing the integral temperature
The most important influences on the integral temperature in addition to design and operating data are various values which are summarized as factors (**34**).

(*a*)   *Effective peripheral force.*   The effective peripheral force related to face width can be calculated from the following equation

$$w_t = K_A K_v K_{B\beta} K_{B\alpha} K_{B\gamma}\ \frac{F_t}{b}\ (N/mm)$$

Here

F_t = peripheral force
b = face width
K_A = application factor
K_v = dynamic factor
$K_{B\beta}$ = width factor = flank stress $K_{H\beta}$
$K_{B\alpha}$ = transverse factor = flank stress $K_{H\alpha}$
$K_{B\gamma}$ = helix factor

The overload factors K_A, K_V, $K_{B\alpha}$ and $K_{B\beta}$ are determined in accordance with DIN 3990, Part 1. The following values apply to the helix factor $K_{B\gamma}$ which depends on the total contact ratio ε_γ

$K_{B\gamma} = 1$ where $\varepsilon_\gamma \leqslant 2$

$K_{B\gamma} = 1 + 0.2\sqrt{\{(\varepsilon_\gamma - 2)(5 - \varepsilon_\gamma)\}}$ where $2 < \varepsilon_\gamma < 3.5$

$K_{B\gamma} = 1.3$ where $\varepsilon_\gamma \geqslant 3.5$

(*b*)   *Average coefficient of friction.*   The following equation can be used to supplement the relationship given in Section 2.4.3 for determining the coefficient of friction

$$\mu_m = 0.12\left(\frac{w_z R_{a0}}{\eta_0 \, v_\Sigma \, \rho_{red}}\right)^{0.25} X_R X_\eta$$

Here

R_a = effective arithmetic mean roughness of pinion and gear R_a
$\frac{1}{2}(R_{a1} + R_{a2})$ (μm).
As a first approximation $R_a \cong R_t/6$ with peak-to-valley height R_t (μm)
R_{a0} = 0.35 μm mean roughness of test gears
w_t = effective peripheral force
η_M = oil viscosity at mass temperature
As a first approximation $\eta_M = \eta_{oil}$ (mPas) with the oil viscosity at operating temperature η_{oil} (mPas).
η_0 = oil viscosity at 100°C
v_Σ = mean total peripheral velocity over the contact.
As a first approximation $v_{\Sigma m} = v_{\Sigma c}$ (m/s) with the total velocity at the pitch point $v_{\Sigma c} = 2v \sin \alpha_{wt}$ with the peripheral velocity v and the transverse pressure angle α_{wt} at the pitch cylinder.

$$p_{red} = \frac{a \times \sin \alpha_{wt}}{\cos \beta_b} \times \frac{(1 + \Gamma_E)(\mu + \Gamma_E)}{(u + 1)^2}$$

with the gear ratio $u = Z_1/Z_2$, the centre distance a and the helix angle β

X_R = roughness factor = $(R_a/R_{a0})^{0.25}$

X_η = viscosity factor = $(\eta_0/\eta_M)^{0.25}$

(c) *Flash factor*. The flash factor X_M takes account of the elastic and thermal characteristics of the gear materials. A first approximation for a steel/steel pair is

$$X_M = 50(K \times N^{-3/4} \times s^{1/2} \text{ m}^{-1/2} \text{ mm})$$

(d) *Geometry factor*. The geometry factor X_{BE} take account of the geometrical dimensions for determining Hertzian stress and sliding velocity at the pinion tooth tip.

$$X_{BE} = 0.51\sqrt{(\mu + 1)} \times \frac{|\sqrt{(1 + \Gamma_E)} - \sqrt{(1 - \Gamma_E/u)}|}{\{(1 + \Gamma_E) \times (u - \Gamma_E)\}^{1/4}}$$

Here Γ_E is a coordinate on the contact path on the pinion tooth tip. It is calculated as follows

$$\Gamma_E = \frac{\sqrt{\{(d_{a_1}/d_{b_1})^2 - 1\}}}{\tan \alpha_{wt}} - 1$$

where

d_{a_1} = tip circle diameter of pinion (mm)

d_{b_1} = base circle diameter of pinion (mm)

(e) *Contact ratio factor*. By means of the contact ratio factor X_ε, the local flash temperature at the pinion tooth tip occurring for an assumed load distribution, e.g., as in Fig. 3.45, for the contact ratio $\varepsilon_\alpha = 1.0$, is to be converted to the flash temperature averaged over the contact path. For $\varepsilon_1, \varepsilon_2 < 1.0$

$$X_\varepsilon = \frac{1}{2\varepsilon_\alpha \times \varepsilon_1} \{0.7(\varepsilon_1^2 + \varepsilon_2^2) - 0.22\varepsilon_\alpha + 0.52 - 0.6\varepsilon_1\varepsilon_2\}$$

with

$$\varepsilon_{1.2} = \frac{z_{1.2}}{2\pi} \{\sqrt{(d_{a1.2}/d_{b1.2})^2 - 1} - \tan \alpha_{wt}\}$$

ε_1 or $\varepsilon_2 > 1.0$ but smaller than 2.0

$$X_\varepsilon = \frac{1}{2\varepsilon_\alpha \varepsilon_1} (0.18\varepsilon_{1,2}^2 + 0.7\varepsilon_{1,2}^2 + 0.82\varepsilon_{1,2} - 0.52\varepsilon_{2,1} - 0.3\varepsilon_1 \varepsilon_2)$$

where the first index applies to $\varepsilon_1 \geqslant 1.0$, and the second to $\varepsilon_2 \geqslant 1.0$. Here

ε_α = transverse contact ratio
ε_1 = tip contact ratio of pinion
ε_2 = tip contact ratio of gear

(*f*) *Contact factor.* The contact factor X_Q takes account of the unfavourable effect of the contact shock at maximum sliding velocity.
Pinion driving
where

$X_Q = 0.6$ for $1.5\varepsilon_1 \leqslant \varepsilon_2$
    $= 1.0$ for $1.5\varepsilon_1 > \varepsilon_2$

Gear driving
where

$X_Q = 1.0$ for $1.5\varepsilon_2 > \varepsilon_1$
    $= 1.6$ for $1.5\varepsilon_2 \leqslant \varepsilon_1$

(*g*) *Tip relief factor.* The following relationship applies

$$X_{C_a} = 1 + 1.55 \times 10^{-2} \varepsilon_{max}^4 C_a$$

with

ε_{max} = maximum tip contact ratio of pinion or gear (ε_1 or ε_2)
C_a = tip relief (μm)

Here
Driving pinion

$$C_a = \min\left(\frac{C_{a1}}{C_{eff}}\right) \text{where } \varepsilon_1 > 1.5\varepsilon_2$$

$$= \min\left(\frac{C_{a2}}{C_{eff}}\right) \text{where } \varepsilon_1 < 1.5\varepsilon_2$$

Driving gear

$$C_a = \min\left(\frac{C_{a2}}{C_{eff}}\right) \quad \text{where } \varepsilon_2 > 1.5\varepsilon_1$$

$$= \min\left(\frac{C_{a1}}{C_{eff}}\right) \quad \text{where } \varepsilon_2 < 1.5\varepsilon_1$$

The maximum effective tip relief is found from

$$C_{eff} = \frac{K_A(F_{t/b})}{C_\gamma}$$

with

C_γ = contact spring stiffness ($N/mm/\mu m$)

For straight gears

$$C_\gamma = C$$

The decisive individual spring stiffness C is calculated as follows

$$1/C = 0.04723 + 0.15551/z_{n_1} + 0.25791/z_{n_2} - 0.00635x_1$$
$$- 0.11654 \times x_1/z_{n_1} - 0.00193x_2 - 0.24188x_2/z_{n_2}$$
$$+ 0.00529x_1^2 + 0.00182x_2^2$$

with

$$z_{n_{1,2}} = \frac{z_{1.2}}{\cos^3 \beta} = \text{number of teeth in normal profile}$$

$$x_{1,2} = \text{addendum modification coefficient}$$

and

$$C_\gamma = C(0.75\varepsilon_\alpha + 0.25)$$

(*h*) *Mass temperature.* The mass temperature ϑ_M depends, as a first approximation, on the oil temperature and the average surface temperature. Then

$$\vartheta_M = (\vartheta_{oil} = C_1\vartheta_{fla\ int})X_s$$

with

ϑ_{oil} = spray or oil sump temperature ($°C$)
C_1 = auxiliary factor = 0.7 (determined empirically)
X_s = lubrication factor

(*i*) *Lubrication factor.* The lubrication factor X_s is intended to take account of the better heat transfer with splash lubrication compared with spray lubrication. The following values have been found in practice

splash lubrication $X_s = 1.0$
spray lubrication $X_s = 1.2$

3.3.5.4.4 *Determination of integral temperature limit*

The following equation has been derived to calculate the integral temperature limit for scoring

$$\vartheta_{\text{int s}} = \vartheta_{\text{MT}} + C_2\, X_{\text{W rel T}}\, \vartheta_{\text{fla int T}}\ (^{\circ}\text{C})$$

here

C_2 = weighting factor = 1.5
ϑ_{MT} = mass temperature in a test gear ($^{\circ}$C)
$\vartheta_{\text{fla int T}}$ = average flash temperature over the contact in a
  test gear ($^{\circ}$C)

$X_{\text{W rel T}}$ = microstructure factor

3.3.5.4.5 *Determination of influences on the integral temperature limit*

For the most important factors affecting the integral temperature limit, relevant data on the test gear set and the damage force achieved in it with the gear oil must be available.

(*a*) *Microstructure factor.* The microstructure factor $X_{\text{W rel T}}$ allows the differences in material and heat treatment between the test and the design gear sets.
Here

$$X_{\text{W rel T}} = X_{\text{W}}/X_{\text{W}_{\text{T}}}$$

with

X_{W} = microstructure factor of design gear
$X_{\text{W}_{\text{T}}}$ = microstructure factor of test gear

Table 3.13 gives information on X_{W}.

(*b*) *Mass temperature in a test gear.* The following relationship serves as an approximation for the mass temperature ϑ_{MT}, which occurs with the gear oil concerned in a test gear set, for example in the FZG test rig

$$\vartheta_{\text{MT}} = 0.23\, T_{1\text{T}} + 80$$

Table 3.13 Reference values for the microstructure factor X_W

Material	Microstructure factor X_W
Austenitic steels	0.45
Steels with above-average austenite content*	0.85
Steels with normal austenite content*	1.0
Steels with below-average austenite content*	1.15
Nitrided steels	1.5
Phosphated steels	1.25
Other cases (e.g., quenched and tempered steels)	1.0

* case-hardened steel

with

T_{1T} = pinion torque at scoring damage force level in the test gear (Nm)

The relationship between the mass temperature and the pinion torque in the FZG test A/8.3/90 as per DIN 51 354 is shown in Fig. 3.46.

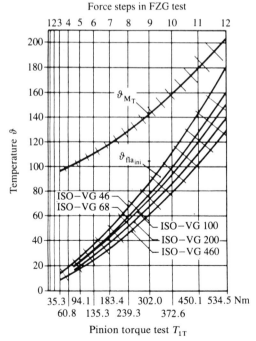

Fig. 3.46 Scoring temperature in the FZG test from Michaelis (34)

(c) *Average flash temperature in a test gear set.* The average flash temperature over the contact in a test gear can be calculated using the following approximation equation

$$\vartheta_{\text{fla int T}} = 0.2 T_{1T}\left(\frac{100}{v_{40}}\right)^{0.02}$$

with

v_{40} = viscosity/density ratio at 40°C (mm^2/s)

The relationship between the average flash temperature and the pinion torque in the FZG test A/8.3/90 to DIN 51 354 is also shown in Fig. 3.46.

3.3.5.4.6   Accuracy of the calculation procedure

Figure 3.47 indicates the accuracy of the calculation procedure on the basis of the integral temperature limit. It can be seen that virtually no scoring damage can be expected with calculated resistances greater than 2.0, while resistances less than 1.0 lead to a high probability of scoring. Gears with calculated resistances between 1.0 and 2.0 can be operated

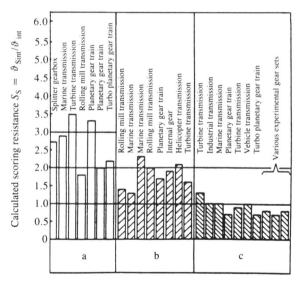

Fig. 3.47   Scoring resistance of in-service gears from Michaelis (34): **(a) no damage; (b) no damage when run-in well; (c) scoring**

without damage, if running-in is performed carefully, manufacturing asperities are worn down sufficiently, and there is satisfactory vertical and horizontal support of the gears.

3.3.5.4.7 Example

The procedure for calculating scoring resistance on the basis of the integral temperature limit is explained below with the aid of an example (**35**).

(a) *Given values.* The calculation is based on the following values:

– *Operating conditions*
  Rated output $P = 13\,000$ kW
  Rotational speed $U_1 = 6850$ min$^{-1} = 114.17$ s^{-1}
  Oil temperature with spray lubrication $\vartheta_{\text{oil}} = 50°C$
  Driving gear

– *Design data*
  Gear ratio $U = z_2/z_1 = 162/35 = 4.629$
  Centre distance $a = 710$ mm
  Normal module $m_n = 7$ mm
  Face width $b = 210$ mm
  Normal pressure angle at reference cylinder $\alpha_n = 20°$
  Helix angle at pitch circle $\beta = 13.8017°$
  Addendum modification coefficient $X_1 = X_2 = 0$
  Tip relief
    pinion $C_{a_1} = 35$ μm
    gear $C_{a_2} = 40$ μm
  Arithmetic mean roughness $R_a = 0.65$ μm (ground tooth flanks)

– *Geometric data*
  Pitch circle diameter
    pinion $d_1 = 252.284$ mm
    gear $d_2 = 1167.715$ mm
  Outside diameter
    pinion $d_{a_1} = 266.280$ mm
  Base circle diameter
    pinion $d_{b_1} = 236.237$ mm

  Transverse contact ratio $\varepsilon_\alpha = 1.7137$
  Overlap ratio $\varepsilon_\beta = 2.3866$

Total contact ratio $\varepsilon_y = 4.1003$
Transverse pressure angle at pitch cylinder $\alpha_{wt} = 20.5456°$

- *Operating data*
  Application factor $K_A = 1.3$
  Dynamic factor $K_V = 1.13$
  Transverse factor (flank) $K_{H\alpha} = 1.0$
  Width factor (flank) $K_{H\beta} = 1.35$
  Nominal peripheral force $F_t = 143.665$ N
  Peripheral velocity $v = 90.488$ m/s

- *Gear oil*
  Additive-free mineral oil ISO VG 46
  Kinematic viscosity at 40°C $v_{40} = 46$ mm^2/s
  Viscosity at 50°C (operating temperature) $\eta_{oil} = 30$ mPas
  Failure force level to DIN 51 354 7
  Torque at failure force level $T'_{1T} = 184$ nm

(b) *Calculation of influencing quantities*
- Helix factor for scoring

$$K_{By} = 1.3 \quad \text{where } \varepsilon_y \geqslant 3.5$$

- Effective peripheral force

$$\omega_t = 1.3 \times 1.13 \times 1.35 \times 1.0 \times 1.3 \times \frac{143\,665}{220}$$

$$\omega_t = 1683.555 \text{ N/mm}$$

- Total velocity at pitch point

$$v_{\Sigma_c} = 2 \times 90.488 \sin 20.5456$$
$$v_{\Sigma_c} = 63.514 \text{ m/s}$$

- Equivalent radius of curvature

$$\rho_c = 710 \times \sin 20.5456 \times \frac{4.629}{(4.629 + 1)^2 + 0.9745}$$

$$\rho_c = 37.355 \text{ mm}$$

with

$$\cos \beta_b = \cos \beta \times \frac{\cos \alpha}{\cos \alpha_t} = 0.9745$$

- Average coefficient of friction

$$\mu_m = 0.12\left(\frac{1683.555 \times 0.65}{30 \times 63.514 \times 37.355}\right)0.25$$

$$\mu_m = 0.0423$$

- Coordinate on contact path at pinion tooth tip

$$\Gamma_E = \frac{\sqrt{\{(266.280/236.237)^2 - 1\}}}{t_g\ 20.5456} - 1$$

$$\Gamma_E = 0.3877$$

- Geometry factor at pinion tooth tip

$$X_{BE} = 0.5\sqrt{(1 + 4.629)} \times \frac{|\sqrt{(1 + 0.3877)} - \sqrt{(1 - 0.3877/4.629)}|}{\{(1 + 0.3877)(|4.629 - 0.3877|)\}^{0.25}}$$

$$X_{BE} = 0.1682$$

- Tip contact ratio of pinion

$$\varepsilon_1 = \frac{35}{2\pi}\left\{\left(\frac{266.280}{236.237}\right)^2 - 1 - t_g\ 20.5456\right\} = 0.8095$$

$$\varepsilon_2 = \varepsilon_\alpha - \varepsilon_1 = 1.7137 - 0.8095 = 0.9042$$

$$\varepsilon_1 \text{ and } \varepsilon_2 < 1.0$$

- Contact ratio factor

$$X_\varepsilon = \frac{1}{21.7137 \times 0.8095}$$
$$\times \{0.7(0.8095^2 + 0.9042^2) - 0.22$$
$$\times 1.7137 - 0.52 - 0.6 \times 0.8092 \times 0.9042\}$$

$$X_\varepsilon = 0.2649$$

- Contact factor

$$X_Q = 1 \quad \text{where } 0.667 < \frac{\varepsilon_1}{\varepsilon_2} = 0.895 < 1.5$$

– Number of teeth in normal profile

$$Z_{n_1} = \frac{35}{\cos^3 13.8017} = 38.216$$

$$Z_{n_2} = \frac{162}{\cos^3 13.8017} = 176.883$$

– Individual spring rate

$1/c = 0.04723 + 0.155\,51/38.216 + 0.257\,91/176.883$
$1/c = 0.05276$
$c = 18.9547 \ N/mm/\mu m$

– Contact spring rate

$C_\gamma = 18.9547(0.75 \times 1.7137 + 0.25)$
$C_\gamma = 29.1007 \ N/mm/\mu m$

– Maximum effective tip relief

$$C_{eff} = \frac{1.3(143\,665/220)}{29.1007}$$

$C_{eff} = 29 \ \mu m$

– Tip relief for driving gear and $\varepsilon_2 < 1.5\varepsilon_1$

$C_a = \min(\frac{35}{29})$
$C_a = 29 \ \mu m$

– Tip relief factor

$X_{C_a} = 1 + 1.55 \times 10^{-2} \times 0.9042^4 \times 29$
$X_{C_a} = 1.3005$

– Microstructure factor for steel with retained austenite content below 10 percent

$$X_{W\,rel} = \frac{1.15}{1.0} = 1.15$$

(c) *Calculation of temperatures*
- Maximum flash temperature at pinion tooth tip

$$\vartheta_{\text{fla } E} = 0.0423 \times 50 \times 0.1682 \times \frac{1683.55^{3/4} \times 90.488^{1/2}}{710^{1/4}} \frac{1}{1.0 \times 13\,005}$$

$$\vartheta_{\text{fla } E} = 132.54°C$$

- Average flash temperature over contact

$$\vartheta_{\text{fla int}} = 132.54 \times 0.2649$$
$$\vartheta_{\text{fla int}} = 35.1°C$$

- Mass temperature for design gear set

$$\vartheta_M = (50 + 0.7 \times 35.1) \times 1.2 \quad \text{with } X_s = 1.2 \text{ with spray lubrication}$$
$$\vartheta_M = 89.5°C$$

- Integral temperature

$$\vartheta_{\text{int}} = 89.5 + 1.5 \times 35.1 \quad \text{with } C_2 = 1.5$$
$$\vartheta_{\text{int}} = 142.2°C$$

- Mass temperature for test gear set

$$\vartheta_{MT} = 0.23 \times 184 + 80$$
$$\vartheta_{MT} = 122.3°C$$

- Average flash temperature over the contact for test gear set

$$\vartheta_{\text{fla int T}} = 0.08 \times 184^{1.2}\left(\frac{100}{46}\right)^{(46-0.4)}$$

$$\vartheta_{\text{fla int T}} = 49.4°C$$

- Integral temperature limit for scoring

$$\vartheta_{\text{int s}} = 122.3 + 1.5 \times 1.15 \times 49.4$$
$$\vartheta_{\text{int s}} = 207.5°C$$

(d) *Calculation and evaluation of scoring resistance*
The following apply to scoring resistance

$$S_{FT} = \frac{\vartheta_{\text{int s}}}{\vartheta_{\text{int}}} = \frac{207.5}{142.2}$$

$$S_{FT} = 1.46$$

Taking account of the practical values of scoring resistance given in Fig. 3.47, it should be emphasized that, with $S_{FT} = 1.46$, the possibility of scoring can only be excluded if care is taken to ensure a good contact pattern by means of careful running-in. In no circumstances should the requirements for the scoring load capacity of the intended additive-free oil be reduced further, perhaps with a lower viscosity.

4 Gear Lubricants

4.1 INTRODUCTION

In principle, liquid, gaseous, consistent, and solid substances can be used as lubricants, so long as they possess the tribological characteristics required by the gear set. Table 4.1 provides a summary of the variety of potential types of lubricant. In this classification system, lubricants are differentiated on the basis of state of aggregation and consistency. Another method is to classify lubricants according to their applications; Table 4.2 contains such a list. A classification list of this sort can, of course, be made as detailed as required.

A lubricant which is ready for use consists usually of a base lubricant, which can be a liquid or a grease, or even a solid lubricant, and certain additives. The base liquids are mineral oils or synthetic liquids, or mixtures of the two.

4.2 REQUIREMENTS AND CHARACTERISTICS

In an extremely simplified form, the characteristics of the gear oil can be divided into selection characteristics and quality characteristics (Table 4.3). The selection characteristics include, for example, the viscosity, which has nothing to do with the quality of an oil. It affects the forces to be transmitted by the contact film between the tooth flanks which is under pressure (elasto-hydrodynamic lubrication), and the operating temperature.

The differentiation between primary and secondary characteristics does not represent an absolute quality ranking. Rather, the primary characteristics are the specific demands made by the gear set on the lubricant, such as wear and scoring behaviour. The secondary characteristics are the general characteristics which are also important. For example, neither the base oil nor the additives of a gear oil should attack the material of the seals.

The action of additives which protect against wear and scoring will be explained using the following scenario. Free surface energy is available as a result of the pressures and temperatures present on the tooth flanks. Thus, the additives in the gear oil can react chemically or chemically and physically with the flank surfaces, causing protective or separating layers

Table 4.1   Variety of possible lubricants

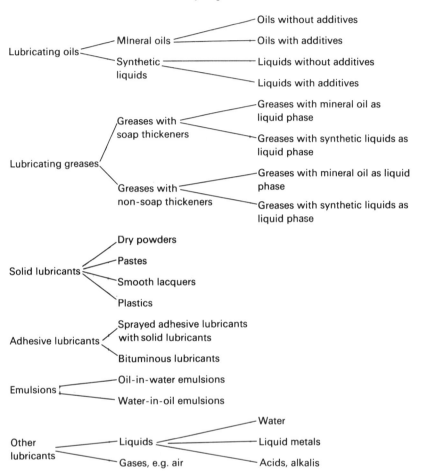

to form. Direct metallic contact between the base metals of the gears, which would otherwise cause increased wear or even scoring, is thus prevented. As the shear strength of these reaction layers is lower than that of the base material, they are continually removed and reformed, without any associated destruction of the flanks. Good lubrication of the mating gears and thus reliability of the gear set are, therefore, maintained by continuous, controlled slight abrasion, without any transition

Table 4.2 Classification of lubricants according to application

Spindle oils	Machine oils
Engine oils	Gear oils
Compressor oils	Turbine oils
Hydraulic oils	Refrigerator oils
Clock oils	Metal machining oils
White oils	Corrosion-inhibiting oils
Insulating oils	Heat conducting oils
Moulding oils	Black oils, axle oils
Greases	

Table 4.3 Characteristics of gear oils

	Quality characteristics	
Selection characteristics	*Secondary characteristics*	*Primary characteristics*
e.g., Viscosity, Pour point, Flash point	e.g., Viscosity/temperature behaviour; Chemical behaviour (corrosion, aggression towards non-ferrous metals); Thermal resistance to oxidation; Foam behaviour; High-temperature behaviour; Low-temperature behaviour; Cold behaviour; Compatibility with sealing components	e.g., Friction performance, Wear performance, Scoring performance, Running-in characteristics

to a state of wear. If the gear oil contains particularly reactive additives, such surface reactions can naturally occur at lower pressures and temperatures, that is, when they are not wanted. Corrosive wear can result. The characteristics of lubricants are determined by the base oil and the additives, and are described in classifications and specifications.

4.3 BASIC TYPES OF GEAR LUBRICANTS

In view of the significance of viscosity and EP characteristics of gear oils, a classification based on these aspects can be very useful for making an appropriate selection.

Lubricants for gears can be roughly divided according to the type and quantity of additives. In small gear sets, in particular if no complex measures can or should be taken to ensure good seals, lubricating greases are used, which usually have a very low consistency and are then called fluid gear greases. Low-speed, open gear sets are often lubricated with so-called adhesive lubricants, which can be applied with a brush or by spraying. However, mineral oils, with and without additives, constitute by far the majority of gear lubricants. As has already been discussed in detail, gear pairs can only achieve great scoring load capacities with additive-free oils if the viscosity is high.

To improve the sliding characteristics, organic fatty oils or fatty acids can be added to the pure mineral oil, which gives better adhesion of the oil to the flank surfaces. Such compounded oils are used in worm drives. For higher demands, the base oil is mixed with additives based on sulphur, chlorine, and phosphorus compounds, which form protective reaction layers on the tooth flanks under pressure and temperature loading. High-pressure gear oils, also known as light EP oils, can be used in highly stressed industrial gears and must be used in vehicle drives, e.g., change-speed gearboxes.

If so-called additives or similar substances, e.g., based on lead naphthenates, are added to the base oil in fairly large quantities, the result is very active EP oils, which are generally called hypoid gear oils. These oils can be aggressive and can sometimes have a deleterious effect on seal materials, so their use must remain restricted to gears subjected to extreme loadings. Therefore, they are not used in industrial gears, but they have made it possible for reliable offset bevel gears (hypoid gears) to be fitted as the final drives in vehicles. The modern trend is to develop gear oils with very balanced combinations of additives, which cover a wide range of applications and have no unwanted side effects. These so-called multi-purpose gear oils are intended to provide adequate protection against wear and satisfactory scoring resistance in hypoid gears, while being suitable for use in less severely loaded gears.

4.4   LIQUID GEAR LUBRICANTS − GEAR OILS

4.4.1   Introduction

In the case of liquid lubricants, we must differentiate between those based on mineral oils and those based on synthetic liquids. It would be technically more correct to distinguish between base oils which are made in a conventional manner (mineral oils) and those which are not (synthetic oils). The following chapters deal with the most important aspects of the structure of liquid lubricants, including the most significant combinations of characteristics to be derived, and the most important stages in their production, especially for lubricating oils based on mineral oils **(36)–(39)**.

4.4.2 Base oils made from mineral oils

4.4.2.1 Composition and structure

4.4.2.1.1 Fundamentals

The initial substance for the processing of the raw oil is the crude oil which has had coarse impurities, petroleum gas, and water removed. The main constituents of crude oil are hydrocarbons; these also contain organic oxygen, sulphur, and nitrogen compounds, as well as compounds with trace elements.

A typical analysis of a crude oil is given in Table 4.4. Of course, the precise composition of the crude oil depends on its provenance i.e., its origins.

4.4.2.1.2 Chemical composition of mineral oil

Mineral oils are, therefore, mixtures of hydrocarbons of various molecular sizes (the molecular weight spectrum ranges from about 50 to about 1000) and molecular structures. A hydrocarbon is a compound of carbon 'C' and hydrogen 'H'. There are numerous possible combinations for the formation of mineral oils, which are, therefore, very complex mixtures and not a precisely defined chemical substance.

There are certain basic types of hydrocarbon mixtures, which permit them to be classified as (a) chain-like or aliphatic hydrocarbons and (b) ring-shaped hydrocarbons. Both of these basic types can occur as saturated or unsaturated compounds, characterized by one or more double bonds. As will be discussed later, unsaturated compounds are particularly reactive.

Table 4.4 Typical analysis of crude oil

Component	Concentration
Carbon	80–85 (wt%)
Hydrogen	10–17 (wt%)
Sulphur	<7 (wt%)
Trace elements	<1 (wt%)
e.g., N, Cl, P, Na,	
Mg, V, O_2, etc.	

$$H- \quad \text{H-atom (Monovalent)}$$

$$-\overset{|}{\underset{|}{C}}- \quad \text{C-atom (Quadrivalent)}$$

$$H-\overset{H}{\underset{H}{\overset{|}{\underset{|}{C}}}}-\overset{H}{\underset{H}{\overset{|}{\underset{|}{C}}}}-\overset{H}{\underset{H}{\overset{|}{\underset{|}{C}}}}\cdots\cdots\overset{H}{\underset{H}{\overset{|}{\underset{|}{C}}}}-H$$

$$H-\overset{}{\underset{H}{\overset{|}{\underset{|}{C}}}} - \overset{H-\overset{|}{\underset{|}{C}}-H}{\underset{H}{\overset{|}{\underset{|}{C}}}}\cdots\cdots\overset{H}{\underset{H}{\overset{|}{\underset{|}{C}}}}-H$$

Basic components of hydrocarbon compound.

Alkane-n- paraffins (normal paraffins) C_nH_{2n+2}. The C atoms are arranged in a chain; free valencies are combined with H atom. The chains are all unbranched.

i-paraffins (iso-paraffins) C_nH_{2n+2}. More or less branched C chains are saturated with hydrogen.

$$H-\overset{H}{\underset{H}{\overset{|}{\underset{|}{C}}}}-\overset{H}{\underset{H}{\overset{|}{\underset{|}{C}}}}-\overset{}{\underset{H}{\overset{|}{\underset{|}{C}}}}\cdots\cdots\overset{H}{\underset{H}{\overset{|}{\underset{|}{C}}}}-H$$

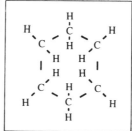

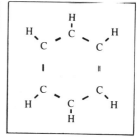

Alkene-olefin C_nH_{2n}. Chain-like or branched arrangement of C atoms with one or more double bonds; the remaining valencies are saturated with H atoms.

Cycloalkane-naphthenes C_nH_{2n}. Cyclic arrangement of 5—7 atoms. The C atoms are saturated with H atoms.

Benzene, basic elements of aromatics. Circular arrangement of 6 C atoms with 3 double bonds. The other 6 free valencies are saturated with H atoms.

Fig. 4.1   General structure of hydrocarbon compounds

Figure 4.1 shows the general structure of the basic hydrocarbons in schematic form. The basis types of hydrocarbons can be described as follows.

(a)   *Chain-like, saturated hydrocarbons* (*chemically unreactive*). The collective term for these hydrocarbons is alkanes or paraffins. The simplest hydrocarbon is, therefore, methane with the empirical formula CH_4.

Hydrocarbons with a chain-like arrangement of the C atoms are called normal or n-paraffins (unbranched chains). Their empirical formula is C_nH_{2n+2}.

If side chains branch off from the main chain, they are called iso-paraffins or i-paraffins rather than normal or n-paraffins. The empirical formula for n- and i-paraffins is the same.

Therefore, different molecules with the same empirical formula but a different structural formula can be composed with the same number of C and H atoms. They are called isomers, and they differ in the most important physical characteristics, such as boiling point or melting point and density. The greater the number of C atoms in a molecule, the greater the possible number of isomers. For butane (C_4H_{10}) there is only one isomer, for pentane (C_5H_{12}) there are two, and for nonane (C_9H_{20}) there are 35. As a hydrocarbon with 25 C atoms would have over 36 million possible isomers, the total number of possibilities increases towards infinity.

Paraffins with 1–4 carbon atoms are gaseous under normal conditions, with 5–17 C atoms they are liquid and with more than 17 C atoms they are solid.

(*b*) *Chain-like, unsaturated hydrocarbons* (*chemically reactive*). The collective name for these hydrocarbons is alkenes or olefins. As they contain at least one double bond, they are not stable chemically, but are reactive. The general empirical formula for the olefins is C_nH_{2n}.

If there are two double bonds, they are called dienes. Their empirical formula is C_nH_{2n-2}.

Compounds with double bonds always try to combine with other atoms, for instance oxygen. The addition of oxygen is called 'ageing'.

Like paraffins, olefins occur in straight chains (n-olefins) and in branched chains (i-olefins). The olefins are recognized by their 'ene' name ending, such as, for example, ethylene or ethene, propene, butene, etc.

(*c*) *Cyclic, saturated hydrocarbons* (*non-reactive*). The collective name for these hydrocarbons is cycloalkanes, cycloparaffins or naphthenes.

The C atoms are arranged in a ring rather than in a chain. Their general empirical formula is C_nH_{2n}.

Naphthenes can also have side chains and be extremely branched.

(*d*) *Cyclic unsaturated hydrocarbons* (*reactive*). The collective name for these hydrocarbons is cycloalkenes, cycloolefins, or unsaturated naphthenes, if there is only one double bond. The empirical formula is C_nH_{2n-2}.

(e)   *Cyclic   unsaturated   hydrocarbons   –   specifically   aromatics*
(*reactive*). The cyclic hydrocarbons in which double bonds alternate
with single bonds are a special case. These compounds are particularly
low in hydrogen and correspondingly chemically reactive. Their empirical formula is C_nH_n. The best known representative of these hydrocarbons is benzene C_6H_6.

Combinations of chain-like and ring-like, of saturated and unsaturated
hydrocarbon compounds are usually met. They are then called paraffin-
based, naphthene-based or aromatic, depending on which type of hydrocarbon determines the physical and chemical characteristics of the whole
compound. Table 4.5 shows, in a simplified form, the classification of a
mineral oil. If it is not possible to assign an oil definitely to one class, the
mineral oil is considered to be of mixed base.

4.4.2.1.3   Relationship between structure and characteristics
The molecular structure and the molecular weight of the hydrocarbons determine their chemical and physical behaviour. In particular the
following characteristics, which are important for base lubricating oils,
are dependent on them:
(a)   viscosity and viscosity–temperature behaviour;
(b)   state of aggregation, particularly the liquid range;
(c)   oxidative and thermal stability.

(*a*)   *Viscosity   and   viscosity–temperature   behaviour.*   Increasing molecular size, increasing chain branching, and increasing intermolecular cohesion forces increases viscosity, while increasing molecular flexibility
reduces viscosity. These relationships can be described by the following
function

$$V = \frac{C_n(D/L + \text{cohesion})}{\text{molecular flexibility}}$$

Here, V means the viscosity, C_n the number of carbon molecules, and
D/L the ratio of the diameter and the length of a molecule as a measure

Table 4.5   Classification of mineral oils

Classification	
Paraffin-based	> 75 percent paraffin
Naphthene-based	> 70 percent naphthene
Aromatic	> 20 percent aromatic

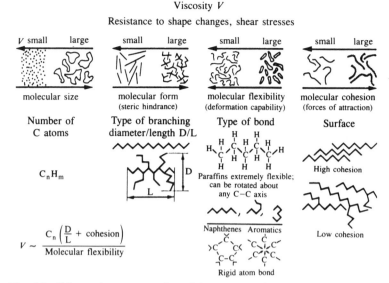

Fig. 4.2   **Schematic representation of the effects of the structure on viscosity**

of the type of branching. Figure 4.2 also illustrates this relationship schematically.

As temperature increases, the relative importance of these influences is displaced, with the exception of the molecular size which remains unchanged. Increasing temperatures lead to lattice expansion with increased molecular flexibility and thus to a decline in the influence of the chain branching and a reduction in the forces of cohesion. As different hydrocarbon structures react differently to the changing influencing factors, the individual hydrocarbon types have differing viscosity–temperature behaviours.

Naphthenes and aromatics demonstrate increasing molecular flexibility with temperature, resulting in a marked viscosity–temperature dependence, whereas the paraffins, which are already fairly flexible, are less affected, so their viscosity – temperature behaviour is not as marked as that of the other types of hydrocarbons.

(*b*)   *State of aggregation and liquid range.*   The dependence of the state of aggregation, primarily of the liquid range, on the molecular weight and the molecular structure is illustrated by Fig. 4.3. It can be seen that, as the molecular weight rises, the melting point and the boiling point

Lubrication of Gearing

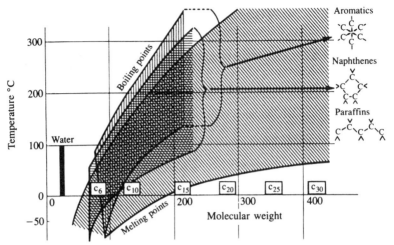

Fig. 4.3   Relationship between the liquid range of hydrocarbons and molecular weight and molecular structure

shift towards higher temperatures. With the same molecular weight, the liquid ranges of paraffins, naphthenes, and aromatics are shifted to higher temperatures in this sequence. The liquid range is the temperature difference between the melting point and the boiling point.

(*c*)   *Oxidative and thermal stability.*   A clear distinction must be made between oxidative and thermal stability. While oxidative stability means the resistance to a reaction with oxygen, thermal stability is the resistance of a molecule to decomposition due to the effect of heat.

Because compounds with double bonds have a greater tendency to react with oxygen than molecules with single bonds, paraffins and naphthenes are most resistant to oxidation than olefins and aromatics. On the other hand, chain-like compounds are made to vibrate more by the effect of heat, causing the molecules to decompose if a certain limit is exceeded, than ring-shaped compounds, which can store more thermal energy before the molecule decomposes. Naphthenes and aromatics, therefore, have greater thermal stability than paraffins. A comparison of paraffins and aromatics thus produces the following interesting findings.

While paraffins are fairly stable oxidatively, they are relatively susceptible to heat. On the other hand, aromatics are oxidatively susceptible, but thermally stable. This has repercussions on the applications of

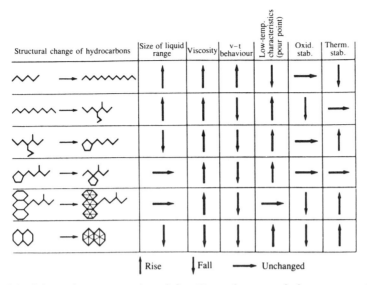

Fig. 4.4 Schematic representation of the effects of structural changes on various characteristics of hydrocarbons

mineral oils. It is always necessary to consider which characteristic is most important.

The effects of structural changes in the hydrocarbon on the liquid range, viscosity, viscosity–temperature behaviour including low-temperature behaviour, oxidative and thermal stability are shown schematically in Fig. 4.4.

4.4.2.2 Production of mineral oil-based base oils

4.4.2.2.1 General

Methods for the processing and treatment of crude oils for producing lubricating oils can be classified under the following basic processes

(a) physical separation
(b) chemical conversion
(c) addition of additives

Using another method of classification, one can differentiate them according to

(a) distillation

(b)   refining
(c)   addition of additives

The most important individual technological stages in modern lubrication oil manufacture are (5):

(a)   Distillation: atmospheric distillation
                       vacuum distillation
(b)   Extraction: sulphuric acid treatment
                       solvent extraction
                       (i)   sulphur dioxide extraction
                       (ii)   duo-sol process
                       (iii)   furfural extraction
                       (iv)   propane deasphalting
(c)   Dewaxing: solvent dewaxing (methyl ethyl ketone (MEK) process)
      Hydrogen treatment: hydrofinishing

Figure 4.5 shows a schematic flow diagram of a lubricating oil refinery (40).

4.4.2.2.2   Distillation

(*a*)   *Atmospheric distillation.*   As already mentioned, mineral oil consists of various hydrocarbons which have their own boiling points (see also Fig. 4.6 as an example). The hydrocarbon mixture which constitutes mineral oil thus has no boiling point, but a boiling range, within which the individual hydrocarbons boil.

During heating to certain temperatures, therefore, hydrocarbons having boiling points at those temperatures vaporize. By collecting these individual vapour components, the mineral oil, i.e., a hydrocarbon mixture, can be separated into its constituent parts. The crude oil is not separated into all the hydrocarbons, however, but only into coarse groupings, which are called fractions. Atmospheric distillation is, therefore, the physical separation of the crude oil by boiling ranges. Figure 4.7 (top) shows a flow diagram of atmospheric distillation.

The crude oil passes from the tank into the tube still, where it is heated to about 360 °C. The hydrocarbons with boiling temperatures below this temperature vaporize and pass up into the fractionating tower. Here they rise, water vapour flowing in the same direction entraining the light components, while the heavy, unvaporized residue remains in the sump. As the hydrocarbon vapours rise, they cool. Then those whose temperatures fall below their boiling points condense again.

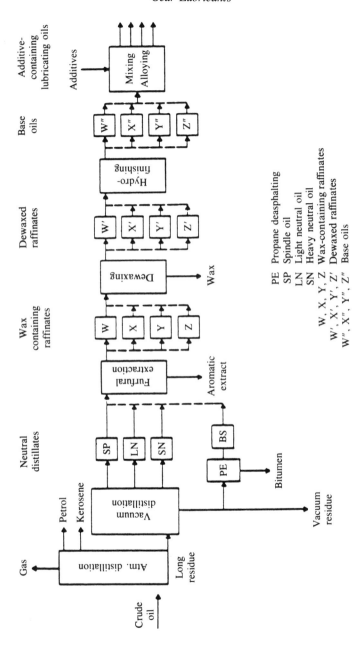

Fig. 4.5 Schematic flow diagram of a lubricating oil refinery (40)

PE   Propane deasphalting
SP   Spindle oil
LN   Light neutral oil
SN   Heavy neutral oil
W , X , Y , Z   Wax-containing raffinates
W' , X' , Y' , Z'   Dewaxed raffinates
W" , X" , Y" , Z"   Base oils

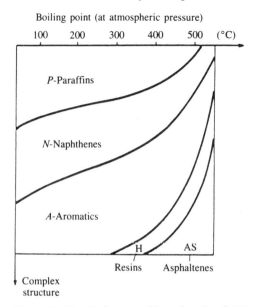

Boiling point (at atmospheric pressure)

Fig. 4.6   Chemical composition of crude oil (40)

In higher temperature zones, the condensate present which is in liquid form, i.e., a distillate fraction, can be drawn off. These temperature zones are defined by so-called bubble trays. The oil vapours rise up through the individual bubblers and must flow through the distillate on the bubble tray. The still vaporous components leave the distillate again and rise further up the fractionating tower, while those fractions whose boiling range corresponds to that of the liquid on the bubble tray remain there, and flow through an overflow back down to the tray underneath (rectification). The rigorousness of the separation of the crude oil thus depends on the number of bubble trays in the fractionating tower. The proportion of lighter boiling hydrocarbons in the remaining mixture increases from bottom to top in the tower.

Because heat is continually introduced into the system by incoming crude oil vapours, the fractionating tower must be cooled from the head down, to maintain the temperatures corresponding to the fractions to be drawn off at each bubble tray. Head cooling is performed by gasoline, which flows counter to the rising oil vapours. In this way the temperature gradient in the tower is controlled. Table 4.6 contains an example of the fractions drawn off in atmospheric distillation.

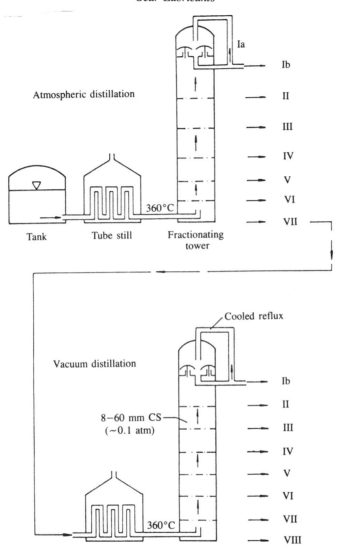

Atmospheric distillation

Ia

Ib

II

III

IV

V

VI

VII

360°C

Tank   Tube still   Fractionating
tower

Vacuum distillation

Cooled reflux

Ib

II

8-60 mm CS
(~0.1 atm)

III

IV

V

VI

VII

360°C

VIII

Fig. 4.7 Flow diagram of atmospheric distillation and vacuum distillation

(*a*) *Vacuum distillation.* As was seen in the above section, no lubricating oil fractions are produced by atmospheric distillation. If they are
to be obtained by distillation from suitable distillation residues – and not
all crude oils are suitable – either the temperature would have to be

Table 4.6   Fractions produced in atmospheric distillation (example)

Product		Boiling range (°C)
Ib	Light gasoline	25–200
II	Heavy gasoline	
III	Kerosene	180–250
IV	Light gas oil	200–360
V	Heavy gas oil	
VI	Unvaporized residue	>360

raised or the boiling temperature of these fractions would have to be lowered by reducing the pressure. The first method is not viable, because, at higher temperatures, the thermal stability of the hydrocarbons is not sufficient to prevent decomposition of the molecules. There would thus be no separation by boiling ranges, only a change in the size of the molecules (thermal cracking), and thus no lubricating oil fractions would be obtained.

Reducing the pressure to about 0.01–0.08 bar reduces the boiling temperature of hydrocarbons in the atmospheric residue, which are to be obtained as lubricating oil fractions, to such an extent that they vaporize at about 360°C. Bubbling in water vapour also reduces the boiling point. The bottom section of Fig. 4.7 gives a flow diagram of vacuum distillation. The oil is reheated to about 360°C. Instead of bubble trays, so-called sieve plates are used in vacuum distillation.

The feedstock used for vacuum distillation is the residue from atmospheric distillation, derived from a crude oil containing suitable lubricating oil fractions. Table 4.7 gives an example of the fractions resulting from vacuum distillation. The fractions produced in vacuum distillation are called distillates. In the past these were used for low-grade lubrication tasks; today they are practically never used as lubricating oils without further processing.

Depending on the type of crude oil used for processing, the vacuum residue can be further processed to produce brightstock or bitumen, or it can be added as a component to heavy heating oil. Figure 4.8 contains a greatly simplified representation of the separation of crude oil into fractions by distillation.

Table 4.7 Fractions produced in vacuum distillation (example)

Product	Viscosity range (mm²/s)
Spindle oil	
II light	12–21 at 20°C
III heavy	37–70 at 20°C
Machine oil	
IV light	21–45 at 50°C
V heavy	53–90 at 50°C
Cylinder oil	
VI light	12–21 at 100°C
VII heavy	21–76 at 100°C
VIII residue	

4.4.2.2.3 Refining

(*a*) *Purpose of refining.* The distillates contain a variety of unwanted constituents which must be removed before the oil can be used as lubricating oil. Table 4.8 lists some of these substances, together with their deleterious effects.

For many areas of application, a chemical conversion, i.e., a change in the hydrocarbon distribution, is necessary to obtain optimal lubricating oils. So aromatics are undesirable in lubricating oils due to their proneness to ageing, whereas they may be useful in heat carrier oils. Paraffins with longer chains should be removed from lubricating oils to be used at low temperatures, to lower the pour point. Removal of unwanted constituents from the distillates and chemical conversion of the composition of the oils is regarded as the actual refining process.

Table 4.8 Undesirable constituents in distillates

Substances	Effects
Acids	Corrosion, poor storage qualities
Resins, asphalts	Poor storage qualities, ageing, sediments
Sulphur compounds	Corrosion, unpleasant odour, poor storage qualities
Unstable compounds	Ageing
Waxes	Poor low-temperature characteristics

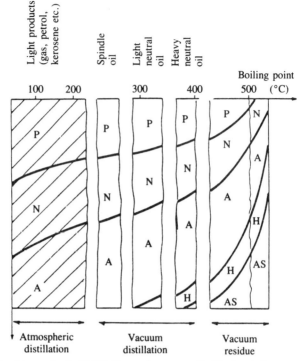

Fig. 4.8 Atmospheric distillation and vacuum distillation (schematic) (40)

(b) *Sulphuric acid refining and treatment with bleaching clay.* One of the oldest and now least used methods of refining is with sulphuric acid. Figure 4.9 shows a flow diagram of this process. In the upper part of the diagram it can be seen that the distillate is mixed with concentrated sulphuric acid. They are then agitated for several hours in tanks; during this process air comes into close contact with the mixture. The sulphuric acid accelerates the polymerization, and anticipates in the refining process events which would otherwise occur during later use of the oil under the influence of oxygen, heat, and light. The acid causes the formation of polymerization products from the resins and asphalts present in the distillate. The acid sludge which is produced is insoluble in oil and can, therefore, be separated from the oil by settling. Some of the sulphur compounds and unstable compounds also react with the sulphuric acid.

Annihilation of the acid sludge or recovering the sulphuric acid from it is expensive. This is probably one of the reasons for the decline in the importance of this method. Aromatics in the distillate also react with the sulphuric acid, producing oil-soluble sulphonic acids which sour the oil. Potassium hydroxide solution is added to neutralize it.

The sulphuric acid treatment is usually followed by treatment with bleaching clay (see Fig. 4.9). The bleaching clay, in this case aluminium hydrosilicate, acts as an adsorbent. This removes other unwanted substances, especially the pigmented constituents, from the oil. As a result the oil becomes lighter in colour.

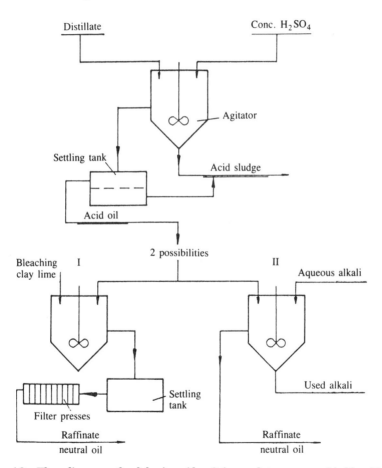

Fig. 4.9 Flow diagram of sulphuric acid refining and treatment with bleaching clay

After the oil is mixed with the bleaching clay at a temperature of about 70–80°C and allowed to settle, the remaining mixture is passed through a filter press. As the bleaching clay which is filtered out contains a certain amount of oil in addition to the unwanted constituents, this process involves a certain loss.

The entire process from the sulphuric acid treatment through the potassium hydroxide stage to the bleaching clay stage is not very selective; it is a fairly severe refining process. It has therefore, proved advantageous in producing very pure oils, including for instance medicinal white oils, transformer oils, and turbine oils. Depending on the degree of refining, a loss of 45–80 percent to be expected with medicinal white oils and a loss of 20–35 percent has to be expected with turbine oils.

In view of the disadvantages of sulphuric acid refining, characterized by the difficulties of disposing of the acid sludge and the cost of the chemicals used, and also by the fact that this method is not very selective, attempts to develop alternative refining processes began some time ago. These include the so-called extraction methods. The basis of these methods is to dissolve the unwanted and/or wanted components in the distillate with solvents and to separate the resulting phases. Some of these methods are described in the following chapters.

(c) *Sulphur dioxide extraction.* A frequently used method is sulphur dioxide extraction, the so-called Edelneau method. First the distillate must be dried because traces of moisture are added with the SO_2, and the mixture is thoroughly agitated. Because sulphur dioxide has a particular solvency for aromatic hydrocarbons, an extract phase forms which mainly contains the aromatics dissolved in the SO_2, and an oil phase forms which also still contains SO_2. Distillation must then be used to dissolve the SO_2 out of both the extract phase and the oil phase. Figure 4.10 shows a flow diagram of sulphur dioxide extraction (41).

The extraction is performed at temperatures between −12 and −50°C, depending on what feed material is used and what purity of raffinate is required. The solubility of aromatic hydrocarbons in SO_2 is particularly marked at low temperatures. The more SO_2 is added to the distillate, the higher is the level of raffination. However, as the level of raffination rises, the raffination losses also naturally increase; these can be between 22 and 55 percent.

Some sulphur compounds are dissolved by the sulphur dioxide and removed from the oil. By adding some benzene it is possible to remove some of the other, unstable, compounds from the distillate.

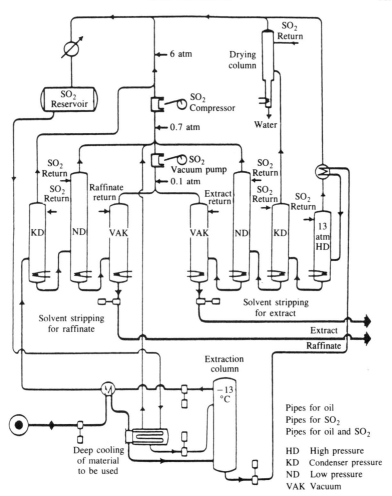

Fig. 4.10   Flow diagram of a sulphur dioxide extraction plant (41)

With some raffinates, the SO_2 extraction is followed by a mild sulphuric acid treatment. This is mainly necessary if the raffinate is to be made lighter in colour. The fact that aromatics are dissolved out with this process means that the viscosity–temperature behaviour of the oil is improved.

(*d*) *Furfural extraction.* Other solvents act very selectively with aromatic hydrocarbons, and also with naphthenic hydrocarbons: for example, furfural is frequently used for solvent extraction. It has a greater solvency with regard to cyclic hydrocarbons than paraffins or isoparaffins. If distillates are mixed with furfural and the heavy furfural extract phase is then separated from the lighter furfural-containing paraffinic oil phase, a raffinate low in aromatics, with a good viscosity–temperature characteristic and an aromatic extract can be obtained after the furfural is distilled off. The selectivity of the process can be increased, at the cost of yield, by raising the temperature during the extraction process and/or by increasing the quantity of furfural in the distillate. The following figures can be given as an example. With an initial viscosity index of 10–20, the VI can be increased to about 60–70 with a yield of about 70 percent. If a VI of 90–100 is required, as is necessary for base oils for engine oils, the yield falls to about 50 percent. Figure 4.11 gives a schematic representation of furfural extraction. Some of the heavy compounds (tar) had already been removed from the feed products for the

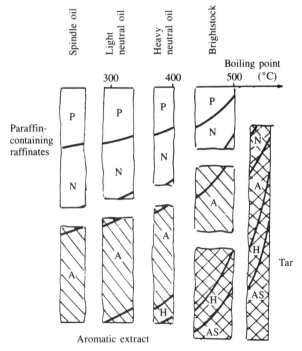

Fig. 4.11   Furfural extraction (schematic) (40). P = paraffins, N = naphthenes, A = aromatics, H = resins, AS = asphalts

furfural extraction by means of another process (e.g., propane extraction).

(*e*) *Duo-sol extraction.* A modern, elegant method is the so-called Duo-sol method, in which two solvents are used. Figure 4.12 shows a flow diagram of this process.

One solvent, e.g., propane, is used to dissolve the constituents of the distillate which are wanted, for example the paraffinic hydrocarbons. The other solvent, e.g., a phenol/cresol mixture, is used to dissolve the unwanted components of the distillate, for example aromatic hydrocarbons or certain unstable compounds. By distillation of both the solvate phase and the extract phase, the solvents are separated again and can be recycled into the process.

(*f*) *Propane de-asphalting.* The viscosity of lubricating oils depends on their boiling range; the higher the boiling range, the more viscous is the oil. There is still, however, an upper limit for the boiling range of distillates containing paraffins; this is the maximum temperature to which the distillate residue itself can be heated under vacuum without the occurrence of cracking reactions. This fact also limits the maximum viscosity of lubricating oils which can be obtained by further processing using distillation methods. If lubricating oil fractions of higher viscosity are required, they must be extracted from the asphalt-containing vacuum residue by means other than distillation.

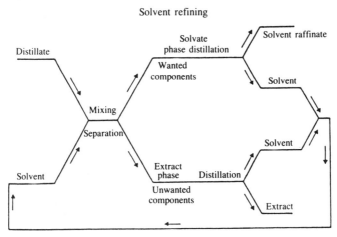

Fig. 4.12   Flow diagram of the Duo-sol process

It is known that liquid propane dissolves the oily and paraffinic components of this residue easily, while the asphalt remains.

This is the basis on which the so-called propane deasphalting process was developed, which is often used to supplement vacuum distillation to produce high-viscosity lubricating oil fractions. Such base oils obtained from the vacuum residue, i.e., not by further processing of distillates, are called brightstock. Figure 4.13 gives a schematic representation of propane deasphalting.

(*g*) *De-waxing.* The paraffin-based distillates, in particular, contain high molecular weight, straight-chain paraffins, which are dissolved in the oil at high temperatures, but increasingly crystallize out as the temperature falls. Then they agglomerate and cause the oil to solidify. If they are removed from the oil, the pour point is lowered. Figure 4.14 shows a flow diagram for solvent dewaxing.

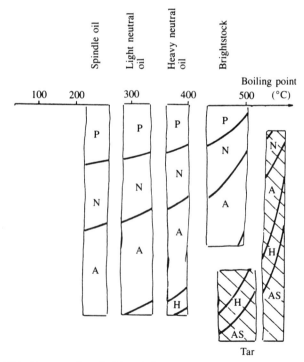

Fig. 4.13   **Propane deasphalting (schematic) (40); (for key see Fig. 4.11)**

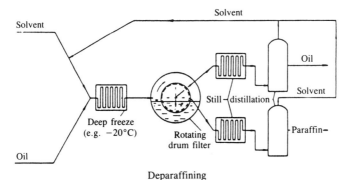

Deparaffining
Fig. 4.14 Flow diagram of solvent dewaxing

The oil is combined with a mixture of methyl ethyl ketone and toluene (of course, other solvents may be used) and then cooled. The wax separates from the crystalline mass and is removed from the oil, for example with a rotary drum filter. The oil/wax mixture is drawn through the filter cloth by vacuum in one of the chambers of the rotary drum, and the wax extract is deposited on the cloth. This extract is then repulsed by higher pressure in one of the chambers. Washing the filtered wax extract with solvent and recrystallization produces an oil-free wax. The solvent from the filtrate, that is from the dewaxed oil, is reclaimed for further use by distillation. What remains is a dewaxed, solvent-free lubricating oil fraction, which can be processed by mixing with other fractions or by redistillation to produce base oils of the required viscosity. Figure 4.15 shows a schematic representation of solvent dewaxing.

(*h*) *Refining by hydrogenation.* Recently, catalytically controlled hydrogenating refining processes have become increasingly important. For 'fining purposes' and for finishing treatment, the mild-acting catalytic hydrogenation processes known collectively as hydrofining or hydrofinishing have displaced acid refining. On the other hand, in addition to solvent refining, the destructive catalytic hydrogenation processes known as hydrocracking, which permit extensive, directed intervention in the chemical structure of the distillate used, are also finding increasing application. Furthermore, they can help to prevent the occurrence of very heavy carbon-rich by-products. From the chemical engineering point of view, hydrocracking represents a transition to the production of synthetic lubricating oil components.

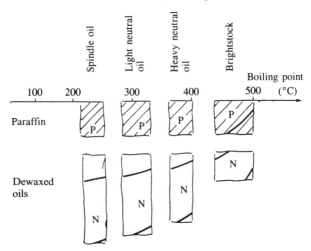

Fig. 4.15   Solvent dewaxing (schematic) (40); (for key see Fig. 4.11)

With excess hydrogen, the sulphur in sulphur compounds is converted to H_2S and thus removed from the oil. Furthermore, nitrogen compounds are converted and certain oxygen compounds are degraded. Diolefins and olefins, that is, unsaturated compounds, are hydrogenated to saturated hydrocarbons, so the oil is stabilized.

4.4.2.2.4   Secondary raffinates

Secondary raffinate oils are produced from used oils, mainly used engine oils. Sometimes it is possible to separate the non-recyclable constituents and impurities such as used additives, sludge, abraded material etc. almost completely from the base oil using fairly simple methods, such as acid refining and distillation, and to obtain base oil of high purity. Secondary raffinates quite often have characteristics which are equally as good as those of fresh oil, and occasionally they are even superior. As a rule they are, therefore, to be regarded as equivalent base oil components.

The main problem in recycling used oils is collecting the oils in the least expensive manner; unfortunately, metal machining oils are often mixed with them, and there is no easy way of removing additive residues from these.

4.4.2.2.5   Available base oils

On the basis of the processing methods described above, the following base oils are available

distillates
acid raffinates
solvent raffinates
secondary raffinates
brightstocks

While distillates are practically no longer used without further processing, the acid raffinates are used in certain applications requiring highly refined, light-coloured oils. An example is white oils, which are used to manufacture pharmaceutical white oils for medicaments and cosmetics or to make lubricating oils for the food industry.

The standard base oils for manufacturing the various lubricating oils are now the solvent raffinates. Brightstocks are needed as components for viscosity modification and to achieve specific characteristics in lubricating oils for certain applications. Figure 4.16 gives a schematic representation of the yield of lubricating oils obtainable from crude oil; it is approximately 10 percent (**40**).

The high-quality base oils available after refining are defined chemically by hydrocarbon distribution and physically by density, viscosity,

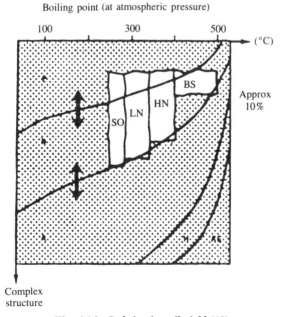

Boiling point (at atmospheric pressure)

Complex
structure

Fig. 4.16   Lubricating oil yield (40)

and flow characteristics including dependence on temperature and pressure, pour point and other characteristics. Raffinates are also designated as additive-free mineral oils or lubricating oils.

4.4.3 Synthetic base oils

4.4.3.1 Reasons for the use of synthetic lubricating oils
Chemical synthesis offers interesting possibilities for manufacturing particularly high-quality lubricating oil components for special requirements or to provide special characteristics. Here it should be borne in mind that lubricants manufactured on the basis of synthetic liquids do not represent an all-encompassing substitute for mineral oil-based lubricants, but are to be regarded as a supplement for certain applications.

The cost factor, as well as other possible disadvantages of synthetic liquids, reinforces this point of view. It makes it preferable to try to solve a lubrication problem in the first instance with mineral oil-based lubricants, with suitable additives or active substances. Only when the limits for the application of mineral oils have been exceeded are synthetic liquids considered as possible problem-solvers.

The reasons for selecting synthetic lubricants can be summarized as follows (**42**):

-   the characteristic required *cannot* be obtained with additives;
-   the demands on the required characteristic *cannot* be met or cannot be met *economically* with additives.

The advantages of synthetic liquids as the basis for lubricants include better performance in respect of the properties shown in Table 4.9.

As already mentioned, there is no synthetic liquid which combines all the listed advantages. When selecting a synthetic lubricant, it is necessary to rank the values of the characteristics required or expected. The advan-

Table 4.9 Advantages of synthetic liquids

Thermal stability
Oxidative stability
Viscosity–temperature characteristic
Flow characteristics at low temperatures
Volatility at high temperatures
Temperature application range
Resistance to radiation
Low flammability

Table 4.10 Disadvantages of synthetic liquids

Hydrolytic behaviour
Corrosion behaviour
Toxic behaviour
Compatibility with other additives
Solubility for additives
Availability:
in principle
in specific viscosity situations
Cost

tages of synthetic liquids are countered by certain disadvantages, however, some of which are listed in Table 4.10.

Of course, one should not expect to find the totality of all disadvantageous properties in every synthetic liquid. Cost is an exception here, however, as all synthetic liquids are more expensive than mineral oil-based lubricants.

4.4.3.2 Types of synthetic liquids

Synthetic liquids can be assigned to certain types or groups on the basis of their chemical composition. Certain classification systems can be used. Table 4.11 shows a comparison of mineral oil-based and synthetic liquids. The main distinction is between liquids obtained by distillation and refining, and those obtained by chemical reactions. In Table 4.12 the

Table 4.11 Classification of mineral oil-based and synthetic liquids

Mineral oil-based lubricants (obtained by distillation and refining)		Synthetic lubricants (obtained by chemical reactions)	
Classical techniques	*Modern techniques*	*Synthetic hydrocarbons**	*Synthetic liquids*
Acid refining	Hydrocracking	Polyalphaolefins (PAO)	Carbonic acid esters
Solvent extraction	Hydrotreating		diesters
Dewaxing		Polyisobutenes (PIB)	polyolesters
		Dialkyl aromatics	(complex esters)
			Polyalkylene oxides
			Phosphate esters
			Silicon oils
			Diphenyl ether
			Halogenated HC

* Contain only C and H

Table 4.12   Classification of synthetic liquids on the basis of their chemical composition

Synthetic liquid	Composition
Synthetic hydrocarbons	
Poly-alpha-olefins (PAO)	
= synth. hydrocarbon fluids (SHC)	C, H
Alkylbenzenes	C, H
monoalkylbenzene	
dialkylbenzene	
Polyglycols (polyethers)	
Copolymers of ethylene oxide, propylene oxide etc.	C, H, O
Carbonic acid esters	
Monocarbonic acid ester	C, H, O
Dicarbonic acid ester	C, H, O
Polyolester	C, H, O
Phosphoric acid esters	
Triaryl-, trialkyl- and alkylarylesters	C, H, O, P
of phosphoric acid ester	
Silicone oil	
Dimethylsilicone oil	C, H, O, Si
Methylphenylsilicone oil	C, H, O, Si
Phenylsilicone oil	C, H, O, Si
Polyphenylether	
Meta- and para-polyphenylether	C, H, O
Polyfluoroalkylether	C, F, O

synthetic liquids are differentiated and classed according to their chemical composition.

4.4.3.3   Manufacture of synthetic liquids

4.4.3.3.1   Principles

The next few chapters deal with some aspects of the manufacture of the most important synthetic fluids (43).

While mineral base oils represent a mixture of different chemical compounds, synthetic fluids are normally obtained from uniformly formed, defined compounds by means of a chemical reaction, which can be controlled by pressure, temperature, and volume ratios. The process leading to the desired end product often involves several stages, which always require cleaning of the intermediate product.

The price of a synthetic liquid is, therefore, composed of the cost of the raw materials and the cost of the individual stages of the reaction.

As in the chemical industry in general, the raw materials are obtained in large part from crude oil and natural gas by thermal decomposition. Among the most important petrochemical compounds are ethylene and its derivatives (Fig. 4.17).

Almost all synthetic lubricants can be manufactured from this basic substance by means of appropriate reactions (Fig. 4.18).

4.4.3.3.2 Synthetic hydrocarbons

This class of compounds contains paraffin-like liquid substances which consist only of carbon and hydrogen, and, as a result of the chain length, degree of branching and the position of the branches, possess a certain viscosity, a high viscosity index and a low pour point (**42**)–(**44**).

(a) *Polyalphaolefins* (*PAO*). Catalytically controlled polymerization and copolymerization of low-molecular, gaseous olefins which are obtained in refineries using cracking processes, can be used to produce polyolefin oils which consist mainly of isoparaffins with a small number of short and terminal side chains.

The individual reactions do not give a one hundred percent yield. Therefore, cleaning processes must be interposed between the individual reaction stages to return unconverted components to the reaction tower and to remove unwanted by-products. Figure 4.19 gives a flow diagram of the manufacture of synthetic hydrocarbons.

(b) *Alkylbenzenes.* The class of synthetic hydrocarbons includes the alkylated aromatics, which can be made by the addition of olefin to a benzene nucleus and which have a wide range of uses in the chemical industry. So ethyl benzene is used as the initial product for making

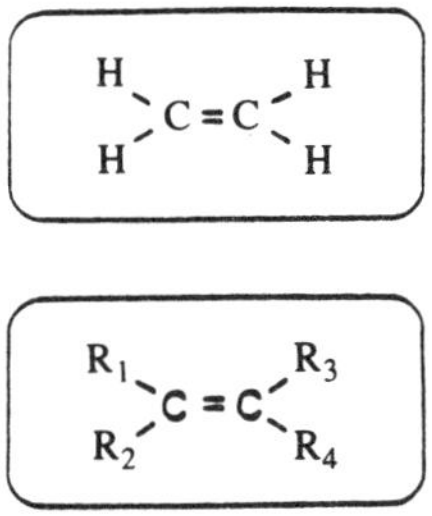

Fig. 4.17 Ethylene and ethylene derivative

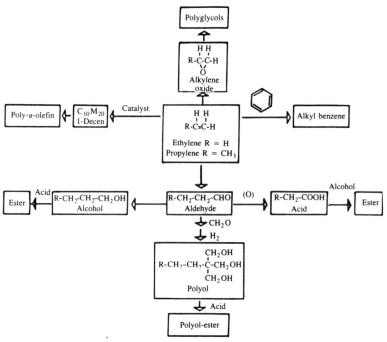

Fig. 4.18   Ethylene derivatives as initial products for manufacturing synthetic liquids (43)

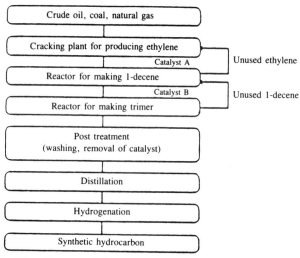

Fig. 4.19   Flow diagram of the manufacture of synthetic hydrocarbons (43)

140

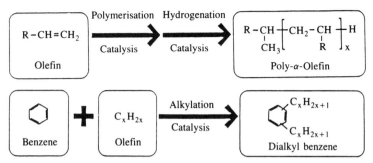

Fig. 4.20   Flow diagram of the manufacture of polyalphaolefins and alkyl-benzenes (43)

styrene, sulphonated alkyl benzenes are used as detergents, and 90 percent of phenol production is via the oxidation of cumene, which is obtained by the reaction of propylene with benzene. Figure 4.20 gives a flow diagram for such a reaction.

4.4.3.3.3   Polyglycols (polyethers)

Polymers of ethylene oxide, propylene oxide or higher alkyl ethers are called polyalkylene glycols. As with the synthesis of polyalphaolefin, the initial material is ethylene or an ethylene derivative, which is oxidized to a cyclic ether by reaction with oxygen.

By base-catalysed polymerization of a cyclic alkylene ether with an alcohol as the initiation molecule, long-chain compounds are formed, the chain length or molecular weight of which can be manipulated to influence the viscosity of the end product. If ethers of the type $R = H$ and $R = CH_3$ are combined, ethylene propylene copolymers are produced; the ratio of ethylene oxide (EO) to propylene oxide (PO) has a decisive influence on the water solubility of the product. Figure 4.21 gives a flow diagram for such a reaction.

4.4.3.3.4   Esters

Esters are organic, oxygen-containing compounds, which can be obtained by the reaction of an alcohol with an organic acid (**43**)(**46**)(**47**).

(*a*)   *Carbonic acid ester.*   If the acid used is a component which has two carboxyl groups

$$-\underset{\underset{O}{\overset{\parallel}{}}}{C}-OH \quad \text{(carboxyl group)}$$

the result is a diester.

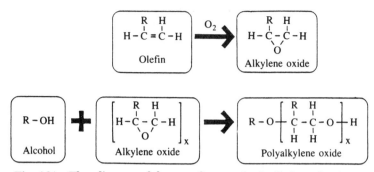

Fig. 4.21   Flow diagram of the manufacture of polyalkylene glycol (43)

The various known diesters differ either in the alcohol or the acid component, Fig. 4.22 gives a flow diagram of such a reaction. To produce them, defined, mainly bivalent carbonic acids (adipic acid, sebacic acid etc.) are esterified with monovalent, higher alcohols, i.e., made to react under dehydration; or monobasic, higher carbonic acids are esterified with multivalent alcohols (neopentanol, pantaerithrol etc.). This gives largely uniform and technically pure esters, whose characteristics can be kept very constant and can be varied within very wide limits due to the variety of synthetic fatty acids and alcohols used for esterification. They can be manufactured in very varied viscosity states and can be made largely resistant to oxidation by the specific introduction of certain features into the chemical structure. In aviation, turbine lubrication, which used to be the largest application of these esters, they have been abandoned due to the higher thermal stability of polyolesters or sterically hindered esters (47). Polyolesters are manufactured by the reaction of a multivalent alcohol such as trimethylolpropane or pentaerythritol with the appropriate number of acid molecules. Figure 4.23 shows such a reaction as a flow diagram.

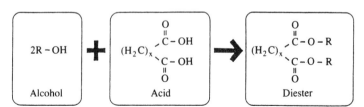

Fig. 4.22   Flow diagram of the manufacture of diesters (43)

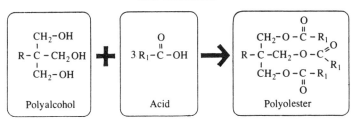

Fig. 4.23 **Flow diagram of the manufacture of polyolesters** (43)

(*b*) *Phosphoric acid esters.* Phosphoric acid esters are generally divided into three groups:

- phosphoric acid triarylesters;
- phosphoric acid trialkylesters;
- phosphoric acid arylalkylesters.

The triarylesters are of particular commercial importance as lubricants, being mainly used today as low-flammability hydraulic fluids in turbine installations (regulation and control equipment, bearings) and other installations susceptible to fire. The esters are manufactured by a catalytic reaction of phenols with phosphoroxytrichloride at 150 to 200°C. Figure 4.24 gives a flow diagram for this reaction.

4.4.3.3.5 Silicone oils

Silicone lubricants are semi-organic polymers and copolymers which contain a repeated lattice of silicon-oxygen units and organic side chains bonded to silicon. They are manufactured as shown in the flow diagram in Fig. 4.25.

4.4.3.3.6 Polyphenyl ethers

Experiments have shown that polyphenyl ethers can be used as high-temperature lubricants with high oxidative stress or high radiation loading (43)(49). Polyphenyl ethers are made by the reaction of phenols with halogen aromatics as shown in the flow diagram in Fig. 4.26.

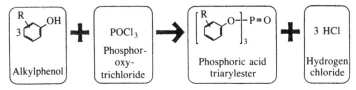

Fig. 4.24 **Flow diagram of the manufacture of phosphoric acid esters** (43)

Lubrication of Gearing

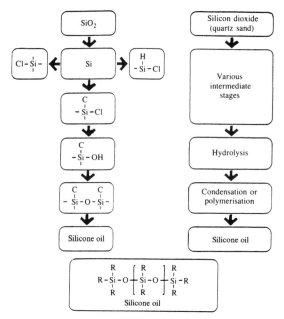

Fig. 4.25 Flow diagram of the manufacture of silicon oils (43)

The wide range of applications of polyphenyl ethers is countered by the relatively high melting point (poor low-temperature characteristics) and the high price. The high price is explained by the difficult chemical manufacture of the meta-substituted initial compounds. The melting point of technically interesting products is over 50°C, so pre-heating devices are necessary at the lubrication points, which adds to the operating costs.

The polyphenyl ethers also include the polyphenyl oxybenzenes, which differ from the ethers only in the chain centre. The high chemical stability of polyphenyl ethers is testified by their inert response to strong acids.

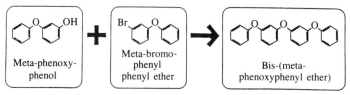

Fig. 4.26 Flow diagram of the manufacture of polyphenyl ethers (43)

Even concentrated hydrobromic acid, which splits alkyl ether, cannot split polyphenyl ethers.

4.4.3.3.7   Other synthetic fluids

In practice, maximum operating temperatures have risen considerably in some cases. For this reason, the requirements for thermal stability, oxidation stability, and high boiling point are in the first rank of necessary characteristics for new lubricants. These compounds, which, it is true, are essential for special applications, but generally are not of great significance in the total volume of lubricants, include the groups of substances in Table 4.13.

The synthesis of these compounds is, in most cases, extremely complex and often can only be performed under inert conditions or with precise temperature control. The yield of finished products is very small compared with the feedstocks, and increases production costs due to the need for additional cleaning processes to regain the initial compounds. These fluids are, therefore, only used in small quantities at high prices in quite special applications.

4.4.3.4   Characteristics of synthetic lubricants

4.4.3.4.1   General

Synthetic lubricants are developed for and used in special applications if mineral oils are incapable of meeting the requirements (e.g., extremely high or low temperatures and loads, unusual ambient conditions) or special demands (e.g., low flammability). The applications of synthetic lubricants may be considered if they can reduce machine downtime and faults, excessively short oil-change intervals, and safety risks, to such an extent that they represent a more cost-effective solution despite their relatively high price.

It is important to recognize that there is no synthetic lubricant which is superior to mineral oil in all its characteristics. Yet synthetic lubricants

Table 4.13   Special synthetic fluids as lubricants

Polyperfluoroalkyl ethers
Tetraalkyl silanes
Ferrocene derivatives
Tetra-substituted urea derivatives
Heterocyclic compounds
Aromatic amines
Hexafluorobenzene

can be so formulated that, while comparing the important and unimportant characteristics, they represent an optimal compromise which the mineral oil cannot provide.

For example, silicone oils offer only moderate wear protection, and this cannot really be improved with additives. Nevertheless, they provide an optimal response to the requirement for viscosity which is independent of temperature for use in precision measuring equipment. Many polyphenyl ethers are solid to 50°C; as, however, they can be used as high-temperature lubricants with a high radiation resistance, their poor behaviour at lower temperatures can be accepted. These two examples show that the synthetic lubricant must be selected according to the special requirements of the application, and may only be useful for this purpose.

The characteristics of synthetic fluids derive from the physical and chemical characteristics of the base fluids and from the effectiveness of added active substances or additives. The physical and chemical characteristics of the base fluids are the viscosity–temperature behaviour, the low-temperature flow properties, the boiling range, compatibility with paint and seals, miscibility with mineral oil, hydrolytic stability, and the capacity to dissolve chemical agents or additives.

The additive-dependent characteristics include oxidation stability, wear and scoring resistance, corrosion resistance etc. As will be discussed below, certain characteristics of a base lubricant cannot be influenced by additives.

4.4.3.4.2 *Limitations of mineral oil*

In order to make a relative assessment of the properties of synthetic fluids and to evaluate the advantages to be gained from their use, it is necessary to be familiar with the limitations of mineral oils.

Some of these characteristics are summarized in Table 4.14. Of course, they are not precise values; they are merely intended to give an indication of the order of magnitude of the values which can be achieved.

The operating temperature of lubricating oils is characterized by their oxidative and thermal resistance. In Fig. 4.27, the oxidation limit for mineral oils without and with oxidation inhibitors and the limit of thermal stability are given as a function of the service period of an oil fill.

4.4.3.4.3 *Temperature limits for synthetic fluids*

The maximum operating temperatures of some synthetic fluids for short-term and long-term use are given in Table 4.15. An indication of

Table 4.14 Some limits for mineral oils

Flash point	100–250°C
Pour point	− 50°C
Resistance to thermal decomposition	300–350°C
Resistance to oxidation	100–150°C
Viscosity/temperature behaviour	0.77
Viscosity index	90–130
Resistance to radiation	10^8–10^9 rad
Temperature range for use:	
continuous	80–120°C
short-term	120–150°C

the temperature limits of some synthetic fluids, shown by oxidation limit and the limit for thermal stability, is given in Fig. 4.28.

In Table 4.16, a high-temperature characteristic of some synthetic fluids, indicated by the flash point, and the pour points of these fluids are compared.

4.4.3.4.4   Viscosity and viscosity–temperature characteristic

Polyglycols and polyalphaolefins can be manufactured in virtually any viscosity state. The viscosity is dependent on the degree of polymerization. Both types have a high viscosity index. Obviously, including

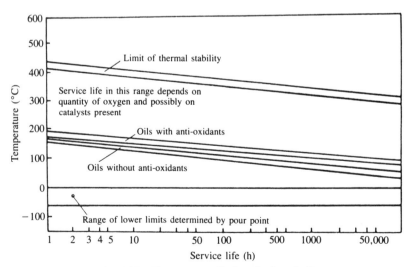

Fig. 4.27   Temperature limits of mineral oils

Table 4.15   Maximum operating temperatures of synthetic fluids

Class of compounds	Continuous	Short term
	($°C$)	
Mineral oils	90–120	130–150
Mineral oil super-raffinates	170–230	310–340
Synthetic hydrocarbons	170–230	310–340
Carbonic acid esters	170–180	220–230
Polyglycols	160–170	200–220
Polyphenyl ethers	310–370	420–480
Phosphoric acid alkyl esters	90–120	120–150
Phosphoric acid aryl esters	150–170	200–230
Silicilic acid esters, polysiloxanes	180–220	260–280
Silicones	220–270	310–340
Silane	170–230	310–340
Halogenated polyphenyls	200–260	280–310
Perfluoro hydrocarbons	280–340	400–450
Perfluoro polyglycols	230–260	280–340

an oxyether in the polymer chain of the polyglycols influences the mobility of the chains. Furthermore, in the Ubbelohde diagram (for the representation of the relationship between viscosity and temperature), the behaviour of mineral oil is divergent.

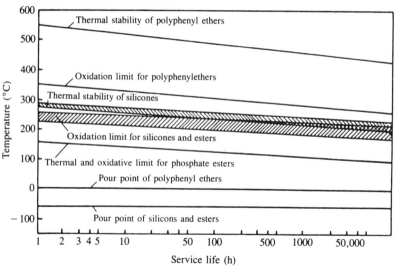

Fig. 4.28   Temperature limits of some synthetic fluids

Table 4.16   Flash point and pour point of synthetic fluids

	Flash point *(°C)*	*Pour point*
Mineral oils	100–250	−50
Phosphate esters	180	−57
Chlorinated hydrocarbons	210	
Diesters	230	−60
Polyglycols	230	0−−50
Polyphenyl ethers	290	−7
Silicate esters	185	−65
Siloxanes	200	−70
Silanes	260	
Silicones	385	−70

Specific branching of the polyalphaolefins produces the naturally high viscosity index. Because the high viscosity index is a natural characteristic of these fluids and does not have to be achieved by adding so-called viscosity index improvers, they have absolute shear stability.

The special advantages of the polyalphaolefins, in addition to their high viscosity index, are their excellent low-temperature performance, their low volatility, and their low coking tendency. In the viscosity range of engine and gear oils, polyalphaolefins have such a good intrinsic viscosity–temperature characteristic that multigrade oils can be formulated without adding VI improvers.

As can be seen from Tables 4.16 and 4.17, polyglycols and polyalphaolefins have a very low pour point in relation to the viscosity, which is not achieved by mineral oils of comparable viscosity. This gives easy starting in winter temperatures in outdoor installations, while the starter motor requires little power and the mixed-lubrication phase, in which high wear occurs, is short.

Alkyl benzenes likewise can be manufactured in almost all viscosity grades. Their viscosity index depends on the length of the alkyl chains in the benzene ring. The longer the side chains are, i.e., the more paraffinic they are, the higher is the VI. The same applies in practice to phosphoric acid triaryl esters.

Dicarbonic acid, polyol esters, and complex esters exhibit a broad spectrum of different viscosity states and viscosity indices, and are, therefore, not included in the system given.

Table 4.17 Viscosity characteristics of some synthetic fluids compared with those of mineral oils

| | Polyalphaolefins | | Polyglycol | Mineral oil | |
	A	B		150 SUS Neutral oil	Brightstock
Visc. 100 (°C mm^2/s)	5.75	40.8	31	5.15	32.7
Visc. 40 (°C mm^2/s)	29.5	405	224	29.7	540
Visc. −17.8 (°C mm^2/s)	1010	39000	–	–	–
Visc. −40 (°C mm^2/s)	7000	–	–	–	–
Viscosity index	140	151	181	100	91
Pour point (°C)	−60	−43	−29	−15	−6
Flash point (°C)	235	271	280	218	300

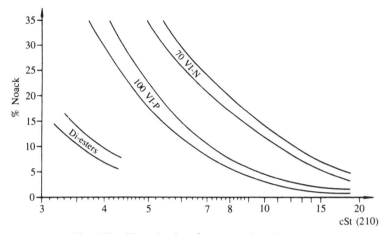

Fig. 4.29   Vaporization characteristics of esters (60)

In addition to good high-temperature and low-temperature flow characteristics, these fluids are distinguished by a low volatility with respect to the viscosity. This is illustrated by Fig. 4.29.

Figure 4.30 shows the viscosity–temperature characteristic of some synthetic fluids compared with mineral oil with and without VI improvers.

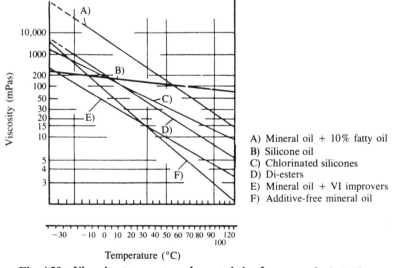

A) Mineral oil + 10% fatty oil
B) Silicone oil
C) Chlorinated silicones
D) Di-esters
E) Mineral oil + VI improvers
F) Additive-free mineral oil

Fig. 4.30   Viscosity–temperature characteristic of some synthetic fluids

4.4.3.4.5   Mineral oil and additive solubility

Mineral oils and additives are not unrestrictedly soluble in all synthetic lubricants. This should be borne in mind when manufacturing so-called part-synthetic lubricants as well as when formulating lubricants with additives, and in particular when unintentionally mixing lubricants and operating fluids derived from different base oils (**43**).

Polyalphaolefins, alkyl benzenes, and esters are unreservedly soluble in mineral oil. Water-soluble and non-water-soluble polyglycols absorb at most 5 percent oil. In addition there are mineral/oil-soluble types, whose solubility is, however, not unrestricted.

Phosphoric acid esters dissolve between 2 and 4 percent mineral oil, their spontaneous ignition temperature being lowered even at these low contents, however.

When using additives or active substances, it is necessary in all cases to examine carefully their solubility in the synthetic fluid concerned, as well as their effectiveness, i.e., their response from within this fluid. There may be applications in which new additive developments are necessary.

4.4.3.4.6   Compatibility with seals and paint

In their behaviour, polyalphaolefins resemble highly refined, paraffinic mineral oils. This means that a change from mineral oils with a viscosity index (VI) of between 80 and 100 to a polyalphaolefin can cause a shrinkage of NBR seals. As products with this base are, however, usually used in high-temperature applications, the seals in the equipment should, in any case, be made of fluorinated rubber. It is, nevertheless, advisable to consult the machine manufacturer.

Polyglycols can be used with fluorinated rubber, polytetra-fluorethylene, and to a certain extent with acrylonitrile butadiene rubber and silicon rubber. Butyl rubber and neoprene should be avoided, as they tend to cause shrinkage and hardening.

Fluorinated rubber and polytetrafluoroethylene can also be recommended for use with dicarbonic acid and phosphoric acid esters. In addition, silicon fluorinated types of rubber and polysulphides can be used in conjunction with dicarbonic acid esters.

All materials containing silicon must be avoided in applications where there must be good air separation performance. This also applies in conjunction with mineral oil.

Table 4.18 (**43**) summarizes the effects of synthetic fluids on the most important seal materials.

Table 4.18 Effects of synthetic fluids on seal materials

Type	Synthetic hydrocarbon	Polyglycol	Diesters	Phosphoric acid esters
NBR	Shrinkage, independent of viscosity	Shrinkage	Swelling	Not resistant
ACM	Only slight changes	Moderate to severe swelling, dependent on viscosity	–	Not resistant
CR	Slight swelling, severe shrinkage dependent on viscosity	Severe shrinkage	–	Not resistant
FKM	Resistant	Moderate swelling	Resistant	Resistant
VMQ	Severe swelling, dependent on viscosity	Moderate swelling	Swelling	Resistant

4.4.3.4.7 Gas solubility

Investigations have shown that the solubility of liquid petroleum gases (LPG) such as propane and butane in polyglycols is lower than in mineral oil and polyalphaolefins.

Polyalphaolefins exhibit a lower solubility of refrigerants such as R12, R22, and R114 than mineral oils. Products with this base experience a smaller fall in viscosity when mixed with the refrigerants, and thus afford greater protection against viscosity-induced wear, and achieve a distinct improvement in effectiveness in oil-sprayed screw-type compressors due to the better sealing effect.

To illustrate the differences in behaviour, Fig. 4.31 shows the miscibility gaps of a polyalphaolefin, a mineral oil, and a semi-synthetic refrigerator oil with R22.

4.4.3.4.8 Extreme pressure and wear-protection performance

Polyglycols and polyalphaolefins as base fluids contain neither sulphur compounds nor chlorine or phosphorus compounds. Nevertheless,

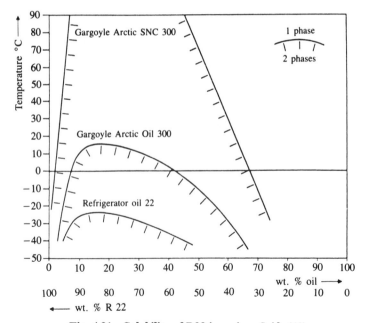

Fig. 4.31   Solubility of R22 in various fluids (43)

Table 4.19 **Load capacity of synthetic fluids in the FZG test as per DIN 51 354**
(43)

		Scoring force level	
Fluid	*Viscosity* *(mm²/s, 100°C)*	*A/8.3/90*	*A/16.6/140*
Mineral oil	27	10	–
Gear oil	29	12	12
Polyglycol	31	12	9
with additives	31	12	12
Polyalphaolefin	10	7	–
with additives	10	12	8–12
Dicarbonic acid ester	8	8	–
Phosphoric acid ester	5	7	–

polyglycols of high viscosity withstand the highest force level in the Test A/8, 3/90 in the FZG test machine without jumping to the wear level. Even at twice the peripheral velocity and with an increased temperature, they reach force level 9 with the tooth profile A which is susceptible to scoring.

Polyalphaolefins, dicarbonic acid esters, and phosphoric acid esters perform as mineral oils of the same viscosity in this test. Table 4.19 contains some results.

The load capacity can be increased with suitable extreme pressure additives if necessary, so even in the FZG test A/16, 6/140, force level 12 can be achieved without notable wear.

Because phosphoric acid esters are frequently used as hydraulic fluids, wear tests in vane pumps are more relevant to the evaluation of their performance. In the V 104 test unit, 10 mg total wear was measured on the ring and vanes after 250 h at 140 bar; this is a good result.

4.4.3.4.9   Friction performance
Table 4.20 contains a comparison of the friction coefficients obtained with doped mineral oils and various synthetic fluids. The good friction performance of the synthetic fluids, especially the polyglycols, is clear.

4.4.3.4.10   Ageing
The essential changes in a lubricant which are a result of ageing are given in Table 4.21. The characteristics marked with a [+] are suitable for determining the ageing state.

Table 4.20   Friction coefficients of various synthetic fluids, measured at 20°C in a modified Tannert apparatus (43)

	Product			
Pair ――――――― *Steel against*	*Mineral oil with lead additives*	*Mineral oil with sulphur/ phosphorus additives*	*Polyglycol*	*Synth. hydrocarbon*
WM 80 (Sn white metal)	0.189	0.160	0.101	0.126
V 738 (Sn white metal)	0.184	0.144	0.100	0.174
V 840 (Sn white metal)	0.232	0.186	0.121	0.193
Thermit (Sn white metal)	0.180	0.171	0.116	0.166
Tego II (Pb bronze)	0.163	0.176	0.130	0.119
Average	0.190	0.167	0.115	0.155

Assuming suitable additives, synthetic lubricants generally have a longer service life in a machine than mineral oil-based lubricants.

4.4.3.4.11   Overview of characteristics of synthetic fluids compared with mineral oils

Table 4.22 summarizes the most important characteristics of the synthetic fluids discussed. It is complemented by the comparative details in Table 4.23.

Table 4.21   Possible changes in a lubricant due to ageing (43)

Rise/fall in viscosity
Increase in neutralization coefficient
Increase in IR ageing bands
Degradation of anti-oxidants
Change in colour
Change in refractive index
Increase in water content
Increase in toluene insolubles

Table 4.22 Some characteristics of synthetic fluids (43)

Characteristics	Mineral oil	Poly-α-olefin	Alkyl-benzene	Diester	Polyolester	Polyglycol	Phosphate ester	Silicone oil
1 VT performance	4	3	4	2	2	2	5	1
2 Low temperature flow properties	4	3	3	3	2	3	4	3
3 Wear protection	4	4	4	4	4	1	2	5
4 Friction performance	3	3	3	3	2	1	2	5
5 High temperature oxidation stability (doped)	4	2	4	3	1	1	4	3
6 Water separation capacity	4	2	2	2	2	5	2	5
7 Air separation capacity	3	2	2	2	2	4	2	–
8 Rust protection (doped)	1	1	1	3	3	3	4	3
9 Miscibility with mineral oil	–	1	1	3	4	5	5	5
10 Compatibility with paint	1	1	1	5	5	3	5	2
11 Compatibility with seals	1	1	1	4	4	3	4	1
12 Hydrostatic stability	1	1	1	4	4	2	4	3
13 Low vaporization loss	4	1	4	1	1	3	3	3

Quality level: 1 = Excellent 2 = Very good 3 = Good 4 = Adequate 5 = Poor

Lubrication of Gearing

Table 4.23 Comparison of most important characteristics of synthetic fluids

	Viscosity–temperature performance	Fluid range	Low temperature performance	Thermal stability	Oxidation stability	Hydrolytic stability
Mineral oils	G	G	G	F	F	E
Mineral oil super-raffinates	E	G	G	G	F	E
Synthetic hydrocarbons	G	G	G	G	F	E
Organic esters	G	E	G	F	F	F
Polyglycols	G	G	G	F	F	G
Polyphenyl esters	F	G	P	E	G	E
Alkyl phosphate esters	G	G	G	F	G	F
Aryl phosphate esters	F	P	P	G	G	F
Silica esters and polysiloxanes	E	E	E	G	F	F
Silicones	E	E	E	G	G	G
Silanes	G	G	G	G	F	E
Halogenated polyaryls	G	G	F	G	G	E
Fluoro-hydrocarbons	F	G	F	G	G	F
Perfluoro-polyglycols	F	G	G	G	G	G

	Fire resistance	Lubricity	Compressibility	Volatility	Resistance to radiation	Density	Handling/storage
Mineral oils	L	G	A	A	H	L	G
Mineral oil super-raffinates	L	G	A	L	H	L	G
Synthetic hydrocarbons	L	G	A	L	H	L	G
Organic esters	L	G	A	A	A	A	G
Polyglycols	L	G	A	L	A	A	G
Polyphenyl esters	L	G	H	A	H	H	G
Alkyl phosphate esters	H	G	H	A	L	H	G
Aryl phosphate esters	H	F	H	L	L	H	F
Silica esters and polysiloxanes	L		A	A	L	A	
Silicones	L	P	L	L	L	A	G
Silanes	L	F	A	H	L	L	G
Halogenated polyaryls	H	G	H	H	L	H	G
Fluoro-hydrocarbons	H	P	L	A	L	H	F
Perfluoro-polyglycols	H	G	L	A	L	H	G

Assessment:

H = High   L = Low   A = Average   P = Poor

E = Excellent   G = Good   F = Fair

4.5 CONSISTENT GEAR LUBRICANTS – GEAR GREASES

4.5.1 Introduction

Lubricating greases can be divided into metal soap and non-soap types. Another system differentiates them into mineral oil-based and synthetic lubricating greases. The so-called adhesive lubricants occupy a special position. Table 4.24 gives a survey of the various types of lubricating grease.

In the following chapters, the concept 'lubricating grease' will be defined, the chemical composition of grease as a suspension of a solid in a fluid will be dealt with, and the structure of greases will be discussed. The most important methods of manufacturing greases will be presented, and finally, some important characteristics of greases will be compared and assessed (50)–(58).

4.5.2 Definition of lubricating greases

Regarded from the physical point of view, lubricating greases are dispersions, or, to be more precise, suspensions of solids in liquids.

Table 4.24 Types of lubricating greases

LUBRICATING GREASES

Greases with metal-soap thickener*   Lubricating greases with non-soap thickener†   Adhesive lubricants

Mineral oil as liquid phase   Synthetic fluid as liquid phase‡

Mineral oil as liquid phase   Synthetic fluid as liquid phase‡

Sprayed adhesive lubricants with solid lubricants   Bitumen-containing lubricants

* Soap thickeners: Simple soap, mixed soap, complex soap with various metal bases, e.g. Ca, Na, Li, Al, etc.

† Non-soap thickeners: Organic and inorganic solids, e.g. clay (bentonite, silicon gel, polyurea, soot, and pigments)

‡ Synthetic fluids: Silicones, esters, polyglycols

There are the following definitions of greases.

- Lubricating greases are consistent lubricants, which consist of mineral oil and/or synthetic oil and a thickener. They may contain agents and/or solid lubricants (DIN 51 825).
- A lubricating grease is a solid or semi-fluid substance, which results from the dispersion of a thickener in a fluid lubricant; it may contain other constituents, which have special characteristics (ASTM D 288-69).
- Lubricating greases are set, i.e., not free-flowing lubricants, and as such are to a certain extent resistant to forces of deformation. Lubricating greases are lubricating oils which are prevented from migrating.
- Regarded physically, lubricating greases are colloidal suspensions of suitable thickeners (solid phase) in mineral oils and/or synthetic oils (fluid phase).

4.5.3 Chemical composition of lubricating greases

4.5.3.1 General

Although lubricating greases appear externally to be uniform, homogeneous substances, in the chemical sense they consist of a variety of individual compounds of a largely organic nature. These may be classed in three groups of components, which generally account for the following proportions of the formulations of lubricating greases:

- base oils – 70–95%
- thickeners – 3–30%   – base grease

- agents – 0–10%   – additives

In the individual case, the chemical composition of a lubricating grease is a result of the proportions and the structures of these component groups.

4.5.3.2 Base oils

The base oils, which represent the main constituent by quantity and the carrier of the greases, must have good lubricating qualities themselves, i.e., they must have the nature of lubricating oils. Furthermore, they must be chemically neutral and non-reactive, as they must not react with other grease components and must not attack the materials at the friction points (corrosion) and in their vicinity (e.g. seal materials).

These demands are generally met by the hydrocarbon mixtures in the higher boiling mineral oil cuts; hence, for qualitative reasons and price reasons, the majority of lubricating greases contain mineral-oil base oils. The selection of the base oil in the individual case is dictated by the quality level and the application of the grease. Particularly high-quality types of grease are made on the basis of dewaxed, hydrofined solvent raffinates, and brightstock. Additional information on mineral oil-based base oils can be found in Section 4.4.2.

More than 95 percent of greases manufactured today have mineral oils as their base oils. In addition, for special applications and extreme operating conditions, characterized by high and low temperatures, high loads, high and low rotational velocities, aggressive media, etc., there are special lubricating greases based on synthetic fluids. The groups of substances in Table 4.25 are used as synthetic fluids in lubricating greases. Further information on synthetic fluids for lubricants can be found in Section 4.4.3.

4.5.3.3   Thickeners

4.5.3.3.1   Overview
The actual constituent which gives a grease its essential characteristic, namely the structure, is the thickening agent. The groups of substances in Table 4.26 may be used as thickeners.

As a rule, the thickeners are solid compounds at normal temperatures, which are colloidally dispersed in the base oil and are precipitated when the lubricating greases are extracted with non-polar solvents. The majority of thickeners for greases are simple metal soaps. The most common

Table 4.25   Synthetic fluids   for lubricating greases
Dicarbonic acid esters
Methylol esters
Silicones, siloxanes
Alkoxyfluorine oils
Polyalkylene glycols
Polyalphaolefins
Alkyl benzenes
Polyphenyl ethers

Table 4.26   Types of thickener for lubricating greases

Simple metal soaps	Complex metal soaps	Non-soaps
Calcium	Calcium	Oleophilic silicon dioxide
Aluminium	Aluminium	Organophilic clay minerals
Sodium	Sodium	Polyureas
Lithium	Lithium	Polymer hydrocarbons
Lead	Barium	Metal oxides, hydroxides, carbonates
Zinc	Zinc	Oleophilic graphite
		Inorganic pigments

thickener is lithium-12-hydroxystearate – this appears in about 65 percent of all greases.

4.5.3.3.2   Simple metal soaps
Metal soaps are the neutral metal salts of naturally occurring fatty acids and their hydrogenated forms with the corresponding metals, preferably those from the group of alkali metals (lithium, sodium) and alkaline earth metals (calcium, barium). The general formula for the soaps is

$$R-COO^-Me^+, RCOO^-Me^{++-}OOCR$$

The fatty acids used differ in the number of their carbon atoms (chain length) and double bonds. The 12-hydroxystearic acid occupies a special position; it is obtained from naturally occurring ricinoleic acid (castor oil) by hydrogenation. Vegetable, animal, and synthetic fatty acids are distinguished according to their origins. Table 4.27 contains those which are most important for manufacturing lubricating greases. Their lithium and calcium salts give lubricating greases which are particularly resistant to working.

The following simple metal soaps are of particular importance

– calcium soaps and calcium-12-hydroxystearate soaps
– sodium stearate and oleate soaps and sodium-12-hydroxystearate soaps
– lithium-12-hydroxystearate soaps
– aluminium soaps
– barium soaps

Table 4.27 Fatty acids for manufacturing lubricating greases

Name	Chain length	Double bonds	Main fatty acids of natural glycerides in
Palmitic acid	C_{16}	0	Palm oil, beef tallow, cotton oil, etc.
Palmitoleic acid	C_{16}	1	Herring oil, sardine oil, beef tallow
Stearic acid	C_{18}	0	Beef tallow, sunflower oil
Oleic acid	C_{18}	1	Olive oil, groundnut oil, beef tallow, sunflower oil, soya oil
Ricinoleic acid	C_{18}	1	Castor oil
Linoleic acid	C_{18}	2	Sunflower oil, soya oil, corn oil, cotton oil
Linolenic acid	C_{18}	3	Linseed oil, rape-seed oil, soya oil
Arachidic acid	C_{20}	0	Obtained by hydrogenating unsaturated C_{20} **acids**
Gadoleic acid	C_{20}	1	Rape-seed oil
Unsaturated fatty acid	C_{20}	2–6	Herring oil, sardine oil
Behenolic acid	C_{22}	0	Obtained by hydrogenating unsaturated C_{22} **acids**
Erucic acid	C_{22}	1	Rape-seed oil
Unsaturated fatty acid	C_{22}	3–6	Herring oil, sardine oil

4.5.3.3.3 Complex metal soaps

These are also neutral metal salts. They differ from the simple metal salts in that the same type of cation is present in addition to anions of different acids (carboxylate anions). These mixed salts arise during the co-neutralization of monoacid or polyacid bases with equivalent mixtures of various acids. In addition to the salts of long-chain fatty acids, such as stearates, the soap structure also contains such short-chain fatty acids as acetates, propinates, or inorganic acids such as borates.

Typical examples are the calcium and aluminium complex soaps; the latter are base salts of aluminium hydroxide, which are formed by the reaction of aluminium alkoxide with two equivalent acids:

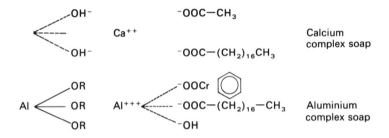

4.5.3.3.4 Oleophilic silicon dioxide

This is the thickener of the so-called gel greases. It consists of spherical, highly disperse SiO_2 agglomerates (average diameter 10–50 cm), which consist, inside the spheres, of larger SiO_2 units and have hydrophobic siloxane groups (Si-OR) and hydrophilic silanol groups (Si-OH) on the surface.

4.5.3.3.5 Organophilic clays

These are the thickeners of the bentonite greases. The individual thickener molecules consist of an inorganic part (clay mineral, such as montmorillonite or hectorite) and an organic part (several fairly long hydrocarbon chains). The two different parts are linked by nitrogen atoms in a true chemical bond.

4.5.3.3.6 Polyureas

These are chain molecules, in which the typical grouping of urea, $-HN-CO-NH-$, occurs several times. They are produced by the reaction of diisocyanates with monovalent and/or bivalent amines in the base oil, and represent the thickeners of the so-called polyurea greases, e.g.

di-urea:

$$R-HN-\underset{\underset{O}{\|}}{C}-NH-(CH_2)_x-HN-\underset{\underset{O}{\|}}{C}-NH-R$$

tetra-urea:

$$R-NH-\underset{\underset{O}{\|}}{C}-HN-(CH_2)_x-HN-\underset{\underset{O}{\|}}{C}-NH(CH_2)_x$$

$$-HN-\underset{\underset{O}{\|}}{C}-HN-(CH_2)_x-NH-\underset{\underset{O}{\|}}{C}-NH-R$$

4.5.4 Structure of the lubricating greases

Lubricating greases differ from liquid lubricants in the existence of a structure. In general, substances with a structure are characterized by the fact that there are such strong correlations between their constituents in the molecular range that they set mutually. The structure of lubricating greases is apparent in their main characteristic, namely consistency.

The carrier of the structure of any lubricating grease is independent of the type of thickener used. In the general model, the thickener molecules possess polar groups which interact with those of neighbouring molecules and thus form three-dimensional molecular aggregates. These thickener aggregates form a more or less regular network throughout the lubricating grease and fix the base oil molecules by addition and inclusion.

Under the electron microscope, the thickener aggregates appear as micelles or crystallites, with diameters in the colloid-disperse range of 10^{-10} to 10^{-7} cm. Figure 4.32 shows an electron microscope image of the micelles of a sodium soap lubricating grease and Fig. 4.33 shows a similar image for a lithium soap lubricating grease. The characteristic differences in the structure of these two different lubricating greases can be seen clearly.

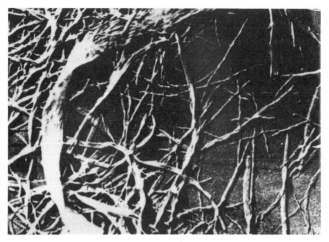

Fig. 4.32 Scanning electron microscope image of a sodium soap lubricating grease

Fig. 4.33 Scanning electron microscope image of a lithium soap lubricating grease

4.5.5 Manufacture of lubricating greases

The manufacture of lubricating greases is described below using the example of thickened greases.

4.5.5.1 Principle and specific problems

The manufacture of lubricating grease involves the task of thickening a suitable liquid lubricant, namely the base oil, with a suitable thickening agent. This thickening process takes place, very generally speaking, by the formation of a spatial network in the base oil, which has the ability to fix the base oil molecules, i.e., to restrict their freedom of movement. This fixing must be durable and must withstand high temperatures and mechanical stresses occurring at the friction points.

Thickening a base oil to form a lubricating grease is not only a question of the type of thickener selected and of the formulation. It depends to a very considerable extent on numerous process parameters, such as:

- pressure and temperature variation (i.e., heating and cooling profile, final temperature);
- shear and convection (i.e., stirring rate, stir geometries);
- concentration conditions;
- type and intensity of homogenization.

With a given formulation, varying these parameters can produce very different thickening effects, which depend ultimately on the formation of different skeletal structures. Herein lie the problems of manufacturing lubricating greases of uniform quality, and also, of course, the numerous possibilities offered by the manufacture of lubricating greases.

4.5.5.2 Chemistry of lubricating grease manufacture

The chemistry of lubricating grease manufacture is concerned primarily with the synthesis of the thickening agents, which are generally produced by reactions in the base oil carrier medium. Here we will deal only with the principle of metal soap formation, as the basis of the manufacture of conventional metal soap lubricating greases (53).

Metal soaps are obtained from the reaction of metal hydroxides (bases) or metal oxides with fatty acids according to the following general principle:

Fatty acid + lye → fatty acid salt (= soap) + water

$$R-COOH + MeOH \rightarrow R-COOMe + H_2O$$

This basic reaction of soap formation is, in the chemical sense, a simple neutralization reaction, in which, in addition to the salt which in this specific case is called 'soap', water of neutralization is produced.

In soap production, it is frequently the case that instead of fatty acids, it is triglycerides which are used, i.e., the corresponding natural raw fat, e.g., tallow instead of tallow fatty acid. It is then a two-stage reaction, in the first of which the triglyceride, which represents an ester of glycerine, is split by reaction, with water. The second stage is the immediate neutralization, or soap formation. See also the following process:

First stage:

$$CH_2-OCO-C_{17}H_{35} + H_2O \quad CH_2-OH + C_{17}H_{35}COOH$$
$$CH-OCO-C_{17}H_{35} + H_2O \rightarrow CH-OH + C_{17}H_{35}COOH$$
$$CH_2-OCO-C_{17}H_{35} + H_2O \quad CH_2-OH + C_{17}H_{35}COOH$$

Glycerol tristearate   + water   glycerol + stearic acid

Second stage:

$$3C_{17}H_{35}-COOH + 3NaOH \rightarrow 3C_{17}H_{35}COONa + 3H_2O$$
$$\text{Na-stearate (Na soap)}$$

This two-stage process is performed in a single reaction. The slower of the two reactions, which determines the speed and the reaction level of the whole reaction, is that of the splitting of the triglyceride (stage 1), which precedes that of soap formation (stage 2). The glycerine released remains to a certain extent in the final lubricating grease, without unduly affecting its properties.

The lyes or metal hydroxides used are called soap bases; the resultant lubricating greases are named after these:

Lithium lye	–	$LiOH \times H_2O$	–	lithium greases
Sodium lye	–	$NaOH$	–	sodium greases
Slaked lime	–	$Ca(OH)_2$	–	calcium greases

The fatty acids used are predominantly saturated and unsaturated, unbranched monocarboxylic acids, which also represent the majority of acids occurring in natural fats and fatty oils. 12-hydroxystearic acid occupies a special position; it is obtained by hydrogenation from ricinoleic acid, the fatty acid of castor oil. Its alkaline and alkaline earth soaps produce lubricating greases of high mechanical stability.

4.5.5.3   Individual stages of the process
The manufacture of lubricating greases does not just consist of the chemical reactions mentioned. A series of physico-chemical processes, controlled by technical process parameters, is equally important. The following three stages can be defined for the manufacture.

(*1*)   *Reaction phase.*   Formation of thickener molecules, e.g., the soaps, by conversion of the relevant components in the base oil. In the case of the soap greases, the water of reaction produced must be removed by evaporation.

(*2*)   *Structure formation phase.*   Formation of a three-dimensional network by aggregation of the thickener molecules produced in the first phase and fixing of the base oil molecules in this system. The form of the structure is determined essentially by the final temperature, the temperature profile of subsequent cooling, and the type of agitation during cooling.

(*3*)   *Mechanical phase.*   Incorporation of additives (from 80°C) and mechanical post-treatment of the grease with the aim of producing a largely homogeneous end product, free of air bubbles and solid contaminants.

These three phases are not always clearly separated. It is also possible to incorporate the thickener in a ready-made form into the base oil, so that the reaction phase is not required. The latter process is used mainly in the manufacture of non-soap-thickened lubricating greases.

4.5.5.4   Discontinuous production
The production of lubricating greases in individual batches is the most widely used method of manufacturing lubricating greases today. The plants used consist of the following linked units (see also Fig. 4.34).

(*1*)   *Reaction phase.*   Open agitator vat for unpressurized reactions or heated autoclaves for pressure reactions. These devices are heated indirectly with steam, thermal oil, or hot air.

(*2*)   *Structure formation phase.*   Agitators which can be heated and cooled and/or pressure coolers.

(*3*)   *Mechanical phase.*   Agitators and mixers for incorporating additives. In many cases it also takes place following cooling in the agitator of phase 2.

Homogenizers and attrition mills, evacuation pumps and units, filters.

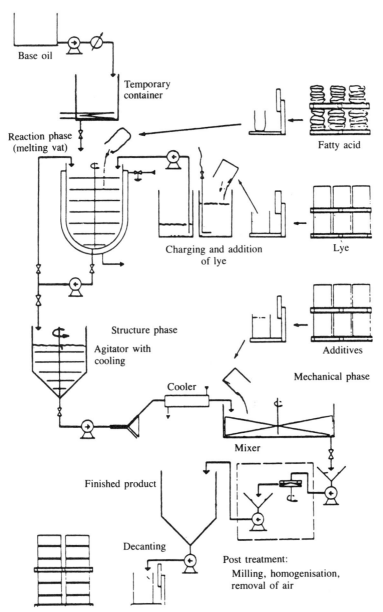

Fig. 4.34 Flow diagram of discontinuous manufacture of lubricating grease (53)

The capacity of the vat is 3–25 t and determines the size of the individual batches. Movement and protective temperature equalization of the product mass without partial overheating by contact with the heater surfaces in the vat walls is a problem. In large-scale plants, extremely heavy agitator designs and drive motors of up to 100 kW are required. Another problem of the discontinuous manufacture of lubricating grease is that the qualitative uniformity of the greases is greatly affected by the subjective capacity of the individual grease melters for precise and uniform work.

The disadvantages of discontinuous manufacture eventually lead to the development of automatically controlled, fully continuous lubricating grease plants. Discontinuous production will, however, continue to be important for special greases which are required in small quantities.

4.5.5.5 Continuous production
There are two fully continuous methods for producing lubricating greases

 – The TEXACO method (TCGP) – Texaco Continuous Grease Process.
 – The OJS method – developed by the BP group.

The two methods are similar in principle and differ only in some details of the process procedure, which are protected by Patent. The plants consist essentially of the following stages:

(*1*) *Reactor section.* At the heart is a continuous flow reactor.

(*2*) *Decompression section.* Water is removed from the reaction mixture by pressure reduction.

(*3*) *Finishing section.* Residual oils and additives are introduced via a continuous flow mixer and the end product is finally homogenized.

Other important parts of these plants are the numerous automatically controlled devices for maintaining temperature and pressure and for metering and removing constituents and end products. The time spent by the constituents in the reaction section is short, i.e., 30–60 s; this keeps thermal stress to a minimum. An additional advantage of fully continuous plants is that they are enclosed, integrated systems, to which oxygen in the air and impurities from the plant atmosphere have no direct access. The development of fully continuous processes represents

Table 4.28   Comparison of some characteristics of calcium, sodium, and lithium soap lubricating greases

	Ca grease	Na grease	Li grease
Dropping point (°C)	80–100	150–200	170–220
Upper application temperature (°C)	40–60	110–120	110–130
Response to water	Resistant	Not-resistant	Some resistance
Mechanical durability	Good	Moderate	Very good
Low temperature performance	Good	Moderate	Very good

an end point in the technology of lubricating grease manufacture for the time being. Figure 4.35 shows a flow diagram of the OJS/BP method as an example.

4.5.5.6   Comparison of important characteristics
The service properties of lubricating greases depend to a great extent on the thickener base and the base oils, and are characterized by the chemical, physical, and technological properties (see Section 4.7).

At this point we will compare some important basic data of the most varied types of lubricating grease. Table 4.28 contains a direct comparison of the most important data of the three classic lubricating greases based on calcium, sodium, and lithium soaps. Table 4.29 contains a comparison of the types of lubricating grease required on a vehicle, which have to meet very varied demands. It can be seen that a multi-purpose grease basically spans the characteristics of the other types of grease.

Table 4.30 contains a comparative survey of the most important service characteristics of the various types of lubricating grease. This information is suitable for the initial selection of a lubricating grease for a given application.

Table 4.29   Requirements for lubricating greases for vehicles

	Lubricating grease	Wheel hub grease	Water pump grease	Multi-purpose grease
Dropping point (°C)	90–100	160–180	100–140	180–200
Temperature range for use (°C)	−30–60	−30–100	<90	−40–120
Water resistant	Yes	No	Yes	No

Table 4.30 Overview of the most important service characteristics of various types of lubricating grease

Type of grease	Drop point (°C)	Low temp. performance	Application temperature Lower (°C)	Application temperature Upper (°C)	Property Water resistance	Mechanical endurance	EP* performance	Anti-corrosion	Cost (Li = 1)	Suitability for Roller bearings	Plain bearings
Metal soap greases:											
Calcium (Ca)	80–100	Good	−35	+50 (+60)	Very resistant	Good	Good	Poor	–	Very limited†	Limited
Sodium (Na)	130–200	Moderate	−30	+120	Not resistant	Moderate	Moderate	Good	–	Good	Good
Lithium (Li)	170–220	Good	−40	+130 (+140)	Resistant	Very good	Slight	Very poor	1	Very good	Good
Aluminium (Al)	120	Good	−35	+100	Swelling	Moderate	Moderate	Very good	3	Very good	Good

Mixed greases:											
Li/Pb	90	Poor	(0)	(+75)	Resistant	Poor	Very good	Good	1.5	Good	Good
Ca/Pb	90	Poor	(0)	(+75)	Resistant	Poor	Very good	Good	1.5	Good	Good
Complex greases:											
Ca-based	>240	Moderate		+120 (+130)	Very resistant	Moderate	Good	Good	0.9–1.2		
Al-based	>230			+160 (+185)	Very resistant	Good	Very good				
Gel greases:											
Silica gel				+130 (+150)							
Bentonite				+150 (+160)	Resistant	OK/good	Slight		3		
Synthetic greases:											
Silicon-based		Good		+320	Resistant	OK/moderate	Poor		30–50		
Ester-based		Very good	−70	+150 (+180)					10–20		

* Can be improved with additives
† Due to high ash content

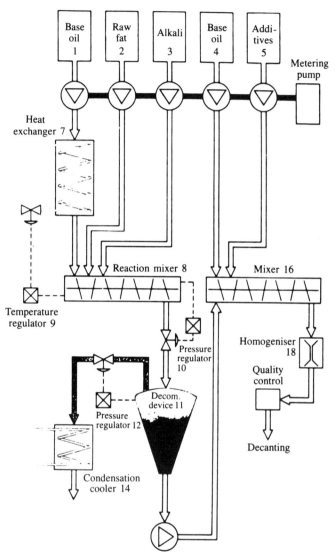

Fig. 4.35   Flow diagram of continuous lubricating grease manufacture using the BP method (53)

4.5.6 Adhesive lubricants and sprayed-on adhesive lubricants

4.5.6.1 General
Adhesive lubricants are used primarily for tooth flank lubrication of so-called open gears. Their use is usually restricted to peripheral speeds of up to about 5 m/s, and in special cases up to 8 m/s.

As shown in Table 4.24, we have to distinguish between products containing bitumen and solvents, the conventional or classical adhesive lubricants, and the products which contain no bitumen and which can mostly be sprayed on. The latter are called spray-on adhesive lubricants (**58**).

4.5.6.2 Bitumen-based adhesive lubricants
As a rule, these are the so-called 'B' lubricating oils as per DIN 51 513, which can have viscosities between 16–36 mm^2/s at 100°C (type BA) and 225–500 mm^2 at 100°C (type BC).

'B' lubricating oils are dark, bitumen-containing mineral oils, in which the bitumen is in a stable solution. 'B' lubricating oils are applied by hand, once-through, and splash lubrication at friction points which require good adhesive properties in the lubricant film, greater than those possessed by lubricating oils without bitumen. They are also used when a particularly high-viscosity lubricating oil is needed.

'B' lubricating oils are used particularly in open gears, guideways, wire ropes, and cores of wire ropes.

'B' lubricating oils may be thinned with a solvent to make them easier to apply. In this connection, it is necessary to observe any regulations concerning the use of solvents in certain situations, e.g., in sealed rooms or in underground mining.

4.5.6.3 Sprayed adhesive lubricants
Adhesive lubricants for spray application are always free of bitumen and often free of solvents. They are complex, expensive special lubricants for lubricating open or simply covered gear sets.

They usually contain a base oil, a thickener, and certain additives. The thickener makes them consistent lubricants, which can be regarded as lubricating greases in a wider sense.

(1) Base oils. Mineral oils, synthetic oils or mixtures of the two are used.

(2) *Thickeners.* Metal soaps or complex metal soaps are used, as well as inorganic non-soap thickeners.

(3) *Additives.* Additives which reduce wear and scoring are used.

The spray-on adhesive lubricants may also contain a solvent to make them easier to apply. The preferred consistency is NLGI Class O.

Great value is placed on the technological property of spray capability. Other important characteristics are adequate corrosion protection, good compatibility with seal materials, and above all, excellent wear and scoring prevention properties, which can be measured, for example, with the aid of the FZG test machine (DIN 51 354).

Spray-on lubricants can be applied by hand with a spatula or with a hand sprayer. However, optimum economy is achieved with automatic spraying equipment.

Further aspects of spray-on adhesive lubricants are dealt with in Section 7.2.

4.6 ADDITIVES

4.6.1 Introduction

The development of modern lubricants and their correct use are of considerable economic significance. Lubricants which are optimally matched to the task produce considerable savings in the form of energy saving, wear reduction, reduced maintenance, and longer service intervals, which can run into billions for an industrialized country (**60**)(**61**).

The properties and qualities of lubricants depend on the one hand on the provenance and viscosity of the base oil and the process parameters used in its production, and, on the other hand, on the type and quantity of the additives used. In modern industry, the use of lubricants whose characteristics have not been improved or altered by additives is scarcely conceivable. The doped lubricant has now reached the rank of a design element, whose characteristics and limitations must be considered by the designer like those of the other components.

The range of additive applications, i.e., the addition of active substances, extends from a few ppm (e.g., foam inhibitors) to concentrations of about 30 percent (e.g., certain engine oils). There is also a great variation in the price of the individual types of additives. The price level ranges from less than 2 DM/kg for cheap petrochemical products and greases to over 50 DM/kg for special additives. The annual turnover of lubricant additives worldwide is about 3 billion DM. This corresponds to production of about 1.5 million tonnes. About 70 000 tonnes of additive concentrates are used annually in the Federal Republic of Germany alone. These concentrates sometimes contain up to 50 percent mineral oil for ease of use.

After a discussion of the most important characteristics of lubricants and the consequent purposes of additives, there will be a brief, systematic overview of the types of additives. Then the classes of chemical compounds, the mechanisms, and the application of oxidation and corrosion inhibitors, of additives for modifying flow properties, of wear and scoring protection additives as well as friction modifiers, of detergent/dispersant additives, and of foam inhibitors and adhesion improvers will be discussed. Information on matching the base oil to additive combinations concludes the chapter.

It should be borne in mind that the lubricant properties required can rarely be achieved with a single type of additive. Mixtures of different additives are mostly used. The constituents of such 'additive packages' can be mutually supportive (synergism) or they may counteract one another (antagonism).

4.6.2 Important properties of lubricants

The most important properties of lubricants are listed in Table 4.31, and are divided into values for selection and those denoting quality.

The selective values, e.g., the viscosity, give no indication of the quality, i.e., the service characteristics of a lubricant, even though during selection account must be taken of the design conditions of the friction point, the operating conditions, and the ambient conditions, in order to ensure problem-free operation and reliability of the friction pair or the machine installation.

The quality values, divided into primary and secondary characteristics, describe the service characteristics and thus the quality of a lubricant. The primary characteristics are those which are associated with a certain sphere of application. For an engine oil, for example, these include the detergent/dispersant characteristics, which are of no significance for a gear oil.

Gear oils, on the other hand, are characterized primarily by their scoring and wear properties, while for applications in which low-flammability hydraulic oils are specified, e.g., in underground mining, their low flammability is one of the primary characteristics.

Table 4.31   Important characteristics of lubricants

	Quality values	
Selective	Secondary characteristics	Primary characteristics
Viscosity	Ash	Friction behaviour
Density	Viscosity/temperature behaviour	Wear behaviour (scoring
Flash point	Viscosity/pressure behaviour	behaviour)
Aniline point	Flow performance at low	Running-in properties
Toxicity	temperatures	Detergent/dispersant
	Flow performance at high	behaviour (foaming
	temperatures	behaviour)
	Low-temperature behaviour	Flammability
	Chemical behaviour (corrosion,	Resistance to radiation
	aggression to NF metals)	
	Stability (thermal, oxidative)	
	Water and air separation	
	capacity	
	Compatibility with seal	
	materials	
	Coking tendency	
	Vaporization behaviour	
	Cloud point/pour point	

All other service characteristics, which are not related to specific applications, are grouped under secondary characteristics. The division into primary and secondary characteristics does not, therefore, indicate a qualitative ranking.

Table 4.32 shows the correlation between some important service characteristics and specific areas of application. It is obvious that some characteristics are important in certain friction locations, but more or less insignificant in others.

4.6.3   Purpose of additives

The additives used in lubricants have to perform the following tasks with regard to modifying their characterstics:

- to impart specific properties which the base lubricant does not have (e.g., detergent/dispersant ability);
- to improve or enhance certain desirable characteristics (e.g., viscosity/temperature behaviour);
- to weaken or repress certain undesirable characteristics (e.g., oxidation behaviour).

Table 4.32 Relative significance of certain characteristics of lubricants for specific applications

Lubricant characteristics	Friction situation								
	Plain bearings	Rolling bearings	Enclosed gears	Open gears, chains, ropes	Precision components	Guideways, interlocks	Vehicle engines	Vehicle trans	Hydraulics
Behaviour in mixed friction	*	**	***	**	**	*	**	***	**
Cooling effect	**	**	***	–	–	–	*	**	**
Friction (low)	*	**	**	–	**	–	*(***)	*(***)	–
Ability to stay at friction point	*	**	–	**	***	*	*	–	–
Ability to exclude impurities	–	**	–	*	–	*	–	–	–
Service temp. range	*	**	**	*	–	*	***	***	**
Corrosion protection	*	**	–	**	–	*	**	**	**
Volatility	*	*	–	**	***	*	***	–	–
Ability to cope with impurities	–	–	–	–	–	–	***	–	*
Foaming tendency	–	–	**	–	–	–	**	**	***
Flammability	–	–	–	–	–	–	–	–	**

Assessment scale: – insignificant; *** very important

Some agents, also called additives, influence not only one but several lubricant characteristics simultaneously. These are called multi-functional additives.

Several additives are often combined in so-called additive packages. Such combinations can be used to influence the chemical and physical properties, and above all the engineering properties of the lubricants, i.e., their service characteristics, and thus to match them to the requirements of the individual case.

It should be borne in mind that certain lubricant characteristics cannot be modified by the use of additives. Some of these are listed in Table 4.33.

4.6.4 Important types of additives

The types of additives and their functions listed in Table 4.34 can be distinguished in a very simplified and rather unsystematic way.

Table 4.35 shows which additive types and packages can be used for the most important areas of application. The scale extends from completely undoped lubricants, e.g., for food machinery, to very highly doped lubricants, e.g., for engine oils.

The characteristics of modern lubricants mentioned earlier show that the diverse requirements of modern lubrication engineering and tribology can no longer be met by natural hydrocarbons, which constitute the base oils derived from mineral oil.

The viscosity–temperature behaviour and oxidation resistance of the base oils can only be improved to a limited extent by modifications to process technology. Anti-corrosion and anti-rust properties, load capacity, and detergent/dispersant behaviour, however, are characteristics

Table 4.33 Characteristics which cannot be influenced by additives

Thermal conductivity
Volatility
Resistance to radiation
Wax separation
Air separation ability
Thermal stability
Gas solubility
Compressibility

Table 4.34 Important additives and their functions in lubricants

Main types	Functions
Neutralizing agents	To neutralize sour compounds, caused by the combustion of sulphurous fuels or, less commonly, by the decomposition of certain EP additives
Foam inhibitors	Reduce surface foam
Oxidation inhibitors	Reduce and delay oxidation. Types: inhibitors, metal deactivators, metal passivators
Rust inhibitors	Reduce rust formation on ferrous surfaces
Corrosion inhibitors	Type A: reduce corrosion of materials containing lead Type B: reduce corrosion of materials containing copper
Anti-wear additives	Reduce abrasive wear in moderately severe conditions, especially with steady-state stresses
Extreme pressure (EP) (anti-scoring) additives	Reduce wear and scoring in severe conditions, especially with shock loading
Friction modifiers	Reduce friction in mixed-friction conditions
Detergent additives	Reduce the occurrence of precipitation at high temperatures, e.g., in internal combustion engines
Dispersant additives	Reduce sludge at low temperatures, e.g., in internal combustion engines
Emulsifiers	To produce water-in-oil or oil-in-water emulsions
Pour point improvers	Lower the pour point of paraffinic oils
Viscosity index improvers	Reduce viscosity/temperature dependence
Adhesion improvers	Reduce tendency of oil to drip off or be thrown off

Table 4.35 Additives and additive packages for specific applications

Type of machinery	Additives used	Special requirements
Food industry	None	Harmless if unintentionally consumed
Oil hydraulics	Oxidation inhibitors Rust inhibitors Anti-wear additives Pour point improvers Viscosity index improvers Foam inhibitors	Smallest possible viscosity change with temperature Least possible steel/steel wear
Steam and gas turbines	Oxidation inhibitors Rust inhibitors	Good water separation ability
Steam engine cylinders	None, possibly fat oils	Retain oil film on hot surfaces; resist washing off of oil by wet steam
Air compressor cylinders	Oxidation inhibitors Rust inhibitors	Least possible tendency to precipitate

Application	Additives	Function
Gears (steel/steel)	Anti-wear additives Anti-scoring additives Oxidation inhibitors Foam inhibitors Pour point improvers	Special protection against abrasion and scoring
Gears (steel/bronze)	Friction improvers Oxidation inhibitors	Reduce friction, temperature rises, wear and oxidation
Guideways and bed tracks (machine tools)	Friction improvers Adhesion improvers	Uniform sliding at low velocities; maintain oil film on vertical surfaces
Encapsulated chillers	None	Good thermal stability; miscibility with refrigerants; low flow point
Internal combustion engines	Detergent/dispersant additives Oxidation inhibitors Neutralizers Foam inhibitors Anti-wear additives Corrosion inhibitors	

which are, at best, present in mineral and synthetic base oils. Only the incorporation of certain additives gives the finished lubricants the desired service characteristics with the necessary reserve of performance to meet the demands of practical service in the intended sphere of application.

Table 4.36 indicates the relationship between additive type, chemical compound, application and functional mechanism.

4.6.5   Description of additives

4.6.5.1   Oxidation inhibitors

4.6.5.1.1   Oxidation process

It is impossible to prevent oxygen being present in lubricating oil circuits: as a result, the oil oxidizes. During the oxidation process, acids form, and, at an advanced stage, oil-insoluble polymerizates, which lead to varnish- and sludge-like deposits. Most crude oils for lubricating oil refining contain natural oxidation inhibitors in the form of polycyclic aromatics or sulphur and nitrogen compounds. As the refining process is designed to produce a high viscosity index, however, these components are largely removed. Therefore, modern uninhibited solvent raffinates exhibit relatively poor ageing stability (Fig. 4.36) (**62**).

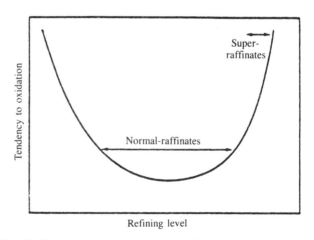

Fig. 4.36   **Oxidation tendency as a function of level of refining (schematic)** (62)

Table 4.36 Chemical compounds, functional mechanisms and applications of the most important additives

Additive	Chemical compound	Application	Mechanism
Oxidation inhibitors	Hindered phenols Amines Organic sulphides Zinc dithiophosphate	Minimize formation of resin-, varnish-, sludge-, acid-, and polymer-like compounds	End oxidation chain reaction by reducing the organic peroxides. Reduce acid formation by reduced oxygen absorption by the oil. Prevent catalytic reactions
Corrosion inhibitors	Zinc dithiophosphate Sulphurized terpenes Phosphorized, sulphurized terpenes Sulphurized olefins	Protect bearing and other metallic surfaces from corrosion	Action as anti-catalysts. Film formation on metallic surfaces as protection against attack by acids and peroxides
Rust inhibitors	Amine phosphates Sodium, calcium and magnesium sulphonates Alkyl succinic acids Fatty acids	To protect ferrous metal surfaces from rust	Polar molecules are preferably adsorbed onto metallic surfaces and serve as a barrier to water. Neutralization of acids
Metal deactivators	Triarylphosphites Sulphur compounds Diamines Dimercaptan thiadizole derivaties	Prevention of catalytic influence on oxidation and corrosion	A protective film is adsorbed onto metal surfaces, preventing contact between the base metal and corrosive substances
Anti-wear additives	Zinc dialkyl dithiophosphates Tricresyl phosphates	Reduce excessive wear between metallic surfaces	Reaction with metallic surfaces produces coatings which deform plastically and improve the contact pattern
Anti-scoring (EP) additives	Sulphurized fats and olefins Chlorinated hydrocarbons Lead salts of organic acids Amine phosphates	Prevent microwelding between metallic surfaces at high pressures and temperatures	Reaction with metallic surfaces produces new compounds with lower shear strength than the base metal. Continuous shearing off and reforming

Table 4.36 *Continued*

Additive	Chemical compound	Application	Mechanism
Friction modifiers	Fatty acids Compound amines Solid lubricants	Reduce friction between metallic surfaces	High polar molecules are adsorbed onto metallic surfaces and separate them. Solid lubricants form a friction-reducing surface film
Detergent additives	Normal or basic calcium, barium or magnesium sulphonates, phenates, or phosphonates	Reduce or prevent deposits in engines at high operating temperatures	Control of formation of varnish and sludge by reaction with oxidation products, during which oil-soluble products or products suspended in the oil occur
Dispersant additives	Polymers such as nitrogen-containing polymethacrylates, alkyl succinimides and succinate esters, high molecular weight amines and amides	To prevent or delay the formation and deposition of sludge at low operating temperatures	Dispersants have a marked affinity for impurities and surround them with oil-soluble molecules which prevent the agglomeration and precipitation of sludge in the engine
Pour-point depressants	Paraffin, alkylated naphthalenes and phenols Polymethacrylates	Lower the pour point of the oil	Prevent agglomeration of paraffin crystals by coating them
Viscosity index improvers	Polyisobutylenes Polymethacrylates Polyacrylates Ethylene propylene Styrol maleic acid ester copolymers Hydrogenated styrol butadiene copolymers	Reduce the dependence of the viscosity on the temperature	Polymer molecules are extremely coiled in poor solvent (cold oil) and assume a greater volume in a good solvent (warm oil) by uncoiling. This makes the oil relatively thicker.

Foam inhibitors	Silicone polymers Tributyl phosphates	Prevent the occurrence of stable foam	Reduce interfacial tension by attacking the oil film surrounding each air bubble. Hence smaller bubbles coalesce to form larger bubbles, which rise to the surface
Adhesion improvers	Soaps, polyisobutylenes and polyacrylate polymers	Increase adhesion of the oil	Increase viscosity. Additives are tenacious and sticky
Emulsifiers	Sodium salts of sulphonic and other organic acids Compounded amine salts	Emulsify oil in water	Reduce interfacial tension by adsorption of emulsifier at oil/water interface. Hence dispersion of one liquid into another.
Bactericides	Phenols Chlorine compounds Formaldehyde derivatives	Increase service life of emulsion. Suppress unpleasant odours	Prevent or delay the growth of micro-organisms

The following parameters play a part in oil oxidation:

- temperatures;
- catalysts, e.g., metal parts, metal abrasion;
- combustion products, ageing products, acids;
- O_2 content.

While oil oxidation proceeds relatively slowly at temperatures below 100°C, it accelerates with rising temperatures.

The oxidation process is very complex, but it can be divided into the following phases (Fig. 4.37):

- induction period/start;
- chain propagation;
- chain branching;
- chain termination.

It can be seen that this reaction, once initiated, is a chain reaction. In successive reactions the peroxides give rise to organic acids, esters, ketones etc., which lead to souring of the oil, and polymerization causes oil-insoluble products and precipitates.

4.6.5.1.2   Types of additives and functional mechanism

The effectiveness of the oxidation inhibitors relies on termination of the oxidation chain, i.e., the exclusion of the peroxides and radicals.

Start

$$RH \rightarrow R^+ \tag{1}$$

Chain propagation

$$R^+ + O_2 \rightarrow ROO^+ \tag{2}$$
$$ROO^+ + RH \rightarrow ROOH + R^+ \tag{3}$$

Chain branching

$$ROOH \rightarrow RO^+ + {}^+OH \tag{4}$$
$$RO^+ + RH \rightarrow ROH + R^+ \tag{5}$$
$${}^+OH + RH \rightarrow H_2O + R^+ \tag{6}$$

Chain termination

$$2\,R^+ \rightarrow R - R \tag{7}$$
$$R^+ + ROO^+ \rightarrow ROOR \tag{8}$$
$$2\,ROO^+ \rightarrow ROOR + O_2 \tag{9}$$

Fig. 4.37   Schematic representation of oxidation process – mechanism of anti-oxidation

Various mechanisms could do this. Due to their high chemical reduction potential, they counteract the formation of organic peroxides, and, as radical traps they lead to the termination of particularly undesirable chain reactions by trapping those radicals which, as intermediate products, are responsible for propagating the ageing reaction. Finally, as deactivators and passivators they prevent reactions which take place under the influence of homogeneous or heterogeneous catalysis, by coating catalytic metallic surfaces or reacting with metallic salts dissolved in the oil.

The following main groups of oxidation inhibitors are of importance:

(a)   Sterically hindered phenols, bisphenols and thiobisphenols;
(b)   Metal dialkyl(diaryl)dithiophosphates and -phosphonates;
(c)   Overbased metal sulphonates, alkylphenolates and alkylphenol sulphides;
(d)   Aromatised amines.

Figure 4.38 surveys the most important types of oxidation inhibitors.

Of course, inhibitors cannot completely prevent oxidative ageing; they can only delay the reaction processes concerned, though to a considerable extent.

Depending on their molecular structure, different inhibitors reveal their full effectiveness in limited temperature ranges; therefore, they are often combined, taking account of synergistic effects, and matched to the base oil mixtures used.

Selection and dosing of oxidation inhibitors, therefore, depends on the type of oil to be formulated. The formulation of modern lubricating oils, therefore, makes special demands. The concentration range for oxidation inhibitors lies between 0.1 and 1.5 wt%.

4.6.5.2   Corrosion inhibitors

4.6.5.2.1   Principles

Corrosion inhibitors in the true sense are those substances which quickly form dense protective films on metallic surfaces and thus stop direct contact between the base material and aggressive media or atmospheres. Corrosion is an electro-chemical process, in which the attacked metal oxidizes and the attacking medium is reduced. A prerequisite for this process is always the presence of oxygen or air. Corrosion inhibitors can usefully be grouped into those for ferrous metals and those for non-ferrous metals (**62**).

Property	Chemical structure	Initial products	Name
Inhibit oxidation	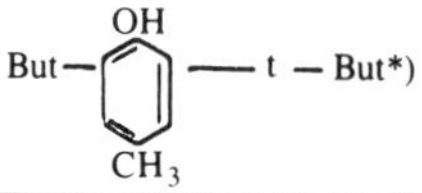 But—⬡(OH)— t — But*) CH₃	Cresol, isobutylene	"Hindered" phenols
Inhibit oxidation	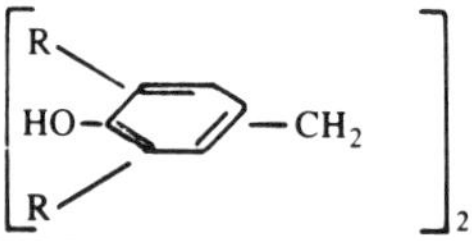 [R＼ HO—⬡—CH₂ R／]₂	Alkylphenol, keton e.g. acetone	"Hindered" bisphenols
Inhibit oxidation	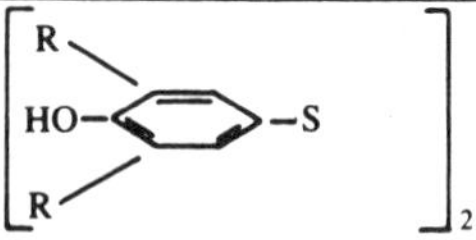 [R＼ HO—⬡—S R／]₂	Alkylphenol, sulphur, s-chloride	"Hindered" thiobisphenols
Inhibit oxidation, metal passivation	[(RO)₂P(S) − S]₂Me	Alcohol, phosphorus pentasulphide, metal oxide (me. mainly zinc)	Metal dialkyl(diaryl) dithiophosphates
Inhibit oxidation, metal passivation	(R)P(S) . . . S . . . R**	Olefins, phosphorus pentasulphide	Alkyl dithiophosphonates
Inhibit oxidation	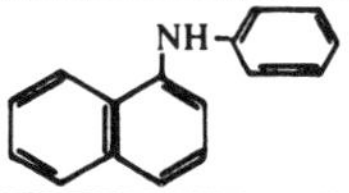 NH—⬡ (naphthalene)	Aniline, 1-naphthol	Phenyl alpha napthlamines (PANA)
Inhibit oxidation	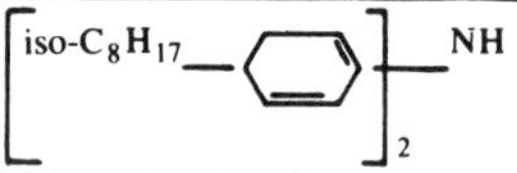 [iso-C₈H₁₇—⬡—]₂ NH	Aniline, phenol, diisobutylene	Alkylated diphenylamines

*Tertiary butyl residue   **Chemical structure uncertain   R = alkyl(aryl) residue

Fig. 4.38   Survey of the most important types of oxidation inhibitor (60)

4.6.5.2.2   *Corrosion inhibitors for ferrous metals – rust inhibitors*

When iron and steel rust, the corrosion takes place in the presence of corrosive substances on the metallic surface at cross-linked and short-circuited galvanic cells. Corrosive substances are water, metal chlorides and bromides, oxides of sulphur and nitrogen, as well as organic and inorganic acids.

The best rust inhibitors are those compounds which carry a strongly polar group on a long-chain alkyl residue. Structures of this type are able to attach themselves to metallic surfaces in an oriented manner and to form dense, hydrophobic protective films.

The rust inhibitors can be divided into ash-producing and ash-free types. The classical ash-producing rust inhibitors belong to the alkaline and alkaline earth metal salts of higher molecular sulphonic acids, the naphthenic acids.

Proven ash-free rust inhibitors are, in addition to the simple fatty acids, primarily dodecenyl succinic acids, in particular the corresponding semi-esters or semi-amides. The fatty acid derivatives of sarcosine are also widely used. These compounds are mainly used in conjunction with imidazoline derivatives, because these combinations have been found to act synergistically. Figure 4.39 contains a survey of the most important rust inhibitors.

When using engine oils, the inhibitor must have a capacity to neutralize as well as to provide effective surface protection. In the case of rust inhibitors for gear oils, protection of the metal surface is of decisive importance, while the neutralization effect is of only subsidiary significance. Widely used rust inhibitors are metal sulphonates (barium, calcium, magnesium) based on sulphonic acids with average molecular weights of 400–450. To increase the anti-rust effect, several additives can be used, which complement each other in their effectiveness synergistically. The inhibitor concentration in gear oils must be kept as low as possible because of the impairment of extreme pressure characteristics by anti-rust additives.

Property	Chemical composition	Initial products	Name
Anti-rust	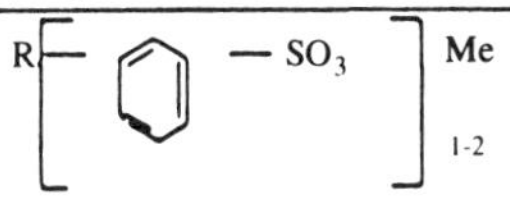 R —⟨ring⟩— SO_3 $\Big]$ Me $_{1-2}$	Alkyl sulphonic acids metal oxide (hydroxide)	Metal sulphates
Anti-rust	$\Big[R - (CH_2)_n COO \Big]_{1-2}$ Me	Fatty acids metal oxide (hydroxide)	Fatty acid salts
Anti-rust	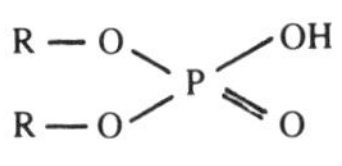 $\begin{array}{c} R - O \\ R - O \end{array} P \begin{array}{c} OH \\ O \end{array}$	Fatty alcohols, ethoxylised alcohols, phosphorous oxychloride	Secondary phosphoric acid esters

Fig. 4.39   Important rust inhibitors (60)

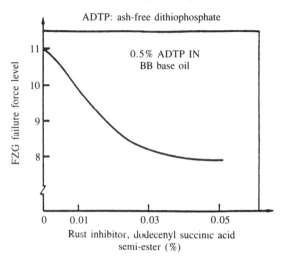

Fig. 4.40   **The antagonistic effect of rust inhibitors in conjunction with EP and AW additives** (62)

Thus, a barium or calcium sulphonate in a concentration of 1 wt% can result in a fall in load capacity (scoring force level) on the FZG test machine of 1–3 units (see Fig. 4.40). For this reason, the concentration of inhibitors in gear oils rarely exceeds values of about 0.25 wt%.

4.6.5.2.3   Corrosion inhibitors for non-ferrous metals
The corrosion of parts made of non-ferrous metals is caused primarily by acids and occurs on bronze and white metal parts. Hence, plain bearings, mainly in engines, but also in gears, are particularly at risk. The most susceptible alloying component is lead, but copper is also attacked.

Protection against corrosion is provided by those substances which form (a) an adsorption layer, (b) a chemosorption layer, or (c) a chemical reaction layer on the surface of the metal. Reaction products of phosphorus pentasulphide and unsaturated hydrocarbons (e.g., terpene) as well as with alcohols and post-treatment with zinc oxide have proved to be effective additives for engine oils. Those produced with zinc oxide include the well-tried zinc dialkyl dithiophosphates. Their effectiveness is due mainly to the formation of chemical reaction layers.

When selecting a corrosion inhibitor for non-ferrous metals, care should be taken to ensure that it is optimally effective but that it does not itself cause corrosion by excessive reactivity. The best known example is sulphide formation on non-ferrous metals by aggressive sulphur compounds. The concentration of corrosion inhibitors is considerably below 1.0 wt%, and in some applications a concentration of only 0.1 wt% is adequate.

4.6.5.3 Pour point improvers

4.6.5.3.1 Wax separation

The flow characteristics of paraffin-based lubricating oils are impaired at low temperatures by so-called setting. This process is caused by stable cross-linking of the paraffins crystallizing out at low temperatures. The resultant wax lattice represents only a part of the lubricating oil, which, however, is able to bind the still-liquid majority of the lubricating oil like a sponge. It is true that the resultant wax lattices have little mechanical strength and can be destroyed by the action of relatively low shear stresses; but they can prevent the lubricating oil from flowing freely.

These paraffins could be virtually totally removed from the mineral oil during dewaxing, but a residue of about 2–5 percent high-setting paraffins is left for reasons of economy and to avoid making too large a sacrifice with regard to the good viscosity–temperature behaviour of paraffin-based base oils. Their effects are counteracted by the addition of small quantities of certain additives. This procedure is also justified by the fact that, as a rule, the paraffin is a disadvantage only in lubricating oils which are used at very low temperatures.

4.6.5.3.2 Additive types and functional mechanisms

The pour point improvers can be classified into the following groups (Fig. 4.41) (63).

(a) *Naphthalene condensation products.* These are the oldest additives used for this purpose; they have been in use since the start of the 1930s. They are obtained by condensing naphthalene with chlorinated paraffin, with aluminium chloride as a catalyst.

(b) *Phenol condensation products.* These are very similar to the naphthalene condensation products and differ from these primarily in the number of alkyl residues on the phenol nucleus.

(c) *Polymethacrylates/polyacrylates.* These are discussed in Section 4.5.6.4. Those used as pour point improvers are mainly polymers with lower molecular weights than those of the VI improvers.

Property	Chemical composition	Initial products	Name
Lower pour point	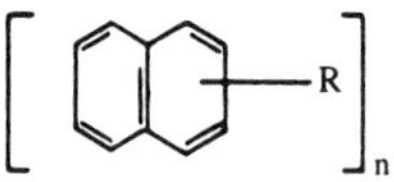 [naphthalene ring]—R]ₙ	Naphthalene, chlorinated paraffin	Naphthalene condensation products
Lower pour point	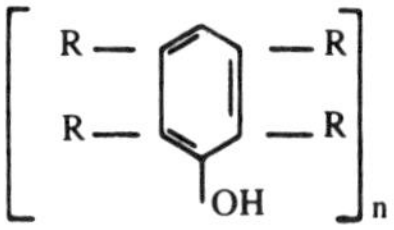 R—[benzene ring with OH]—R, R——R]ₙ	Alkylated phenols chlorinated paraffin	Penol phenol condensation products
Lower pour point	CH_3 | CH—$[CH_2]_n$— | CO | O—R]ₙ	Vinyl acetate olefin *	Vinylacetate olefin copolymers
Lower pour point Raise viscosity index	CH_3 | —CH_2—C— | OCOR]ₙ	Methyl acrylic acid ester, acrylic acid ester *	Polymethacrylates polyacrylates
Lower pour point Raise viscosity index	CH_3 OCOR | | —C———C— | | O-Acetat CH_2 | OCOR]ₙ	Vinyl acetates Fumaric acid ester *	Polyester

*Also a polymerisation catalyst

Fig. 4.41   Survey of the most important pour point improvers (63)

(d) *Vinyl acetal/fumaric ester copolymers.* These additives are also discussed in Section 4.6.5.4.

(e) *Olefin/ester copolymers.* Certain copolymers of unsaturated esters with olefins (ethylene, propylene) are effective pour point improvers.

These additives are not able to stop wax crystallizing out of the lubricating oil. However, at the start of the separation of wax crystals, they attach them adsorbently, and thus prevent them cross-linking into dense and stable wax lattices; they have no effect on the cloud point, i.e., the temperatures at which crystallizing out begins.

The effectiveness of pour point improvers depends not only on their chemical composition, but also to a considerable extent on the type of base oil.

Multigrade oils, which contain polymethacrylates as viscosity index improvers, do not normally require additional pour point improvers. Otherwise, between 0.1 and 1.0 percent of a pour point improver is added. Higher concentrations rarely produce a further lowering of the pour point.

4.6.5.4 Viscosity index improvers

4.6.5.4.1 Functional mechanism
The viscosity–temperature relationship of base lubricating oils is such that, for some applications, the viscosity is too high in the low temperature range and too low at operating temperature. This applies mainly to engine oils, but can also be of significance for gear oils and hydraulic oils. While the viscosity–temperature relationship suitable for monograde oils can also be achieved from mineral oils by suitable refining processes, this is either not possible for formulating multigrade oils, or is only possible with certain restrictions. To improve the viscosity–temperature behaviour, that is, to raise the viscosity index (VI), additives are used which have better solubility at high temperatures than at low temperatures. This means that, at a given concentration, they increase the viscosity of the base oil relatively less at low temperatures than at high temperatures.

This makes the viscosity–temperature curve flat. This function is performed by oil-soluble polymers with high molecular weights. The extent of their influence depends on the structure of the polymer molecules, their average molecular weight, and the molecular weight spectrum. The mechanism is shown schematically in Fig. 4.42.

At low temperatures the long-chain molecules of the polymer are tightly coiled and increase the viscosity of the base oil acting only slightly as a solvent. As the temperature rises the molecules of the polymer gradually uncoil, their spatial expansion increases at an ever greater rate than that of the solvent molecules, and this causes progressive flow impedance. The thickening effect of such polymeric additives is several times greater at high temperatures than at low temperatures, and thus causes a reduction in the temperature-dependence of the viscosity (**63**).

4.6.5.4.2 Types of VI improver
As the additives used as VI improvers have other properties in addition to that of influencing the viscosity–temperature behaviour, they can

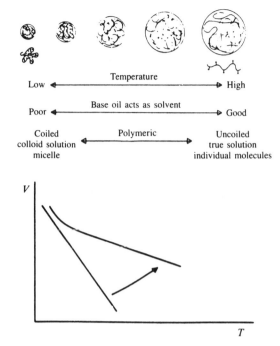

Fig. 4.42   Mechanism of polymeric additives used as viscosity index improvers (schematic)

be defined as follows (**65**)–(**67**):

(a) monofunctional   –   raise viscosity index;
(b) bifunctional   –   raise viscosity index;
                       –   lower pour point;
(c) trifunctional   –   raise viscosity index;
                       –   lower pour point;
                       –   dispersant/detergent effect.

The description of the types of polymers in this list is purely function-based and gives no indication of their chemical composition. From the chemical point of view, the following types of polymers which are used as viscosity index improvers can be differentiated (Fig. 4.43):

–   polyisobutylenes;
–   olefin copolymers;

Type	Spectrum of activity	Chemical composition	Initial products	Name
I	Mono-functional	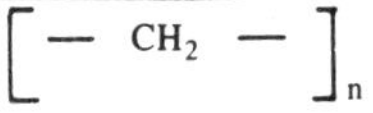 $\left[- CH_2 - \right]_n$	n-olefins, (butene, propene, ethylene, etc.)	Polyolefins (e.g. polybutenes), olefin copolymers ('OCP')
I	Mono-functional	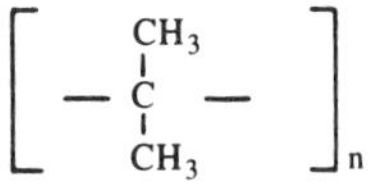 $\left[-\overset{CH_3}{\underset{CH_3}{C}}- \right]_n$	Isobutene	Polyisobutene
I	Mono-functional	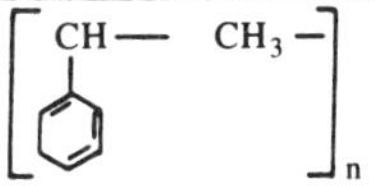 $\left[CH- CH_3 - \right]_n$ (with phenyl ring)	Styrol, olefins	Styrene copolymers (SBC)
II	Bi-functional	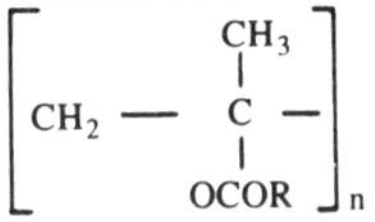 $\left[CH_2 - \overset{CH_3}{\underset{OCOR}{C}} - \right]_n$	Methacryl (acryl) acid*	Polymethacrylates (PMA) (polyacrylates)
II	Bi-functional	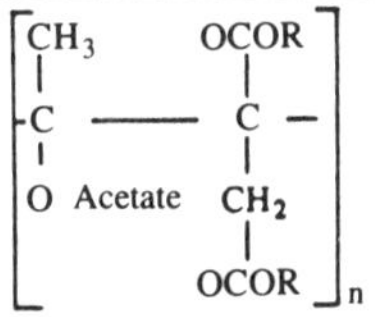 $\left[\overset{CH_3}{\underset{O\ Acetate}{C}} - \overset{OCOR}{\underset{CH_2,\ OCOR}{C}} - \right]_n$	Vinyl acetate, fumaric acid ester*	Vinyl acetate/ fumaric acid ester copolymers
III	Tri-functional	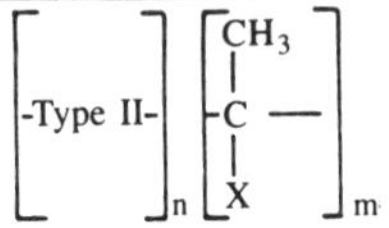 $\left[-\text{Type II}- \right]_n \left[\overset{CH_3}{\underset{X}{C}} - \right]_m$ (simplified representation) X—hydrophilic component with dispersant properties n > m	Large possible variety of components for selection. Dispersant effect usually obtained by incorporating nitrogen compounds	Copolymers of complex composition

*Plus a polymerisation catalyst

Fig. 4.43   Survey of the most important types of viscosity index improvers (60)

– styrene copolymers;
– polymethacrylates, polyacrylates;
– polyesters.

By incorporating polar groups, the VI improvers can also produce dispersant additives. Viscosity index improvers, due to the high molecular structure of the polymers they contain as additives, are more or less sensitive to thermal stresses, and also to mechanical loading by shear forces. The quality selection criterion to be observed for viscosity index improvers, in addition to improving the viscosity–temperature behaviour and the thickening caused by these additives, is, therefore, their resistance to thermal and mechanical influences. Furthermore, it may also be necessary to consider their dispersant capacity and their pour-point-improving properties. Polymeric additives are sometimes also considered to reduce wear at highly stressed friction points.

The demands to be made on polymers as viscosity index improvers may be summarized as follows.

(a) Satisfactory improvement of the viscosity–temperature behaviour in acceptable concentrations (favourable cost/benefit ratio).

(b) Adequate mechanical stability (shear stability), i.e., maintain the SAE viscosity class until the next oil change.

(c) Least possible undesirable side effects, e.g., due to thermal and oxidative influences, which may cause deposits at hot spots.

4.6.5.5 Types of EP and AW additives

4.6.5.5.1 Compounds

Figure 4.44 shows a survey of the most important anti-wear and anti-scoring additives. A distinction must be made between physically adsorbent and chemically reactive compounds (**68**)(**69**).

(a) Chemically active additives
Sulphur compounds

These include sulphurated fat oils, sulphurated terpenes, olefins and dibenzyl disulphides, as well as, in the past, sulphurated sperm oil. The more active the sulphur is, i.e., the more loosely it is bound in the compound, the less energy created by friction is necessary for it to be effective. There are, therefore, limits on the use of large quantities of sulphur compounds, as there may possibly be deleterious effects on non-ferrous metals and some seal materials.

Phosphorus compounds

The most familiar representative of this group of additives is tricresyl phosphate, which can be regarded as a mild additive. Compared with the sulphur compounds, phosphorus-containing compounds have greater anti-wear properties, but poorer anti-scoring properties.

Chemical composition	Initial products	Name
$R-(S)_x-R$	Unsaturated fatty acid esters, cyclic and aliphatic olefins, sulphur, hydrogen sulphide	Sulphurated fat oils and sperm oil, sulphurated terpenes and olefins
$(RO)_3PO$, $(RO)_3P$ $(RO)_2P(O)OH$	Alcohols, phosphorus oxychloride, phosphorus trichloride	Tertiary and secondary esters of phosphoric or phosphorous acid
$R \dots (CHCl)_x \dots CH_2Cl$	Aliphatic hydrocarbons, chlorine	Chloroparaffins
$[(RO)_2P(S) \ S]_2Me$	Alcohol, phosphorus pentasulphide, metal oxide (metal usually zinc)	Metal dialkyl(diaryl) dithiophosphates
$(RO)_2P(S)_2-S_x$ $x = 1$ or 2	Alcohol, phosphorus pentasulphide	Dialkyl(diaryl) dithiophosphate sulphides
$(R)P(S) \dots S \dots R*)$	Olefins, phosphorus pentasulphide	Alkyl dithiophosphonates

R — alkyl(aryl) residue
*Chemical composition largely unknown. P/S ratio depends on manufacturing process

Fig. 4.44   Survey of most important anti-wear and anti-scoring additives

Unlike the sulphur compounds, phosphoric acid esters can cause rusting of ferrous materials due to hydrolysis of the ester in the presence of water.

Chlorine compounds

Whereas, in the past, chlorine compounds had a firm place as anti-scoring and anti-wear additives, today they are only used in isolated applications, for example in metal machining oils. The reason for this is the danger of corrosion, because hydrochloric acid can be separated out at higher temperatures. Chloroparaffin is one of the most important chlorine compounds.

Lead compounds

These include lead naphthenates and lead oleates, which used to have considerable importance primarily as additives for gear oils. They were also useful in running-in oils. Due to problems associated with the use of lead, these additives have now been replaced with other compounds all over the world.

Combined compounds

The most effective compounds of this type include most of the metal dialkyl dithiophosphates based on zinc. The occurrence of certain pro-

ducts of decomposition is not a problem in engine oils due to the
presence of detergent and dispersant additives. In the case of gear and
hydraulic oils, however, it is advisable to use a combination of stable
dialkyl dithiophosphates (without metal), reaction products of phosphoric acid esters with amines and suitable sulphur compounds, such
as, for example, sulphurated terpenes, sulphurated fatty oils or sulphurated olefins. Such combinations meet the demands for thermal
stability, corrosion prevention, and compatibility with seal materials.

(b) *Physically active additives*
The second group of surface-active additives acts by adsorption with the
metal surfaces. These include metal soaps, esters, fatty oils, and organic
acids. These additives react relatively easily with metals or metal oxides
and form intermediate films. There is also another effect, however. Due
to their polarity, the molecules orientate themselves in a brush-like
fashion on the surface of the metal. The strength of the adsorption
depends on the van der Waals forces of the hydrocarbon groups and the
dipole moment of the polar group with respect to the metal surfaces. The
load capacity is increased by layer formation and polarity. The shear
strength of the adsorption layers, however, is low. The layers sheared off
during mixed friction are continuously replaced.

The solid lubricants can also be classed with the physically active
additives. These include primarily molybdenum disulphide (MoS_2) and
graphite, which are used dispersed in oil. Even though it has now been
found, at least for MoS_2, that they can also react chemically, the actual
mechanism of their effectiveness lies in physical attachment to the surfaces. Their anti-wear action is more pronounced than their anti-scoring
effect. As they are solids, particular attention must be paid to their dispersion to ensure that they do not settle out of the lubricant. Often they
are offered as so-called supplementary additives, to be mixed with ready
formulated oils and greases by the lubricant user.

4.6.5.5.2 Formulation of lubricants
The concentration of these additives depends on the application. A
concentration of 1 wt% is sufficient for the formulation of so-called light
gear oils or hydraulic oils, while more than 6 wt% may be necessary to
meet the demands made on rear axle gear oils. For special gear oils, for
example, locking differential oils or running-in oils, concentrations of up
to 10, and in exceptional cases up to 30 wt% may be necessary.

When formulating lubricants with anti-wear and anti-scoring additives, care must be taken to ensure that these additives meet certain other requirements. These include a neutral response to seal materials and non-ferrous metals, adequate rust prevention where there is water ingress, low foaming tendency, and adequate thermal and oxidative stability for satisfactory service life.

4.6.5.6 Friction modifiers

4.6.5.6.1 General

As a rule these are friction-reducing additives (70). This group of substances consists of mild extreme-pressure additives, which correspond approximately to the polymers used in automatic transmission fluids (ATFs) with functional groups and extreme pressure characteristics based on sulphur/phosphorus incorporated into the organic residue. These additives are required to diminish friction losses in engines and gears and thus reduce fuel consumption.

This requirement is problematic, as the most important friction points in an engine are characterized by liquid friction, for which viscosity is the decisive factor. It is, therefore, only possible to influence friction directly at friction points subject to mixed friction conditions.

4.6.5.6.2 Types of friction modifier

It is possible to reduce friction to raise mechanical efficiency in engines and gears with the following measures, which can be used individually or in combination:

- lower viscosity;
- use of friction modifier (FM);
- replace additives which increase friction.

The friction modifiers used in practice today can be classified according to the following criteria:

- ash-producing FM;
- ash-free FM;
- oil-soluble FM;
- non-oil-soluble FM;
- FM containing molybdenum;
- FM containing no molybdenum.

Modern 'fuel economy' engine and gear oils can contain the following groups of substances as friction modifiers:

- oil-soluble, long-chain carbonic acids and their derivatives such as alcohols, esters, and salts;
- oil-soluble, long-chain phosphoric acids or phosphorus acids and their derivatives;
- oil-soluble, long-chain amines, amides and imides, and their derivatives;
- oil-soluble molybdenum compounds, e.g., molybdenum dithiophosphate, complex amine-molybdenum compounds;
- insoluble solid lubricant dispersions, e.g., graphite, molybdenum disulphide, polytetrafluoroethylene.

4.6.5.6.3   Mechanism of friction reduction

A friction modifier molecule dissolved in oil is attracted to metal by means of adsorption forces. It is actually the polar end of the molecule which is orientated towards the metal, while the long hydrocarbon chain is 'attracted' by the oil, which also consists of hydrocarbons.

As a result, the molecules are anchored to the metal surface by their polar ends, the long chains being arranged vertically to the metal surface. Figure 4.45 gives a schematic representation of this process and mechanism.

The mechanism of the action of oil-insoluble solid lubricants is characterized by mechanical coating of the surfaces. The resultant film reduces the friction between the surfaces which move relative to one another.

Only mechanical losses can be reduced by friction reduction by means of measures applied to the oil. It must be remembered that in the range of liquid friction, only reducing the viscosity achieves the desired effect, while the action of friction-reducing additives, the so-called friction modifiers, is restricted to the mixed friction range. When formulating friction-reducing lubricating oils, it must be remembered that reducing the viscosity (reducing friction in liquid friction) enlarges the mixed friction area (higher friction). This disadvantage must be equalized with additives which reduce friction and wear.

Figure 4.46 gives a schematic representation of the combined effects of reducing viscosity and using friction modifiers to formulate modern friction-reducing oils. Starting with oil A, reducing the viscosity produces oil B, which reduces friction in liquid friction, but, due to the greater mixed friction component, leads to more wear and friction in these con-

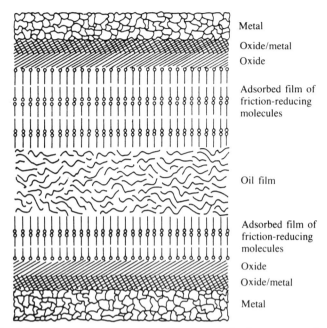

Metal
Oxide/metal
Oxide

Adsorbed film of
friction-reducing
molecules

Oil film

Adsorbed film of
friction-reducing
molecules

Oxide
Oxide/metal

Metal

Fig. 4.45 Adsorption and orientation of polar molecules on metal surfaces (schematic)

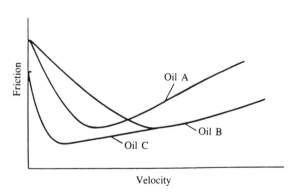

Fig. 4.46 Combined effects of viscosity reduction and use of friction modifiers: A – reference oil; B – viscosity reduction; C – viscosity reduction + friction reducer

ditions. Adding friction modifiers, i.e., substances which reduce friction and wear, gives the optimal oil C.

4.6.5.7   Detergents and dispersants

4.6.5.7.1   General

The detergent/dispersant additives are intended to suspend impurities and thus prevent precipitation and deposits, stop the formation of sludge-like products from solid and liquid impurities and neutralize acid substances. They are intended to re-dissolve existing deposits. The occurrence of such solid and liquid impurities is characteristic of the thickening of engine oils during operation in internal combustion engines. As engine oils are also used for gear lubrication in special cases, the detergent/dispersant additives will be described (**71**) even where they have no direct relevance to gear lubrication.

4.6.5.7.2   Types of DD additives

The detergent additives are oil-soluble or finely dispersed metal salts of organic acids (organometallic compounds), which are used primarily to prevent deposits on hot engine parts. This high-temperature effectiveness of these ash-producing additives is ascribed to their cleaning or dissolving effect.

In the narrower sense, dispersant additives are ash-free organic compounds, which stop sludge formation from solid and liquid impurities at temperatures below operating temperatures, so-called cold sludge. This effect is ascribed to the dispersant capacity of these products, which prevents flocculation and coagulation of colloid products.

The transition between the actions of the detergents and dispersants is fluid and not clearly defined.

As the ability to suspend or disperse is undoubtedly the most important factor, both classes of additives are sometimes called dispersants.

The following products can be used as ash-producing, i.e., organometallic detergent additives

- naphthenates and stearates;
- sulphonates;
- phenolates, phenolate sulphides and salicylates;
- phosphates, thiophosphates, phosphonates and thiophosphonates;
- carbamates and thiocarbamates.

Figure 4.47 gives a comparison of the structural formulae of some important types.

Sulphontes

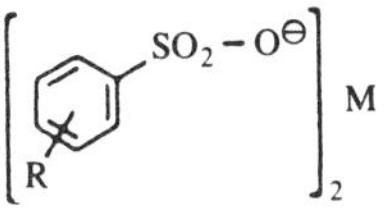

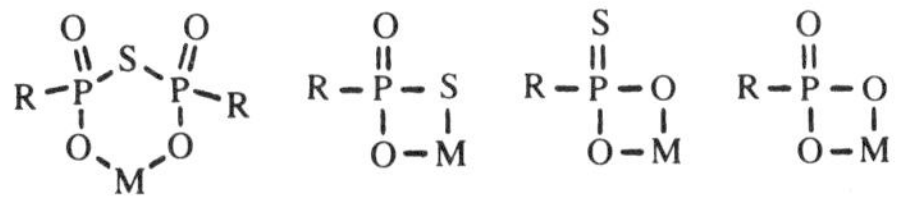

$$\left[\underset{R}{\bigcirc}\!\!\!-\!\!SO_2-O^{\ominus}\right]_2 M$$

$$\underset{\underset{O}{\overset{O}{\underset{\|}{\overset{\|}{R-P}}}}\underset{M}{\overset{S}{\diagdown}}\underset{O}{\overset{O}{\underset{\|}{\overset{\|}{P-R}}}}}{}
\qquad
\underset{\overset{O-M}{}}{\overset{O}{\underset{\|}{\overset{\|}{R-P-S}}}}
\qquad
\underset{\overset{O-M}{}}{\overset{S}{\underset{\|}{\overset{\|}{R-P-O}}}}
\qquad
\underset{\overset{O-M}{}}{\overset{O}{\underset{\|}{\overset{\|}{R-P-O}}}}$$

Thiopyrophosphonates   Thiophosphonates   Phosphonates

R = polyisobutene group, molecular weight: range 500–1200

Salts of alkyl salicylic acids   Phenolates

(MCO₃)

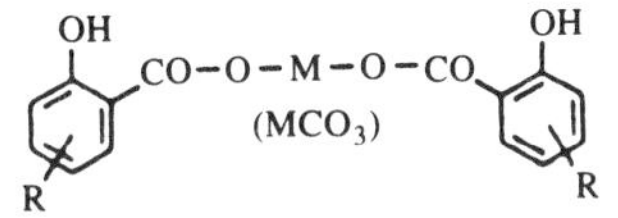

R = long straight-chain alkyl residue

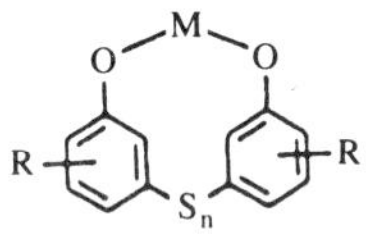

Fig. 4.47 Important types of detergent additives (71)

The following product groups can be used as ash-free, i.e., organic dispersant additives, (a) polyisobutenyl succinic acid derivatives and (b) methacrylate copolymers and fumarates. Figure 4.48 gives the structural formulae of these types in the form of a comparison.

4.6.5.7.3 Purposes and mechanisms of detergent/dispersant additives

The purposes of detergents and dispersants can be summarized as follows

– prevention of sludge formation (cold sludge);
– prevent deposits forming on the piston, primarily in the grooves;

Polyisobutenyl succinic acid derivatives

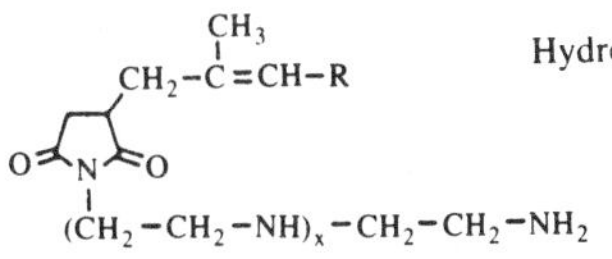

Hydrocarbon part

$x = 3$ as a rule

Polar head

Methyacrylate copolymers

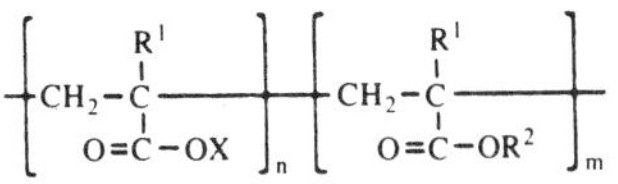

$$\left\{\!\!\underset{\underset{O=C-OX}{}}{\overset{R^1}{\underset{|}{\overset{|}{CH_2-C}}}}\!\!\right\}_n\!\!\left\{\!\!\underset{\underset{O=C-OR^2}{}}{\overset{R^1}{\underset{|}{\overset{|}{CH_2-C}}}}\!\!\right\}_m$$

$R^1 = CH_3$ generally, $R^2 = $ alkyl (chain with C_{22}),

$x = $ amine, amide, imide, OH, polar groups

Fig. 4.48 Structural formulae for dispersant additives (71)

- prevent the formation of varnish on hot metal surfaces (e.g. on piston skirt);
- prevent piston ring sticking (piston sticking);
- prevent wear (mechanical and corrosive) on pistons and cylinders, on the camshaft and the bearings;
- prevent rust;
- prevent the formation of acid;
- prevent the agglomeration of combustion products;
- prevent oil oxidation.

The function and the action can generally be divided into three groups:

- peptization;
- solubilization;
- neutralization.

Peptization means coating solid dirt particles with ashless dispersants or metal detergents (Fig. 4.49). While both metal detergents and ashless dispersants can peptize solid particles, solubilization can only be carried out by ash-free dispersants (Fig. 4.50). Here, oil-soluble foreign substances, such as acids or fuel particles, are coated by the dispersants and their (mainly negative) effects are hindered or diminished. This coated cell (the so-called micelle) looks almost the same as with peptization, but the nucleus is liquid and not solid. Neutralization means the chemical neutralization of acid constituents in the oil (Fig. 4.51).

Particle geometry	Molar ratio of dispersed substance/ dispersant additive	Mechanism of deposit reduction
Variable	Variable	Rate of deposition slowed

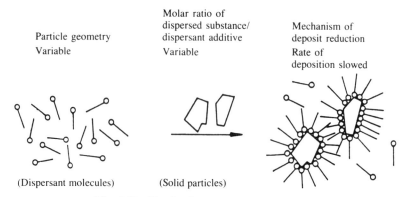

(Dispersant molecules)   (Solid particles)

Fig. 4.49   Peptization process (schematic) (71)

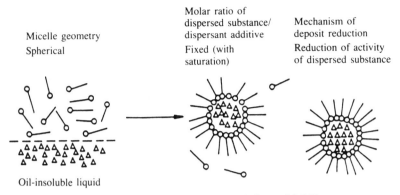

Micelle geometry	Molar ratio of dispersed substance/ dispersant additive	Mechanism of deposit reduction
Spherical	Fixed (with saturation)	Reduction of activity of dispersed substance

Oil-insoluble liquid

Fig. 4.50 Solubilization process (schematic) (71)

4.6.5.8 Foam inhibitors

Foam is a system consisting of two liquid substances which are only partially miscible. The dispersed component consists of relatively large droplets (**63**).

In lubricating oils, foam formation can impair oil flow if the quantity and the viscosity of the surface foam become too great. We can distinguish between surface foam, which can be suppressed with defoaming agents, and 'aero-emulsions', which may be influenced by additives.

To disperse liquid droplets or gas bubbles in the oil, the surface of the dispersed phase must be increased. This requires work which depends, among other things, on the surface tension of the lubricant. The greater

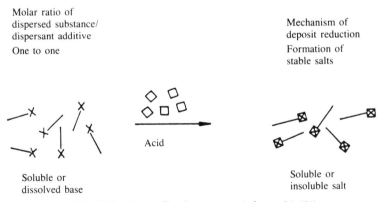

Molar ratio of dispersed substance/ dispersant additive	Mechanism of deposit reduction
One to one	Formation of stable salts

Acid

Soluble or dissolved base	Soluble or insoluble salt

Fig. 4.51 Neutralization process (schematic) (71)

the surface tension of the oil, the greater is the work required, and the lower is the stability of a foam. As detergents reduce surface tension, they can be the cause of foam formation. Correction, therefore, requires an additive which increases the surface tension of the lubricating oil. Foam inhibitors, or antifoaming agents as they are also called, are such additives.

To act as a foam inhibitor, an additive must possess the following characteristics:

(a)   it must be practically insoluble in oil;
(b)   its surface tension must be lower than that of the oil;
(c)   it must be so well dispersed in the oil that it retains its activity even after long storage.

The most familiar additives of this type are the polyalkyl siloxanes or silicones. Figure 4.52 (left) shows their general structure.

The main chain consists of alternating silicon and oxygen atoms. The side chains consist of hydrocarbon chains which are bonded to the silicon atoms. For the polymethyl siloxanes which are mostly used, the substitutes are methyl groups. Cyclical structures are also possible (Fig. 4.52 – right).

The silicones are very active as foam inhibitors, and as little as 10–100 ppm is sufficient for use in lubricating oils. When manufacturing the formulations they are generally added after the detergents/dispersants.

4.6.5.9   Adhesion improvers

Adhesion improvers are additives which are used in lubricants to make the film on machine parts thicker or more stable, for example, at points where fast motion or design factors mean that the oil could too easily be thrown off, and oil losses could be high. The action of such additives is mainly to increase the viscosity of the lubricant, especially at lower shear rates, and to improve the adhesion of oil films to metal surfaces (**62**).

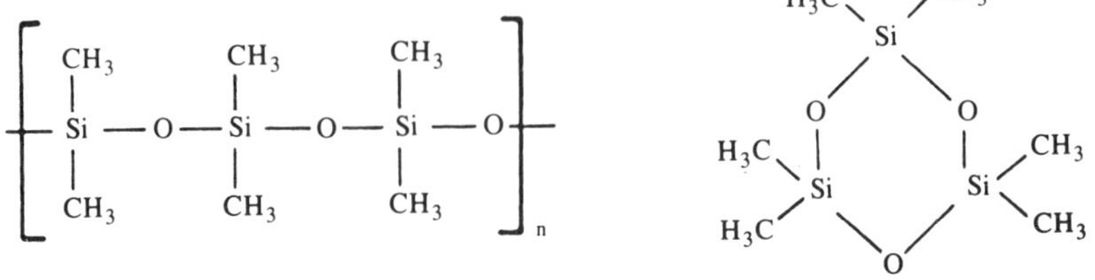

Fig. 4.52   Polyalkyl siloxanes (63)

Two types of adhesion improvers can be distinguished on the basis of chemical structure.

- *Polymeric products*, which effect a considerable viscosity increase at low concentrations due to their high molecular weight.
- *Polar low-molecular weight compounds*, which are surface-active due to their polar structure and, therefore, adhere more strongly to machine parts, and may also be able to form superstructures which greatly increase viscosity.

Polyisobutylene is the main polymeric compound used. Since a product with a high molecular weight (e.g., 500 000–1 000 000) is generally used, lubricating oils with polyisobutylene as the adhesion improver often exhibit a marked structural viscosity (non-Newtonian flow characteristics). Depending on molecular weight, the polyisobutylenes are very viscous to rubber-like substances, which, after pre-mixing in mineral oil, are easily dissolved in lubricating oil.

The polar low-molecular compounds which act as adhesion improvers include metal soaps and unsaturated fatty acids. Due to their polar structure they adsorb easily on metal surfaces and thus promote the formation of stable lubricant films on these surfaces.

4.6.5.10 Additives for metal machining liquids
The tasks of cooling and lubricating are combined in the term 'cooling lubricants'. An additional important function of these products is to wash the swarf away from the machining area.

The cooling lubricants are divided into water-miscible and non-water-miscible products as per DIN 51 385. The different physical mixing conditions require a further subdivision of the water-miscible products into cooling lubricant emulsions and cooling lubricant solutions. The emphasis of the action of these products is on their cooling effect.

In the case of the water-immiscible cooling lubricants, it is the lubricating effect which is most important. Cutting oils are water-immiscible cooling lubricants. Table 4.37 summarizes the types of additives which can be used in water-miscible and water-immiscible cooling lubricants (72)(73). As these additives are not relevant to gear lubrication, they will not be discussed further.

4.6.5.11 Additives for lubricating greases
Virtually the same classes of substances are used as additives for lubricating greases as for liquid lubricants: chemical compounds, mainly of

Table 4.37   Types of additives for cooling lubricants

Water-immiscible cooling lubricants	*Water-miscible cooling lubricants*
High-pressure additives	High-pressure additives
Anti-wear additives	Anti-wear additives
Anti-corrosion additives	Anti-corrosion additives
Polar additives	Polar additives
Solid lubricants	Emulsifiers
	Biocides
	Foam inhibitors

an organic nature, with the most varied functional groups (**60**)(**61**). In detail, these are:

- fatty acids, naphthenic acids, sulphonic acids and their alkaline, alkaline earth and heavy metal salts (lead and zinc);
- esters, semi-esters and amides of organic acids and fatty acids;
- esters and salts of phosphoric acid and thiophosphoric acid;
- aromatic and aliphatic amines;
- substituted phenols, bis- and tetrakisphenols;
- disulphides, polysulphides and thioethers;
- organic halides;
- various heterocyclenes (mainly nitrogenous).

The individual additives contain compounds from the classes of substances listed, in a uniform, defined form, or as mixtures of technical homologues. A doped lubricating grease often contains several additives of different types.

Most additives are easily soluble in the base oil and are also completely dissolved in the thickened base oil of the lubricating grease. In extractive decomposition of lubricating greases they are retrieved together with the base oil in the extraction medium.

4.6.5.12   Interrelationship between base oils, additives, and other materials

Under the heading of the interrelationship between base oil and additives, we will consider, on the one hand, the response of additives, and, on the other, the compatibility of lubricants (**74**).

4.6.5.12.1   Response

By the response of additives, we mean the effect achievable in improving the characteristics of the base oil. The response is different for each additive type and for the base oil under consideration and must be determined experimentally for each separate case. The oil structure, refining level, purity, etc., play an important part. It is interesting that additives can influence one another in their response. A positive influence is known by the name 'synergism', in particular with regard to combinations of oxidation inhibitors, but it is naturally observed with other additive combinations as well.

For mineral base oils, the example of solvent refining in Fig. 4.53 shows the response of oxidation inhibitors as a function of the refining level. Increasing selective removal (shown by the curve distillate-optimal raffinate-white oil) of the oxidation-labile hydrocarbons, primarily aromatic and heterocyclic, as well as compounds with the heteroatoms, sulphur, oxygen, and nitrogen, results firstly in a so-called optimal raffinate, in which the natural oxidation inhibitors remaining in the base oil, primarily sulphur compounds, can produce the maximum effect. Further removal of these compounds clearly reduces the ageing stability; base oils of such an extraction level are considered to be superraffinates. This behaviour can also be applied to other types of additives, e.g., anti-corrosion and anti-wear additives.

The response of most synthetic substances which protect against oxidation, corrosion, wear, and foaming and which improve the dissolution or infiltration of layers of dirt and moisture on sliding surfaces and their

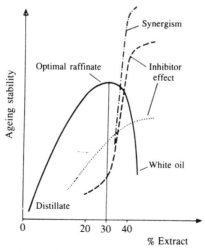

Fig. 4.53 Level of extraction against ageing stability (74)

Table 4.38   Additive susceptibility (response) as a function of base oil quality level (74)

	Additives to counteract:				*Viscosity improvers (relative viscosity*
Quality level	*Oxidation*	*Corrosion*	*Wear*	*Foam*	*increase)*
Distillates	0	+	+	+	
Raffinates (acid, earth, hydrogenation raffinates)	+	+ +	+ +	+ + +	+ + +
Solvates hydrogenation raffinates (VI ≈ 100)	+ +	+ + +	+ + +	+ + +	+ + +
Hydrocracked oils (VI > 120)	+ + +	+ + +	+ + +	*	+ +
Polyolefin oils	+ + +	+ + +	+ + +	*	+ +

0	No effect
+	Noticeable
+ +	Good
+ + +	Very good
*	Oils do not form

dispersion in the oil can generally be regarded as good in high-quality hydrocarbon base oils (Table 4.38); it increases with the refining level and is most marked with pure mineral oils. High-quality base oils derived from crude oil are, therefore, no longer produced as optimal raffinates, but as superraffinates.

The response of VI improvers is relatively better in base oils with a poor VT characteristic than in those which already have a very low viscosity–temperature dependence. Pour point improvers only respond in base oils before the viscosity setting point is reached, and so are mainly used in base oils with high residual paraffin contents or high pour points.

When using combinations of additives, the response of individual types of additives can be very considerably hindered by others, especially in those cases where the additive molecules are to be active at interfaces and certain concentrations must be supplied to the interface concerned for that purpose. Such antagonistic mechanisms and competition for space in the interface are familiar in the case of anti-corrosion and anti-wear substances, and may also occur with other combinations of additives.

4.6.5.12.2 Compatibility

Under this collective term we include the mutual compatibility of mineral oil base oils, and their compatibility with

- additives, synthetic oil components;
- seal materials, metals;
- paints and coatings.

Base oils of paraffinic and naphthenic structure, even when they have different levels of refining, are compatible and miscible with one another. Separation has sometimes been observed, e.g., in the case of mixtures of low-viscosity base oils with high-viscosity cylinder oils or mixtures of components with very different structures, e.g., aromatic extracts and paraffin oils.

Synthetic hydrocarbon oils and organic esters are compatible with one another and with paraffinic and naphthenic solvates, which can be exploited in oil formulations to achieve special characteristics, e.g., low-temperature and VT performance, among others.

Table 4.39 gives a simplified summary of the compatibility of synthetic oils and solvent raffinates. From the example of the polyglycols it can be seen that the polarity of the polyglycol – expressed by the level of hydroxylation – has a decisive influence on the mutual compatibility; the smaller the number of OH groups, the better is the solubility in the crude oil product. In the case of other synthetic oils, the type and size of the so-called organic residues are responsible for the solubility in the solvate; the larger and more numerous in the molecule, the better the solubility.

Table 4.39 **Solubility of crude oil base oils when mixed with synthetic oils (74)**

Synthetic oils	Solvent raffinates
Polyolefin oil	Soluble
Alkyl benzene	Soluble
Bicarbonic acid ester	Soluble
Polyol ester	Soluble
Polyglycol (water soluble)	Insoluble
Polyglycol (water insoluble)	Soluble
Phosphoric acid ester	Some solubility
Silicone oils	Insoluble
Silicate ester	Soluble
Halogenated HC (F, Cl)	Some solubility

In some cases, however, the opposite is required – lower solubility of hydrocarbons in synthetic oils – e.g., for fire-resistant lubricating oils for reasons of flammability.

To summarize on the subject of the compatibility and the mutual solubility of HC base oils and synthetic oils (non-hydrocarbons), it can be stated that, in the case of synthetic oils, the substance class, molecule size and form, the type and size of the organic residues, the polarity of the molecule or the substituents, among other things, are of decisive importance. It is not possible to draw up comprehensive rules; rather, it is advisable to perform solubility tests as a function of temperature and time in each case, especially if one intends to use additives.

Oils and their components are compatible with seal materials if swelling and hardening do not exceed the tolerances given in various product standards. In the case of many machine paints which claim to be oil-resistant, there are problems with oils and liquids etching and dissolving them, because compatibility testing was only carried out with regard to compatibility with the mineral oil components, and not with other substances which may be mixed with it or contained in it, or even with synthetic oils.

The effect of hydrocarbon base oils on seal materials depends greatly on the aromatic content. Antagonism increases from paraffinic oils to naphthenic and aromatic oils; the action is greater with thin oils and higher temperatures. The effect of structure and viscosity can be calculated or estimated for many elastomers. Paraffinic HC oils can also cause seals to shrink or harden. This can be prevented with ester additives, however. We should not ignore the effect of additives or synthetic oil components on seal materials. They can increase swelling or shrinkage and hardening, they can cause vulcanization (sulphur and amine compounds, etc.) on the surface, and can form deposits (Zn dithiophosphate etc.) and thus cause impairment of the sealing function.

Reactions of the hydrocarbon base oils with metallic materials are mainly associated with the sulphur content, which reacts with copper, silver and non-ferrous metals and can give them black or dark discoloration.

4.7 SOLID LUBRICANTS

4.7.1 Introduction

After a description of the areas of use and applications of solid lubricants, including an indication of advantages and disadvantages, the possible groups of substances which may be used as solid lubricants will be given. Descriptions of the crystal structure and some other characteristics will be given for the most important solid lubricants, molybdenum disulphide (MoS_2) and graphite. Using the example of MoS_2, the widely varying functional mechanisms will be indicated and the various factors influencing them will be discussed. In the case of forms which permit direct film formation, their effectiveness is influenced by moisture, oxidation, temperature, load and sliding speed, ambient pressure and radiation as well as materials combinations.

The effectiveness of solid lubricants in forms which provide indirect film formation depends on the concentration, the particle size, and the presence of other additives. Finally, the most important forms are discussed and information is given on surface treatment of materials and the application of the solid lubricants (**75**).

4.7.2 Reasons for using solid lubricants
Solid lubricants are useful only for dry lubrication and in mixed friction, not for liquid friction.

Certain operating conditions and ambient conditions virtually exclude the use of liquid lubricants, even though the tribological problems concerned could be solved better with them. The range of applications of solid lubricants will now be considered (**76**).

4.7.2.1 Areas of use and applications
The fields of use and applications of solid lubricants can be described with the following two broad features.

- Operating conditions in which no liquid or paste lubricants can be used. These include, for example:
  - low sliding speeds, small or oscillating movements;
  - high specific loads;
  - very high or very low ambient temperatures;
  - very low ambient pressures (vacuum);
  - unusual (aggressive) ambient atmospheres.

- Improvement of certain characteristics of liquid and paste lubricants, for example, use as an agent or additive for:
  - reduction of friction;
  - reduction of wear.

4.7.2.2   Advantages and disadvantages of solid lubricants
The advantages of solid lubricants are usually countered by certain disadvantages. The advantages include:

- very wide service temperature range;
- may be used in a vacuum (applies to MoS_2, but only to a limited extent to graphite);
- often very resistant to aggressive media;
- simpler design is possible, e.g., of seals, due to absence of liquid or paste lubricants.

The disadvantages include:

- continuous removal of the lubricant film in the case of direct film formation and consequent dimensional change;
- problematic 'relubrication' in the case of direct film formation;
- greater friction losses with dry friction compared with liquid friction; for use as an additive (not oil-soluble) a suspension is necessary;
- corrosion protection is sometimes problematic, e.g., when using MoS_2.

4.7.3   Groups of substances suitable for use as solid lubricants, and basic forms in which they are used

4.7.3.1   Groups of substances suitable for use as solid lubricants
The following groups of substances may be suitable for use as solid lubricants:

- certain glasses;
- surface reaction layers, e.g., gold, silver, lead, copper, indium;
- certain soft non-metals, e.g., lead and iron sulphide, lead oxide, silver iodide;
- certain organic substances, e.g., polytetrafluoroethylene (PTFE), amides, imides;
- certain substances with a layer lattice structure, e.g., sulphides or selenides of certain heavy metals, e.g., WS_2, WSE_2, MoS_2, graphite.

Within the scope of this work, only the organic substances and those with a layer lattice structure are of significance.

4.7.3.2   Basic forms for direct and indirect lubricant film formation
Solid lubricants can be used as additives (indirect lubricant film formation) or directly as lubricants (direct lubricant film formation). Figure 4.54 illustrates this dual function.

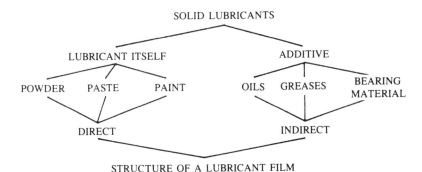

Fig. 4.54 **Solid lubricants for direct and indirect lubricant film formation**

The solid lubricants are available in various forms for direct and indirect lubricant film formation. Tables 4.40(a) and (b) contain a survey of these solid lubricants.

The forms which lead to indirect film formation are of significance for gear lubrication. There are gear oils and greases which contain solid lubricants as wear-reducing additives. The following observations relate particularly to the forms used in practical application.

4.7.4 Molybdenum disulphide (MoS₂) and graphite

MoS_2 and graphite belong to the most important solid lubricants with a layer lattice structure which are widely used.

4.7.4.1 Structure of molybdenum disulphide (MoS₂) and graphite

Layer lattices are those crystal structures in which the atoms or ions are arranged in layers (lamellae), usually in such a way that the individual layers consist only of atoms or ions of the same type and charge (Fig. 4.55) (77). As a result of this structure, such substances can absorb very

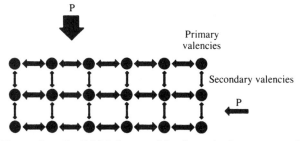

Fig. 4.55 **Formation of solid lubricants with a layer lattice structure (schematic)** (77)

Table 4.40(a)   Application forms of solid lubricants for direct and indirect lubricant film formation – direct film formation

Powder	Varnishes	Pastes
Several types of varying purity and particle size	Solid lubricants as pigments in various binders, with volatile carrier liquids in the as-supplied state	Pastes and dispersions in mineral and synthetic oils with other additives
c. 100%	10–50%	30–70%

Table 4.40(b)   Application forms of solid lubricants for direct and indirect lubricant film formation – indirect film formation

Oils	Greases	Bearing materials
Colloid suspensions: (1) in highly concentrated form as an additive to lubricating oils (2) in low concentration in ready-made lubricating oils	Powder or colloid suspensions in low concentration in ready-made lubricating greases	Additive as lubricating pigment to sliding materials made of various plastics and sintered materials
(1) 5–20% (2) c. 1%	1–10%	5–50%

high pressures vertically to the layer and at an angle to the lamella they are easily displaced. This results in good sliding and lubrication properties.

Molybdenum disulphide (MoS_2) naturally occurs as a mineral (molybdenite). It forms blue-grey, shiny, very thin, soft and elastic flakes. The layer lattice structure of molybdenum disulphide (MoS_2) consists of three planes; there is a molybdenum plane between two sulphur planes (**78**).

Between the atoms of a layer, including between the S and Mo planes, there are strong valency bonds (atom bonds), while the individual layers, i.e., every two S planes, are linked by weaker van der Waals forces. The different bonding forces between the planes, and above all the relatively smaller shear forces for displacing the individual layers, appear to be the key to understanding the physical and mechanical mechanism of molybdenum disulphide. Figure 4.56 gives a schematic illustration of the crystal structure of natural molybdenum disulphide. Perfect crystals are rarely found. Normally they have various lattice imperfections, which cause mechanical weakening of the structure. Thus the shear forces necessary to displace the layers can be reduced.

Graphite is one of the oldest and, due to its relatively low price compared with MoS_2, one of the most widely used solid lubricants. It is obtained both as a mineral and a synthetic. The layer lattice structure of graphite is illustrated schematically in Fig. 4.57 (**77**).

4.7.4.2 Comparison of characteristics

In Table 4.41 some of the characteristics of molybdenum disulphide and graphite are compared with those of polytetrafluoroethylene (PTFE).

The coefficient of friction of graphite is about 0.1. The maximum temperature at which it can be used is 450°C, if oxygen has unrestricted access. Above this temperature, graphite oxidizes. The pressure resistance of graphite is high and increases with rising temperature. Graphite also has high electrical and thermal conductivity. It lubricates very well in moist air and in a carbon dioxide atmosphere. However, its lubricating effect is only moderate in oxygen – in nitrogen it is poor.

The use of graphite is limited by the fact that absorbed gas, oil or water films on the surface are necessary to maintain its good lubricating qualities. In a high vacuum these are removed. They also lose their effectiveness at low temperatures, so the overall effectiveness of graphite falls in these conditions.

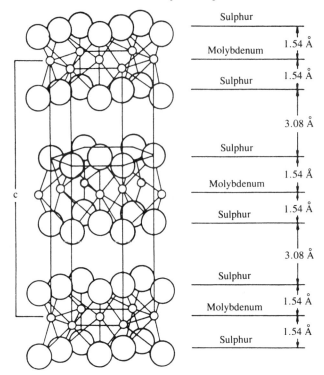

Fig. 4.56   Crystal structure of molybdenum disulphide (schematic) (78)

The coefficient of friction of molybdenum disulphide (MoS_2) lies between 0.03 and 0.06. In humid air it is higher than in dry air, but it falls with increasing sliding velocities and loads, due to a type of 'drying effect'.

The load capacity of molybdenum disulphide (MoS_2), according to recent studies, extends beyond 3000 N/mm^2 i.e., beyond the yield point of the metals and alloys currently used as bearing materials. MoS_2 oxidizes from about 360°C to molybdenum trioxide or molybdenum dioxide. If molybdenum disulphide is used in inert atmospheres, e.g., inert gases or nitrogen, the temperature range for use can be extended considerably.

Unlike graphite, MoS_2 remains fully effective as a lubricant even in an ultra-high vacuum, but it possesses virtually no electrical conductivity. There is the disadvantage that MoS_2 powder causes electro-chemical

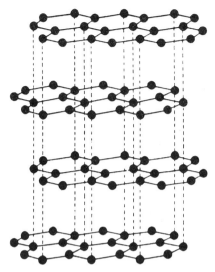

Fig. 4.57 **Crystal structure of graphite (schematic)** (77)

Table 4.41 **Important characteristics of graphite, molybdenum disulphide and polytetrafluoroethylene**

	Graphite	MoS_2	PTFE
Density at 20°C (g/cm^2)	1.4–2.4	4.8–4.9	2.1–2.3
Mohs hardness	1–2	1–2	–
Molecular weight	12.01	160.7	70 000–250 000
Friction coefficient	0.08–0.10	0.03–0.06	0.04–0.09
Start of oxidization in air (°C)	450	360	
Maximum operating temperature (°C) (without oxygen ingress)	550	420	260
Minimum operating temperature (°C)	−20	−180	−250
Products of oxidation, decomposition	CO, CO_2	MoO_3, MoO_2	C_2F_4
Resistance to:			
chemicals	Very good	Good	Good
corrosion	Good	Poor	Good
radiation	Good	Good	Poor

corrosion of iron and steel surfaces in the presence of moisture. This disadvantage must be considered when using it.

4.7.5 Functional mechanism, illustrated with the example of molybdenum disulphide

The effectiveness of solid lubricants with a layer lattice structure is primarily ascribed to the ease with which the crystals can be displaced within the planes (layers). This mechanism and a number of others are described in the following sections (75).

4.7.5.1 *Physico-mechanical mechanism*

The greater spacings between the sulphur planes of molybdenum disulphide, i.e., between the individual layers, and the smaller spacings between the molybdenum and sulphur planes are equated with weaker bonding forces between the layers and stronger forces within the layers.

If a shear stress is applied, the individual MoS_2 layers are, therefore, displaced relative to one another. If they are between two sliding surfaces, the sliding motion is subject to lower friction than without the MoS_2 film. Figure 4.58 gives a schematic representation of this process, which is often called the 'pack of cards effect'.

4.7.5.2 *Adsorbed gas layers*

The good lubricating effect of MoS_2 could also be due to weak bonds between the crystals, caused by adsorption layers. This would mean that, with relative motion, individual crystals would be displaced with respect to one another, and not individual layers or lamellas within the crystal.

The slight amount of sliding within the molybdenum disulphide film must, therefore, also be due to adsorption films, i.e., to gas or water vapour layers. Such foreign layers on the crystal surfaces of the molybdenum disulphide could weaken the bonds between the individual particles.

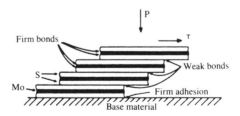

Fig. 4.58   Schematic illustration of the relative motion between MoS_2 layers

Examination of this effect has shown that the lubrication effect of MoS_2 is not, as with graphite, improved by such adsorbed gas or water vapour layers, but impaired. As a result of this mechanism it can be maintained that MoS_2 is superior to graphite in a vacuum, where there are no such layers.

4.7.5.3 Intercrystalline sliding

Another theory explains the low friction with lubrication by molybdenum disulphide (MoS_2) by means of intercrystalline sliding processes. The good friction properties of layer lattice residual lubricants thus depend clearly on the free surface energies. The high-energy lateral faces of the crystals react quickly, so stable oxides form, which have little interaction with the low-energy sliding surfaces. Thus the bonds between the MoS_2 layers are weakened. This assumption may be more accurate for graphite than for molybdenum disulphide, in which the bonds between the S planes are already so weak that the boundary energies must play a smaller part.

4.7.5.4 Separation effect

The functional mechanisms of molybdenum disulphide (MoS_2) discussed so far allow separation of the friction pair sliding against one another. A prerequisite is good adhesion or anchoring of the MoS_2 particles on the surfaces, which is ensured by the strong adhesion of the sulphur atoms in the S boundary planes to metals.

As the MoS_2, on the one hand, must prevent direct metal contact between the friction pair, and, on the other, possesses lower shear strength than the base material, it can prevent excessive wear and permit smoothing of the surface asperites by plastic deformation. This is closely linked to the filling of the surface depressions with MoS_2. This increases the contact surface, which is equivalent to reducing the specific surface pressure.

If the molybdenum disulphide (MoS_2) is suspended in a liquid or paste-type lubricant, it is conceivable that the MoS_2 particles may be orientated in the gap between the friction pair.

It can then act as a buffer and absorb shock forces. Figure 4.59 gives a schematic illustration of these effects of the mechanical separation as an explanation of the mechanism of molybdenum disulphide (MoS_2).

Another effect is worthy of mention in this connection. The MoS_2 particles suspended in a liquid or paste lubricant can alter the flow behaviour. The possible viscosity increase is desirable even in conditions

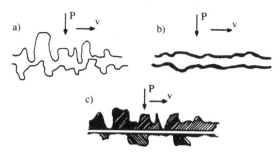

Fig. 4.59 Schematic representation of the functional mechanism of molybdenum disulphide: (a) original surface; (b) surface smoothed by plastic and elastic deformation after being coated with a MoS_2 film; (c) increase in load component as surface depressions are filled with MoS_2

of mixed friction at local points with hydrodynamic or elastohydrodynamic pressures. These two effects naturally apply also to graphite and other solid lubricant particles.

4.7.5.5 Chemical action

The physico-mechanical mechanism of molybdenum disulphide (MoS_2) discussed so far does not indicate its good performance as a lubricant in all cases, so we will also discuss the process of chemical reactions between the MoS_2 and the iron of the friction pair surfaces.

It has been demonstrated (79), that the following reaction takes place

$$MoS_2 + 2Fe = 2FeS + Mo$$

As Fig. 4.60 shows, the conversion of MoS_2 with Fe to FeS and Mo accelerates considerably above a temperature of 720°C. Such temperatures are quite realistic in local areas of tribological contact.

During the chemical reaction between the MoS_2 and the metal of the friction pair, metallic molybdenum (Mo) is formed, which can 'temper' the surface. Diffusion of the molybdenum into the base metal also cannot be ruled out. At the very least, there could be an alteration in the material characteristics in the uppermost layers, for instance an increase in fatigue strength.

4.7.6 Factors influencing effectiveness

Factors influencing the effectiveness of solid lubricants will also be described using the example of molybdenum disulphide with indirect and direct film formation.

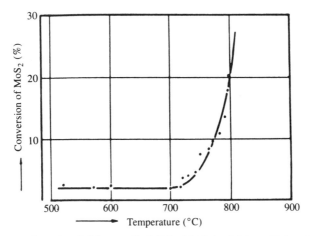

Fig. 4.60 **Reaction of molybdenum disulphide (MoS₂) with iron (Fe) to form iron sulphide (FeS) as a function of temperature**

4.7.6.1   Factors influencing effectiveness with direct film formation
When using solid lubricants for direct film formation, that is, in the form of powders, varnishes and pastes, the effectiveness can be influenced by the following parameters:

– water vapour (moisture) and other gases
– oxidation
– load and velocity
– temperature
– ambient pressure
– high-energy radiation
– material pair
– impurities

As solid lubricants are mostly used in indirect film formation, we will not consider the factors affecting efficiency with direct film formation any further. The reader is referred to the extensive literature on the subject (75).

4.7.6.2   Factors influencing effectiveness with indirect film formation
By indirect film formation we mean that the solid lubricant leaves a carrier medium and is deposited on the surfaces of the friction pair where it must be anchored. Solid lubricant particles can be suspended in liquid

carrier media, e.g., mineral and synthetic oils, and in paste-like media, e.g., lubricating greases.

Their tribological effectiveness then depends on the concentration and the particle size as well as the presence of other, chemically active agents, in addition to the factors already discussed.

4.7.6.2.1 Effect of particle size

Wear tests performed on simulation test equipment, such as, e.g., the four-ball apparatus and the FZG machine, have shown that colloid MoS$_2$ suspensions with fairly large particles (5 μm and larger) produce higher wear than suspensions with small particles (1 μm and smaller) **(80)**.

This has been confirmed in more extensive tests **(81)**. Figure 4.61 gives the wear values as a function of operating time for two different particle sizes. Figure 4.62 shows the relationship between wear values and load. It is quite obvious that the larger particles tend to produce greater wear. Tests performed on the FZG machine produced a contradictory result. As Fig. 4.63 shows, the finer particles permit somewhat greater wear than the suspension with the larger solid lubricant particles. The reason for this contradictory behaviour is probably the relationship between surface roughness and particle size. While the peak-to-valley height of the test element in the four-ball apparatus is about 0.1 μm, that of the flanks of the FZG gear is 2–5 μm. It would be reasonable to assume that

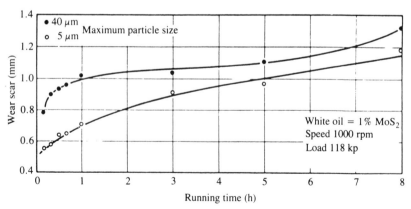

Fig. 4.61   Wear as a function of operating time for MoS$_2$ suspensions with various particle sizes (VKA tests)

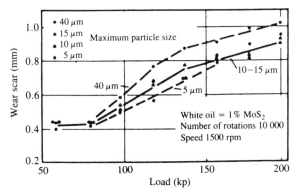

Fig. 4.62 **Wear as a function of load for MoS₂ suspensions with various particle sizes (VKA tests)**

larger particles provide better protection against wear on rougher surfaces, because they enlarge the contact surface by 'filling in' the depressions. On the other hand, smaller particles may be deposited more easily on very smooth surfaces, forming an anti-wear film.

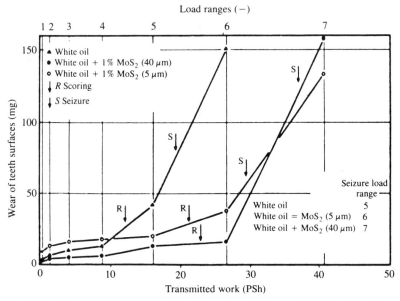

Fig. 4.63 **Tooth flank wear as a function of work transmitted for oils with and without addition of MoS₂ of various particle sizes (FZG tests)**

4.7.6.2.2 Effect of concentration

We also have results of tests conducted on simulation apparatus available for the consideration of the effect of the concentration of solid lubricants in the carrier substance. Figure 4.64 shows test specimen wear on the four-ball apparatus as a function of MoS_2 content in an undoped base oil for two different operating times and loads (81). While, at higher loads, wear is reduced as the concentration increases, at lower loads there seems to be a sort of optimal concentration, above which wear increases again.

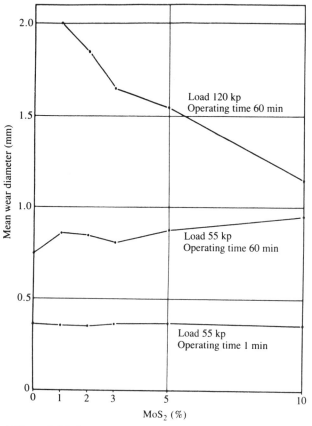

Fig. 4.64   Effect of MoS_2 concentration on wear under various test conditions (VKA tests)

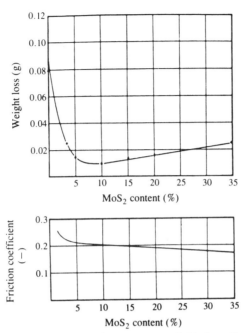

Fig. 4.65 **Effect of MoS$_2$ concentration on wear and friction of solid bodies containing silver and copper (sliding velocity 25.4 m/s, load 0.52 kp, operating time 1 h)** (82)

The results shown in Fig. 4.65, also obtained on the four-ball apparatus, confirm the existence of an optimal concentration for anti-wear performance (**81**). Furthermore, the coefficient of friction falls as the MoS$_2$ concentration rises.

Figure 4.66 shows the wear on the delivery plunger of a lubricating grease pump as a function of operating time for lubricating greases of varying graphite content. It is quite obvious that the highest wear values correspond to the highest graphite concentration. One explanation for these results could be the mutual interference among the particles during orientation on the surface when higher concentrations are present. These results were confirmed during the tests on the FZG test machine. It can be seen in Fig. 4.67 that in durability tests, the grease with the highest graphite content produced the greatest wear, and that grease with the lowest graphite content produced the least wear on the tooth surfaces.

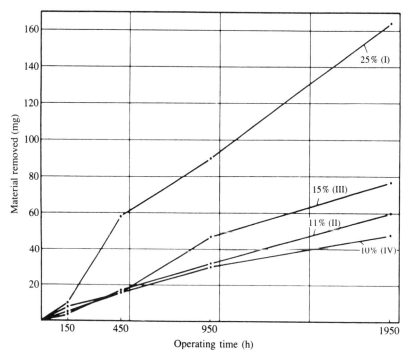

Fig. 4.66   Effect of graphite concentration in lubricating greases on delivery plunger wear in a lubricating grease pump

Interestingly, the damage force level also fell with the graphite content, while the specific wear increased (Fig. 4.68).

4.7.6.2.3   Effect of chemically active additives

When molybdenum disulphide in the form of a colloid suspension is used for lubrication, the effectiveness of the solid lubricant depends on its ability to produce a cohesive protective film on the surfaces of the friction pair. Other agents contained in the oil can be useful or deleterious here. If one assumes chemical reactions on the surfaces in which the MoS_2 also takes part, the question arises as to whether the various additives can 'displace' one another from the surfaces, whether the positive characteristics of different agents are added or multiplied (synergistic behaviour), or whether these agents counteract one another (antagonistic behaviour). If surface energy is released on the friction partners due to the friction conditions, it is possible that, with a combination of substances, components would react one after the other as the relevant free

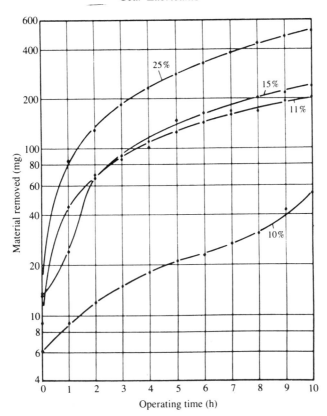

Fig. 4.67 Effect of graphite concentration on tooth surface wear. FZG tests at force level 5

energy level is reached. It is still questionable, however, whether the MoS_2 fits directly into this scenario, as it does adhere to the surfaces and acts physically and mechanically. The surfaces can be coated in such a way that chemical reactions can be delayed or even halted. The reverse process is also possible, however, with the other substances 'occupying' the free surfaces and thus suppressing the effect of the molybdenum disulphide.

This phenomenon can be observed particularly clearly if additives with greater forces of adsorption with respect to the metal surfaces are present in addition to the MoS_2 **(83)(84)**.

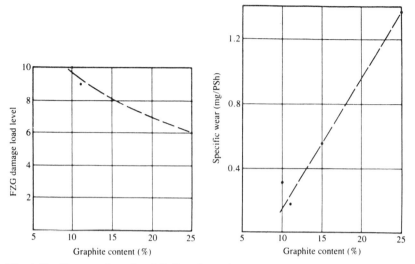

Fig. 4.68   **Damage force level (left) and specific wear (right) in the FZG test as a function of graphite concentration**

There is no disputing that the situation discussed is of particular practical importance when MoS_2 is added to 'ready-made', i.e., fully formulated, lubricants, especially if they contain detergent and dispersant additives, or anti-wear and anti-scoring agents. In an extensive study, the author has demonstrated these effects with a simulation device, the four-ball apparatus **(85)(86)**.

Among other things, various zinc dialkyl dithiophosphates were studied, used in a concentration of 1.3×10^{-5} Mol/kg. The addition of 0.3% MoS_2 has no effect on the wear behaviour. Increasing the MoS_2 concentration to 1 percent gives a surprising result. Particularly in the upper load range, it is quite marked that there is greater wear if molybdenum disulphide and zinc dialkyl dithiophosphate are present simultaneously than if the oil contains only one of the additives (Fig. 4.69). This antagonistic effect means that the lubricant allows the same wear to occur as the base oil alone would. Such antagonistic effects can also occur with other substances in conjunction with MoS_2.

Tricresyl phosphate in the base oil only reduces wear in the lower load range, while molybdenum disulphide reduces wear primarily at higher loads. If the two additives are present simultaneously, these relationships are not changed in any way, i.e., in these test conditions no correlation

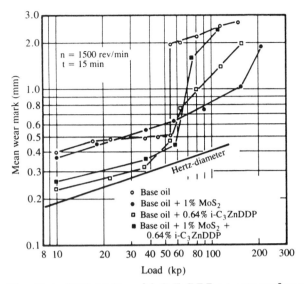

Fig. 4.69   The effect of 1% MoS₂ and i–C₃ZnDDP on wear performance (VKA tests)

can be found between the two additives which would cause an enhancement of their characteristics. As Fig. 4.70 shows for a MoS₂ concentration of 1 percent, the wear curve at low loads is as it would be if there were no MoS₂ in the base oil, and at higher loads it is as it would be if there were no tricresyl phosphate in the base oil.

To obtain information on the effectiveness of molybdenum disulphide in oils with commercial EP additives, a series of tests was carried out with several additives.

The load selected on the four-ball apparatus was 55 kp, the rotational speed was 1480 min⁻¹ and the test time was 1 h.

Figure 4.71 gives some of the results. It can be seen that without MoS₂, wear is considerably reduced by the sulphur/phosphorus additive (additive A), while a trend towards somewhat higher wear values with increasing additive concentration can be detected. This phenomenon also occurs in the presence of molybdenum disulphide, and is even more marked with the sulphur chlorine/lead additive (additive B). The addition of MoS₂ to the base oil reduces wear as expected, even if not so much as the addition of the EP additive, but concentrations of over 5 percent MoS₂ have no further wear-reducing effect. If molybdenum di-

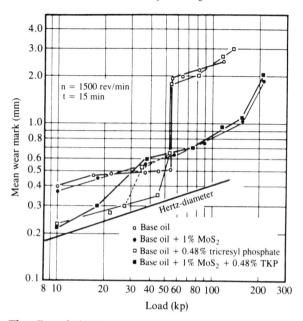

Fig. 4.70　The effect of 1% MoS₂ and tricresyl phosphate on wear (VKA tests)

sulphide and an EP additive are present simultaneously, it is not possible for MoS₂ to give any further improvement in wear protection in these test conditions. The action of the EP additives clearly masks the effect of the molybdenum disulphide.

When detergent additives are present it is likewise not possible to exclude the possibility of impairment of the effectiveness of the MoS₂, as one of the purposes of such additives is to prevent solid particles being deposited on surfaces. Figure 4.72 shows the wear curve for the tests with calcium sulphonate and molybdenum disulphide.

In the lower load range, one can detect a tendency for the wear to be somewhat higher in the presence of both additives than when the base oil contains only calcium sulphonate. The results for higher loads are more interesting and more marked, however. It is true that MoS₂ reduces wear even in the presence of calcium sulphonate, but the wear is still greater than for molybdenum disulphide alone. The dispersant function of the calcium sulphonate clearly impedes the full effectiveness of the MoS₂, so this antagonistic behaviour occurs. It must be noted, however,

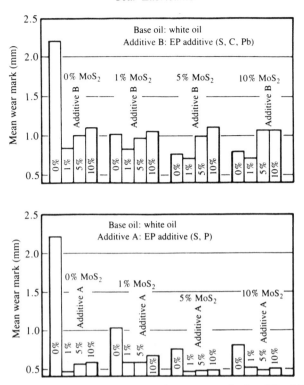

Fig. 4.71 The effect of MoS₂ on wear in the presence of other EP additives (VKA tests)

that this does not alter the fact that MoS_2 reduces wear compared with a lubricating oil consisting of the base oil with calcium sulphonate.

During the course of these investigations, the relationship between molybdenum disulphide and commercial detergent additives was also studied.

As Fig. 4.73 shows for a detergent additive, wear is distinctly lower in the simultaneous presence of MoS_2 and a detergent additive (additive A) than when there is no MoS_2 present, but always higher than if there is no detergent additive in the base oil. The dispersant function of this additive, to which MoS_2 particles are obviously 'foreign bodies', partially suppresses the effectiveness of the molybdenum disulphide. The same effect is sometimes more marked for a highly alkaline detergent additive (additive B).

Lubrication of Gearing

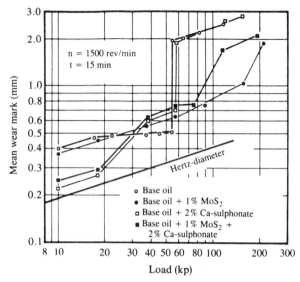

Fig. 4.72   The effect of MoS$_2$ and a calcium sulphonate on wear (VKA tests)

If one considers that the dispersion or suspension of MoS$_2$ by the detergent additives destroys part of its effectiveness, which is equivalent to reducing the concentration of the active substance, it must remain questionable whether it is advisable to add molybdenum disulphide to an engine oil, despite the possible reduction in wear.

4.7.7   Forms in which solid lubricants are used

4.7.7.1   Forms for direct lubricant film formation
The following forms are important for direct lubricant film formation:

- 'In-situ' formation of an MoS$_2$ film
- Solid lubricant powder
- Pastes
- Bonded lubricants (coatings)

4.7.7.1.1   'In situ' formation of an MoS$_2$ film
Recently attempts have been made to form molybdenum disulphide 'in situ' with soluble compounds in lubricating oils, which requires substances containing molybdenum and sulphur to be present. Activated by the heat of friction, MoS$_2$ is then formed by chemical reaction.

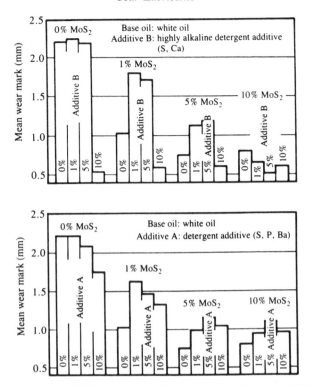

Fig. 4.73 The effect of MoS₂ on wear in the presence of detergent additives (VKA tests)

4.7.7.1.2 Solid lubricant powder

The solid lubricants PTFE, graphite and MoS₂ are also available to users in powder form. PTFE powder is mainly manufactured in a specific particle size, whereas MoS₂ and graphite powders are available in various particle sizes (77). Solid lubricant powders are used to coat small parts.

4.7.7.1.3 Pastes

Pastes may be assembly pastes or high-temperature pastes (77).

(a) *Assembly pastes.* Assembly pastes are pastes of solid lubricants in oils which do not vaporize without residue. They act in a synergetically matched structure. This means that the base oil, the solid lubricant, and other additives act up to a certain upper operating temperature, which is usually about 180°C. Above this temperature, rapid ageing and vaporization of the base oil lead to thickening of the paste. The solid content of assembly pastes can be up to 75 wt%.

(*b*) *High-temperature pastes.* High-temperature pastes, unlike assembly pastes, are pastes of solid lubricants in oils which vaporize more or less without residue above 200°C. The purpose of the oil is simply to distribute the solid lubricants at the friction points and to fix them to the surfaces. The oils must then vaporize, leaving a film of dry lubricant.

4.7.7.1.4 Bonded solid lubricants (coatings)

Coating lubricants are varnishes which contain particles of solid lubricant instead of the conventional colour or metallic pigments. They may be organic or inorganic coatings, depending on the binder. An essential advantage of solid bonded lubricants is that they deposit a much larger supply of lubricant on the surfaces than solid lubricant powders, and thus extend the life of the dry lubricant film significantly. Solid bonded films are often used in dusty environments, where oils or pastes would not be advisable due to the danger of thickening and associated increased wear.

(*a*) *Organic solid bonded lubricants.* There are both air-drying and heat-hardening bonded lubricants. The binders used for air-drying bonded lubricants are cellulose, alkyd, acrylic, and urethane resins. For heat-hardening lubricants they are phenol, epoxide, silicon, and, recently, polyimide resins. The advantage of organic coatings is that they provide a certain amount of corrosion protection for the parts treated. This is generally greater with the stoving coatings than with the air-dried coatings.

(*b*) *Inorganic solid bonded lubricants.* The binders in inorganic coatings are inorganic salts, such as alkali silicates. The advantage of these coatings is that they can be subjected to greater thermal loading than the organic coatings and do not de-gas in a vacuum. Furthermore, they are generally resistant to radiation. Their disadvantage is that they do not provide as much corrosion protection as organic bonded coatings. If more protection is required, the bearing material must be either rustless steel or galvanically coated.

The inorganic bonded coatings also include the enamels, which contain CaF_2/BaF_2, CaF_2, or PbO as the solid lubricants. At temperatures below 250°C these coatings do not lubricate as well as organic bonded coatings. Above this temperature, however, they are superior, and can be used up to 1100°C.

4.7.7.1.5 Application of solid lubricants
To achieve maximum effectiveness of the solid lubricants, it is important that the particles are securely attached to the surface. The surfaces are pre-treated mechanically, e.g., by sanding or wet-blasting, or chemically (see Table 4.42). The coating is applied by: (a) mechanical deposition; (b) tumbling; or (c) spraying.

4.7.7.2 Forms for indirect lubricant film formation
The following forms are significant for indirect lubricant film formation

- self-lubricating contact materials;
- suspensions;
- lubricating greases;
- aerosols.

4.7.7.2.1 Self-lubricating contact materials
Replacing a metallic contact material with a self-lubricating plastic produces a few disadvantages as well as a series of advantages (Table

Table 4.42   Pre-treatment of surfaces of materials to be coated

Material	Pre-treatment
Steel	Degreasing
	Wet- or sand-blasting or acid etching
	Phosphatizing
Stainless	Degreasing
	Wet- or sand-blasting or acid etching
	Oxalic acid treatment
Aluminium and its alloys	Degreasing
	Wet- or sand-blasting or acid etching
	Anodizing
Copper and its alloys	Degreasing
	Wet- or sand-blasting or acid etching
Plastics	Gentle sand-blasting or treatment with emery paper
	Soaking in suitable solvents
Elastomers and rubber	Soaking in suitable solvents
Wood	Removal of surface with emery paper

Table 4.43   Advantages and disadvantages of self-lubricating plastics, particularly thermoplastics

Advantages	Disadvantages
Lubricant-free operation	Low thermal conductivity
Wear particles and foreign bodies can be embedded	Electrostatic charge
	High coefficient of expansion
Resistant to corrosion	Low thermal stability
Relatively good resistance to chemicals	Cannot withstand high pressure loading
Noise damping	
May be very inexpensive	

4.43). The disadvantages can generally be overcome with appropriate selection of the composition, fillers, and reinforcement materials.

The plastics are divided into two groups: linear and network polymers. Linear polymers have bifunctional entities in their chemical structure. Because they can be thermoplastically deformed, they are also called thermoplastics. They can be softened by the heat of friction and harden again when they cool. The most important self-lubricating thermoplastics are:

- polyacetals;
- polyamides;
- polyterephthalates;
- polytetrafluoroethylene;
- polyethylene.

They are used as lubricant materials in the pure form or with suitable fillers.

Compounds with graphite, MoS_2, metal powder, or glass fibre can generally be subjected to higher loads and can be tailored to the most disparate applications by using suitable mixing ratios. In network polymers, the components can be combined to form a two- or three-dimensional network with the aid of at least three functional chemical cross-linking points.

This cross-linking reduces the mobility of the molecules. Therefore, these plastics flow and melt very little even at high temperatures. The most important network polymers, which are also called thermosetting plastics, are: (a) phenol formaldehyde; (b) urea melamine formaldehyde; and (c) unsaturated polyester and polyimide resins. To improve sliding characteristics and strength these are mixed with suitable fillers such as graphite, MoS_2, PTFE, or woven glass fibre, as thermoplastics are. These self-lubricating materials permit maintenance-free operation. To improve running-in properties it has proved useful to provide additional lubrication with liquid lubricants or lubricating greases. These are used extensively with simultaneous oil or grease lubrication.

4.7.7.2.2 Suspensions

Both mineral oils and synthetic oils can be used as carrier liquids for solid lubricant particles. Studies of the lubricating performance of MoS_2 suspensions have shown that, under boundary friction conditions, a layer of MoS_2 from the suspension forms on the surfaces, preventing direct metal contact between the friction pair. Investigations have shown that the suspensions must contain a minimum of 3 percent solid lubricant for high loads, to produce a definite improvement in friction and wear characteristics. Higher concentrations do not bring about any further improvement.

4.7.7.2.3 Lubricating greases

The lubricating greases also belong to the suspensions, but they are dealt with separately here. To improve the load capacity of conventional lubricating greases, it is possible to add up to 10 percent, but usually only about 3 percent of solid lubricant particles.

4.7.7.2.4 Aerosols

Aerosols are sprays of solid lubricant powders, suspensions, greases, pastes, and also bonded coatings in organic solvents with a carrier gas. These are used for application to surfaces which are difficult of access or to large areas and when high operating temperatures are involved.

4.8 CHARACTERISTICS OF GEAR LUBRICANTS

4.8.1 General

The properties of gear lubricants are determined by the formulation – base oil (mixture) and additives – and described by standards, classifications, and specifications. A standard, or, in a narrower sense, a test standard, contains the definition of an individual characteristic and the rules for measuring it. By a specification we mean the description of a product by means of several characteristic features, and, in addition to rules for measuring them, limits are given for meeting the requirements.

Table 4.44   Characteristics of gear oils

Selection values	Quality values	
	Secondary characteristics	Primary characteristics

e.g.
Viscosity
Pour point
Flash point
	e.g.
	Viscosity/
	   temperature behaviour
	Chemical behaviour
	   (corrosion, attack on
	   NF metals)
	Resistance (thermal, to
	   oxidation)
	Foaming performance
	High-temperature
	   performance
	Low-temperature
	   performance
	Cold performance
	Compatibility with
	   sealing elements
		e.g.
		Friction performance
		Wear performance
		Scoring performance
		Running-in performance

A classification can be a specification without limits, but it can also represent simply a type of organizational system.

The individual characteristics of gear oils can be divided into selection values and quality values (Table 4.44). The selection values include those characteristics which must be matched to the specific application, but which do not provide an indication of the quality or the service performance, such as viscosity. Qualitative and service characteristics are indicated by the quality values.

The differentiation between primary and secondary characteristics is not intended to represent a quality ranking according to an absolute scale. Rather, by primary characteristics we mean the specific demands made on the lubricant by the gear set, such as wear and scoring behaviour. The secondary characteristics are the general characteristics which are of interest otherwise. For instance, it may be demanded of a gear oil that the base oil or the additives should not attack the material of the seals.

4.8.2   Characteristics of gear oils

The characteristics of gear lubricants are divided into physical, chemical and technical characteristics (87). The measurement of them is standardized by DIN/ISO procedures. For details on the procedures and the apparatus, the following DIN handbooks should be consulted:

- DIN Handbook 192 – Lubricants
- DIN Handbook 20 – Mineral oils and fuels 1: characteristics and requirements
- DIN Handbook 32 – Mineral oils and fuels 2: test procedures
- DIN Handbook 57 – Mineral oils and fuels 3: test procedures
- DIN Handbook 58 – Mineral oils and fuels 4: test procedures
- DIN Handbook 228 – Mineral oils and fuels 5: test procedures

Latest editions.

4.8.2.1 Physical characteristics
The following physical characteristics and their measurement in standardized procedures will be described briefly:

- viscosity and flow behaviour;
- colour;
- density;
- cloud point;
- pour point;
- flash point and fire point;
- vaporization loss;
- aniline point;
- water content.

4.8.2.1.1 Viscosity
(a) *Definition and dimensions.* Viscosity is a measure of the internal friction of a medium during flow. The term viscosity was defined by Newton (1687). A surface A is displaced with velocity u at a distance h parallel to another surface. In the gap between the two surfaces is a viscous medium. Newton found that the force F necessary to maintain a uniform velocity is proportional to the surface A and the velocity gradient u/h. Therefore, $F \sim A \times u/h$.

The proportionality constant is the viscosity η of the medium between the surfaces. This gives the Newtonian law of friction $F = \eta A \times u/h$ or, with the shear stress $\tau = F/A$ and the velocity gradient $G = u/h$, $\eta = \tau/G$.

This definition gives the following dimensions for the viscosity:

physical system of units: dyne s/cm^2
technical system of units: kp s/m^2
international system of units: N s/cm^2

Furthermore:

1 dyne s/cm^2 = 1 Poise (P) = 100 cP
1 kp s/m^2 = 98.1 P = 9810 cP
1 N s/m^2 = 10 P = 1 daP = 1 Pas

1 cP = 1 mPas

Often, the viscosity related to the density v is wrongly used as the viscosity. This viscosity/density ratio is indicated by v, so $\eta = \eta/\rho$ with the following dimensions:

physical system of units: cm^2/s
technical system of units: m^2/s
international system of units: m^2/s

1 cm^2/s = 1 Stoke (St) = 100 cSt.

When including viscosity in the analysis and design of machine components, the actual viscosity η must be used, and not the viscosity/density ratio v. The temperature dependence and sometimes the pressure dependence of the viscosity must be taken into account. In addition to the physically defined values and the legal units, the following so-called conventional, non-legal designations of viscosity are often used in commerce:

(a) relative flow times E_t, in Engler numbers (E), measured on the Engler apparatus;
(b) flow time in seconds, measured in the Redwood apparatus, in Redwood seconds;
(c) flow time in seconds, measured in the Saybolt apparatus, in Saybolt Universal seconds.

All these conventional designations of viscosity are dependent on the special measuring instruments used and are, therefore, not suitable as a basis for calculations as absolute viscosity values.

(b) *Measurement.* Table 4.45 contains the most important types of viscometers, related to the basic principles of viscosity measurement. Table 4.46 contains the most important viscometers, the values determined with them, and an indication of standard test procedures. The most widely used is the Ubbelohde viscometer as per DIN 51 562 (Fig. 4.74). The measuring principle is based on laminar flow of the liquid to be studied through a capillary tube.

Table 4.45 Principles of viscosity measurement

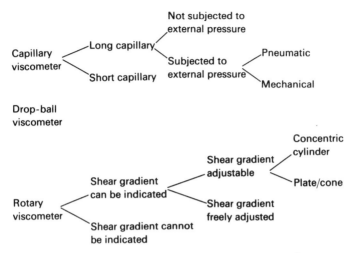

The measured value is the viscosity/density ratio in mm²/s. The reproducibility of the test is about 0.3 percent, and the comparability about 0.7 percent. If the viscosity in mPas is required, it is necessary to convert it, and the density of the liquid at the same temperature is required.

Table 4.46 Types of viscometer

Instrument	Measuring method	Measured value	Standard
Ubbelohde	Long capillary without external pressure	Viscosity/ density ratio	DIN 51 562 ASTM D 2515 IP 71
Umstätter	Long capillary with external pressure	Viscosity	–
Klein/Müller	Long capillary with external pressure	Viscosity	–
Engler	Short capillary	Relative flow time	DIN 51 560
Saybolt	Short capillary	Flow time	ASTM D 88
Höppler	Drop-ball principle	Viscosity	DIN 53 015
Rotovisco	Rotary principle	Viscosity	–
Cold cranking simulator (CCS)	Rotary principle	Viscosity	ASTM D 2602 DIN 51 377
Brookfield	Rotary principle (Immersion principle)	Viscosity	ASTM D 2983 CEC L–32–T–82
Mini-rotary	Rotary principle	Viscosity	ASTM D 3829
Tapered bearing viscometer (TBV)	Rotary principle	Viscosity	–
Ravenfield	Rotary principle	Viscosity	–

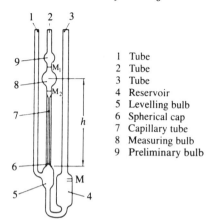

1	Tube
2	Tube
3	Tube
4	Reservoir
5	Levelling bulb
6	Spherical cap
7	Capillary tube
8	Measuring bulb
9	Preliminary bulb

Fig. 4.74   Ubbelohde viscometer as per DIN 51 562

The simplest instrument for determining the viscosity in cP or in mPas is the Höppler drop-ball device (DIN 53 015). The measuring principle is based on the rolling and sliding motion of a ball in an inclined, cylindrical tube, which is filled with the liquid to be examined (Fig. 4.75). The repeatability of the test is between ± 0.5 and ± 1.5 percent, and the comparability is between ± 1.0 and ± 3.0 percent.

Figure 4.76 shows the Umstätter viscometer. Various velocity gradients can be set by means of variable pressures and capillaries with various diameters. In this way the so-called apparent viscosity of non-Newtonian fluids can be determined and then flow curves can be recorded. The velocity can be set by means of the magnitudes of the external pressure. The HSTU viscometer developed by Klein and Müller operates on the same principle, but is suitable for higher velocity gradients (up to 3×10^6 s^{-1}) and higher temperatures (up to 165°C).

In the case of rotary viscometers, the substance to be examined is in the gap between a rotor and a stator. The resistance to motion created by the internal friction of the substance can be measured either at the rotor or the stator and indicated as viscosity. Varying the rotational speed produces a change in the velocity gradient, so flow curves can also be recorded. There are designs with cylindrical or conical stator/rotor systems or cone/plate arrangements (Fig. 4.77).

The cold cranking simulator (CCS) was developed to measure the low-temperature viscosity of engine oils and also of gear oils (Fig. 4.78). The rotational speed of the drive motor, which is operated with constant

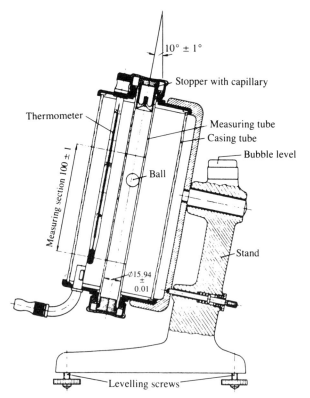

Fig. 4.75 Höppler drop-ball viscometer as per DIN 53 015

voltage, falls due to the internal friction of the liquid, i.e., its viscosity. Thus the rotational velocity is a measure of the apparent viscosity of the liquid. The instrument is calibrated using reference oils of known viscosity. The calibration curve is obtained by tracing this viscosity against rotational speed. The repeatability of the test is 6 percent and the comparability 12 percent. This instrument was developed especially for low temperatures in conjunction with high velocity gradients (ASTM D 2606, DIN 51 377).

So-called immersion viscometers, which are also rotary viscometers, can be used for the combination of low temperatures and low velocity gradients. Figure 4.79 shows the measuring head of the Brookfield viscometer (ASTM D 2983, CEC–L–32–T–82) and Fig. 4.80 that of the Mini Rotary Viscometer (ASTM D 3829).

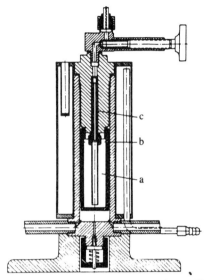

Fig. 4.76   Umstätter viscometer: (a) substance container; (b) air bores; (c) capillary

Rotary viscometers have also been developed for viscosity measurements at high temperatures with high velocity gradients. Figure 4.81 shows the design of the test head of the Tapered Bearing Viscometer (TBV).

(c)   *Factors on which viscosity is dependent.*   Viscosity is dependent on temperature and pressure, and, in the case of non-Newtonian lubricants, on the velocity gradient as well.

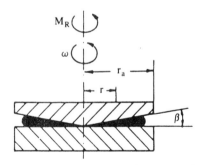

Fig. 4.77   Schematic representation of the plate/cone arrangement of a rotary viscometer

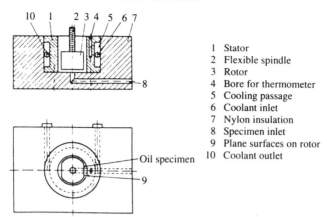

1 Stator
2 Flexible spindle
3 Rotor
4 Bore for thermometer
5 Cooling passage
6 Coolant inlet
7 Nylon insulation
8 Specimen inlet
9 Plane surfaces on rotor
10 Coolant outlet

Fig. 4.78 Cold Cranking Simulator (CCS) as per DIN 51 377

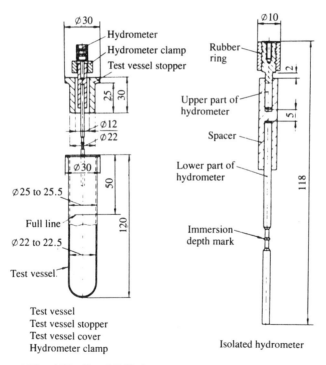

Test vessel
Test vessel stopper
Test vessel cover
Hydrometer clamp

Isolated hydrometer

Fig. 4.79 Brookfield viscometer as per ASTM D 2983

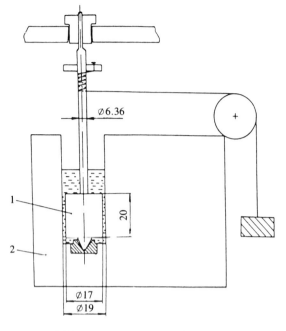

1 Rotor
2 Temperature-equalisation block

Fig. 4.80 Mini rotary viscometer

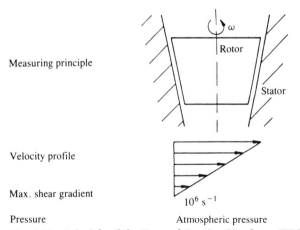

Measuring principle

Velocity profile

Max. shear gradient

Pressure   Atmospheric pressure

Fig. 4.81 Principle of the Tapered Bearing Simulator (TBS)

(i) *Temperature dependence*   The viscosity of lubricating oils falls as the temperature rises. Walther's approximation equation

$$\lg \lg(v^+ + 0.8) = m(\lg T_1^+ - \lg T^+) + \lg \lg(v_1^+ + 0.8)$$

is widely used for linear representation of the dependence of the kinematic viscosity on temperature. Here, $v^+ = v/v_0$ and $v_1^+ = v_1/v_0$ indicate the relative kinematic viscosities at the relative absolute temperatures $T^+ = T/T_0$ and $T_1^+ = T_1/T_0$ (with $v_0 = 1$ cSt and $T_0 = 1°$K as reference value. v and v_1 are the kinematic viscosities at the absolute temperatures T and T_1 and the constant m is the slope of the vT curve. Figure 4.82 shows the conventional diagram for the vT function. The viscosity–temperature behaviour of some lubricating oils is included.

The linear representation is reliable only for the temperature range from 0 to 100°C. In particular, extrapolation to lower temperatures can lead to considerable errors in determining the viscosity. To find the hydrodynamic and elasto-hydrodynamic load capacity, it is necessary to know not v, but η and its dependence on temperature. The relationship

$$\ln \eta^* = \ln k + \frac{c_1}{T + c_2}$$

proposed by Vogel for representing the ηT function, wth $\eta^* = \eta/\eta_0$ as the relative dynamic viscosity at the temperature T in °C and $\eta_0 = 1$ cP as the reference viscosity, k as a dimensionless constant and c_1 and c_2 as two constants with the dimension of a temperature, has proved useful. Measurements have shown that $c_2 = 95°$C can be selected. If the viscosity values of an oil are entered on the diagram, the ordinate of which has a logarithmic division, and on the abscissa of which the values $1/(T + 95°$C$)$ are entered from right to left, the result is a straight line.

(ii) *Pressure dependence*   The viscosity of lubricating oils increases as the pressure rises. As a first approximation, this relationship is expressed by the equation

$$\eta_p = \eta_0 e^{\alpha p}$$

Here, η_0 and η_p indicate the viscosities at atmospheric pressure or at the desired pressure p and α indicates the viscosity–pressure coefficient

$$\alpha = \frac{1}{\eta_p}\left(\frac{\delta n_p}{\delta_p}\right)_T = \text{const.}; \quad \alpha = \frac{\ln \eta_p - \ln \eta_0}{p}$$

at a constant temperature T.

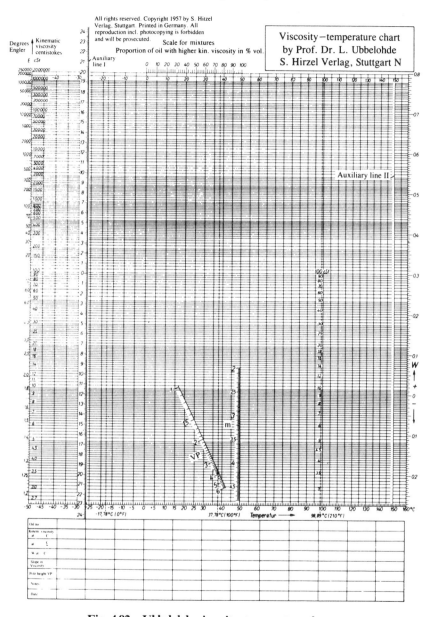

Fig. 4.82   Ubbelohde viscosity–temperature chart

The index p for the viscosity η is only given below if it is necessary for differentiation. The determination of the η_p function is not absolutely simple, as α depends on the structure of the liquid. The following must be observed:

- as the aromatic content C_A increases, the pressure dependence of the viscosity rises;
- as the temperature rises, the pressure dependence of the viscosity falls;
- the viscosity of mineral oils depends to a considerable extent on the pressure; the factor η_{2000}/η_1 (with η_1 and η_{2000} as the viscosity at a pressure of 1 bar or 2000 bar) can lie between 20 and 20 000;
- at higher pressures, the pressure dependence of the viscosity is lower.

Figure 4.83 shows the dependence of viscosity on pressure for two mineral oils at different temperatures. The different behaviour is clear.

(*iii*) *Dependence on velocity gradient* In the case of non-Newtonian substances, the viscosity depends on the viscosity gradient as well as on the pressure and temperature. As the viscosity concept, by its very nature, is defined by the Newtonian law as a proportionality factor, the value of this factor which applies only to a specific shear velocity is called the 'apparent viscosity'. To indicate the flow performance of such substances, it is not enough to measure the viscosity at a specific temperature; a flow curve corresponding to

$$\tau = f(G)$$

must be determined. The effective local viscosity is then

$$\eta = \tau/G$$

for each point of the flow curve.

Figure 4.84 shows possible forms of the flow curve. The intrinsic viscosity is of particular importance. All lubricating oils with macropolymers as viscosity index improvers exhibit such behaviour. Apart from the dependence of the viscosity on shear stress, most non-Newtonian substances exhibit behaviour in which their apparent viscosity varies with the duration of the shear stress. This phenomenon is called rheopexy (curve b in Fig. 4.85) or thixotropy (curve c in Fig. 4.85), depending on whether the viscosity is rising or falling.

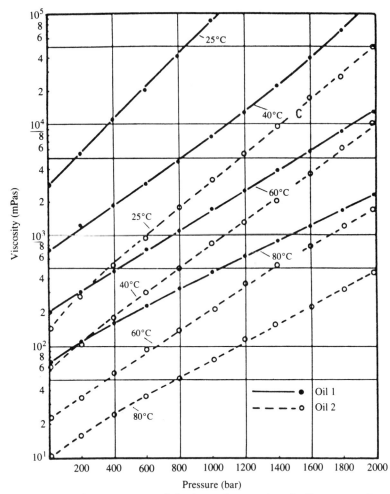

Fig. 4.83 Viscosity–temperature behaviour of two mineral oils at several temperatures

4.8.2.1.2 Colour as per ISO 2049 (substitute for DIN 51 578)
Definition
Colour number according to chromaticity coordinates red, green, and blue as level of spectral transmission.
Unit of measurement
Colour number as per ISO 2049.

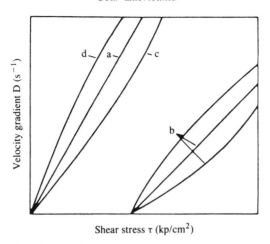

Shear stress τ (kp/cm^2)

Fig. 4.84 Possible shapes of flow curves (schematic): (a) Newtonian oils; (b) Bingham bodies; (c) Intrinsically viscous oils; (d) Dilatant oils

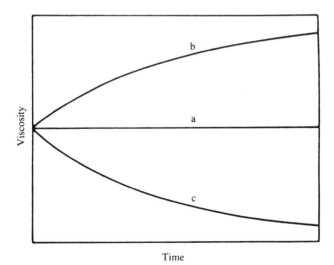

Time

Fig. 4.85 Schematic representation of the relationship between the viscosity and the duration of the shear stress

Measuring method

Comparison of the colour of the specimen with the colour of coloured test discs.

Measurement equipment

Specimen holder of clear, colourless glass and colour comparison device, consisting of light source, coloured test discs, recording glass holder with screening hood, and eyepiece.

Conclusion

With the same viscosity, the colour of undoped oils permits conclusions to be drawn concerning the refining level.

4.8.2.1.3 Density as per DIN 51 757

Definition

Quotient of mass and volume of the specimen.

$\rho = m/V$

where

ρ = density

m = mass

V = volume

Relative density at reference temperature 60°F

$d_{60/60}$ = density of specimen at 60°F, divided by the density of water at 60°F

USA: API gravity $= 141.5/d_{60/60} - 131.5$

Unit of measurement

kg/m^3 (kg/l, g/ml) at 15°C or 20°C.

Measuring method

Process A

Determination with areometer.

Areometers are floats with a density scale.

Process B

Determination by hydrostatic weighing.

Use of a beam balance (Mohr balance).

Measurement of buoyancy.

Process C

Determination with a pyknometer.

Pyknometers are glass vessels of known volume.

Measurement of weight.

Conclusion
The density can give an indication of the chemical composition of lubricating oils. It is lower for paraffinic oils than for naphthenic and aromatic oils. It is a number which indicates how many times heavier a body is in a vacuum chamber than the same quantity of a normal substance of equal volume, e.g., water at 4°C. There is a difference between density and specific gravity. The density is the ratio of weight to volume. Unlike weight, mass is independent of external factors such as temperature, buoyancy in air and acceleration due to gravity.

4.8.2.1.4   Cloud point as per ISO 3015 (substitute for DIN 51 597)
Definition
That temperature at which the specimen becomes opaque due to the precipitation of wax crystals.
Unit of measurement
°C.
Measuring method
Heat to at least 14°C above the cloud point to be expected. Remove moisture, e.g., with filter cloth. Cool in steps of 1°C and examine visually for opacity.
Measuring equipment
Instrument for determining cloud point as per ISO 3015 (see Fig. 4.86).
Conclusion
None.

4.8.2.1.5   Pour point to ISO 3016 (substitute for DIN 51 597)
Definition
The lowest temperature at which an oil just flows, when cooled under specified conditions.
Unit of measurement
°C.
Measuring method
Oils with pour points down to −33°C: heat specimen to 45°C and cool in a water bath at 25°C to 36°C.
Oils with pour points below −33°C: heat to 45°C and cool in a water bath at 7°C to 15°C.
Cool the specimen in a jacketed vessel in temperature steps of 3°C, and check whether oil still moves within 5 s.
3°C are added to the last observation temperature.
Measuring equipment
Instrument for determining pour point as per ISO 3016 (see Fig. 4.86).

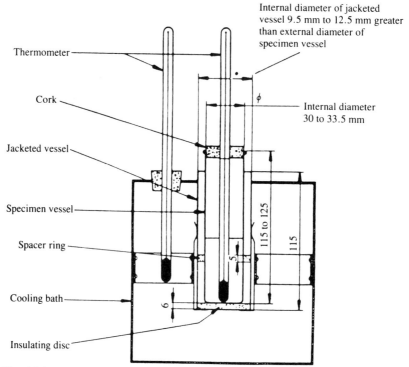

Fig. 4.86   Device for determining cloud point and pour point as per ISO 3015 and ISO 3016

Conclusion

Measure of the flow capability of an oil at low temperatures.

4.8.2.1.6   Flash point and fire point to ISO 2592 (substitute for DIN 51 376) (in open Cleveland crucible)

Definition

Lowest temperature at which vapours form in such quantity that they form a flammable mixture with the air above the surface of the liquid, which briefly ignites (flash point) or burns for at least 5 s (fire point).

Unit of measurement

°C.

Measuring method

Heat specimen initially at 14–17°C per minute, then at 5–6°C per minute. At the latest at 28°C below the expected flash point actuate the ignition flame in temperature steps of 2°C. Heat initially to the flash point, then to the fire point.

Measuring equipment
Cleveland flash point test device in open crucible (see Fig. 4.87).

Conclusion
Indirect measure of the fuel dilution of a lubricating oil. Classification of fuels into risk categories by measurement in a closed crucible, as Abel–Pensky (DIN 51 755) or as Pensky–Martens (DIN 51 758).

4.8.2.1.7 Vaporization loss as per DIN 51 581 (Noack)

Definition
Reduction in mass of doped and undoped oils on heating.

Unit of measurement
Mass component w (V) in percent.

$$w \text{ (V)} = \frac{m_E - m_A}{m_E} \times 100$$

m_E = mass of specimen in g before test
m_A = mass of specimen in g after test

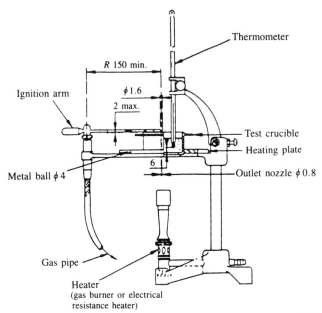

Fig. 4.87   Device for determining flash point and fire point as per ISO 2592

Measuring method

Heat the specimen to 250°C and hold this temperature for one hour. Remove the resultant oil vapours. Weigh the crucible with the specimen before and after heating.

Measuring equipment

Test device for determining the vaporization loss (see Fig. 4.88).

Conclusion

Indirect measure of:

- oil consumption in the case of engine oils;
- change in oil characteristics during use;
- fuel dilution in the case of engine oils.

4.8.2.1.8   Aniline point as per DIN 51 775

Definition

That temperature at which a homogeneous mixture of equal parts by volume of a pale mineral oil and aniline separates on cooling. There

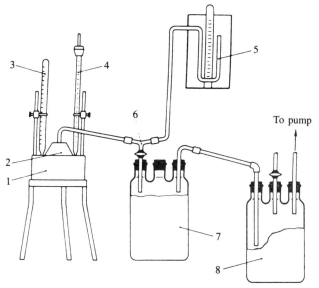

1 Heater block	5 'U' tube manometer
2 Vaporisation crucible	6 'Y' piece with tap
3 Measuring thermometer	7 1st Woulfe bottle (collection vessel)
4 Contact thermometer	8 2nd Woulfe bottle (buffer vessel)

Fig. 4.88   Test device for determining the vaporization loss (Noack) as per DIN 51 581

are the following reference values:
Aromatics: below 0°C
Naphthenes: between 30 and 50°C
Paraffins: above 50°C.

Unit of measurement
°C.

Measuring method
Mix mineral oil and aniline while stirring and heat to form a homogeneous solution; then cool slowly while stirring until it becomes distinctly opaque.

Measuring equipment
Glass specimen container with air jacket, agitator, heating and cooling baths, aniline thermometer.

Conclusion
The aniline point allows conclusions to be drawn as to the composition of mineral oils, i.e., the hydrocarbon distribution.

4.8.2.1.9 Water content to DIN 51 777, Parts 1 and 2

Definition
Water content of new oils (direct and indirect method) and of used oils (indirect method).

Unit of measurement
mg/kg (mass content), (ppm).

Measuring method
Direct method
The specimen, dissolved and dispersed in a titration medium, has Karl Fischer solution (KFS) added to it until the titration point is reached. The water content of the specimen is calculated from the volume of KFS and titre.

Indirect method
The water is driven from the specimen at 120°C with superpure nitrogen, condensed in methanol at room temperature, and subsequently titrated and determined as in the direct method.

Measuring equipment
Figures 4.89 and 4.90 show the test equipment for determining water content as per DIN 51 777.

Conclusion
Water content of new and used mineral oils for description and evaluation of lubricating oils.

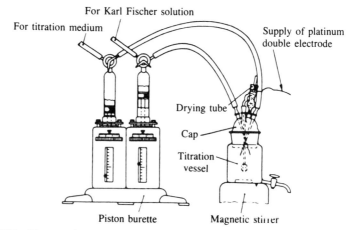

For Karl Fischer solution

For titration medium

Supply of platinum double electrode

Drying tube

Cap

Titration vessel

Piston burette

Magnetic stirrer

Fig. 4.89 Test equipment for determining water content as per DIN 51 777, Part 1

4.8.2.2 Chemical properties

The following chemical characteristics and standard methods for determining them will be described briefly:

- ash, sulphate ash;
- asphaltene content;
- coking tendency;
- neutralization value;
- saponification value;
- total base number, strong base number;
- total acid number, strong acid number;
- oxidation resistance, ageing characteristics.

4.8.2.2.1 Ash as per EN 7 (substitute for DIN 51 575)

Definition

Combustion residue of undoped lubricating oils, which contain no ash-forming additives.

Unit of measurement

wt%.

Measuring method

The specimen is burnt in a suitable vessel until only ash and carbon remain. The residue containing carbon is then incinerated by heating to 775°C.

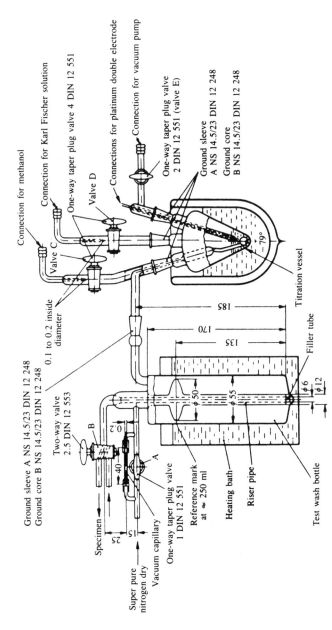

Fig. 4.90 Test equipment for determining water content as per DIN 51 777, Part 2

Measuring equipment

Incineration crucible made of platinum, quartz glass or porcelain, and an electrical muffle furnace.

Conclusion

In the case of undoped fresh oils (new oils), the ash is a measure of ash-forming residues, which are to be regarded as unwanted impurities. In the case of undoped used oils, the ash gives an indication of the quantity of ash-forming impurities which have been produced during use or which have been introduced, if the ash content of the fresh oil (new oil) is known.

4.8.2.2.2   Sulphate ash as per DIN 51 575

Definition

Mineral residue which remains in the form of sulphate on combustion (incineration) of lubricating oils and treatment with sulphuric acid. Combustion residue of doped lubricating oils with ash-forming additives.

Unit of measurement

g/100 g or wt%.

Measuring method

Combustion of the organic components of the lubricating oil in a dry or wet manner. Incineration of the residue after addition of sulphuric acid by calcining.

Measuring equipment

Incineration crucible and dish of porcelain or sintered aluminium oxide and regulated electrical muffle furnace.

Conclusion

In the case of undoped fresh oils (new oil), the sulphate ash is a measure of the purity of the oil. In the case of new oils with organometallic additives, the sulphate ash gives an indication of the total volume of additives. In the case of used oils, the increase in ash compared with the ash of the new oil gives an indication of the solid impurities, metallic abrasion, and chemical reactions of ageing products with metals.

4.8.2.2.3   Asphaltene content as per DIN 51 595

Definition

Asphaltenes are crude oil constituents which are precipitated when the oil forms a solution with thirty times the volume of n-heptane at room temperature, and are soluble in pure toluene.

Unit of measurement

wt%.

Measuring method

Dissolve the specimen in n-heptane and burn off the precipitated and undissolved constituents by filtration. Remove remaining paraffins by extraction with n-heptane. Dissolve the asphaltenes remaining on the filter paper in pure toluene and determine gravimetrically.

Measuring equipment

Conventional laboratory equipment and extraction device with extractor (Fig. 4.91) as per DIN 51 595.

Conclusion

The asphaltene content gives an indication of the behaviour of lubricating oils during operation. Too great a content can cause unwanted bitumen- and tar-like precipitates. This test is carried out with undoped lubricating oils (base oils), as additives can falsify the results.

4.8.2.2.4   Carbon residue as per DIN 51 551

Definition

The Conradson coke residue is determined as a measure of the coking tendency of mineral oils. The coke residue of doped and undoped lubricating oils is the amount of residue remaining after low-temperature carbonization.

Unit of measurement

wt% (mass content in percent).

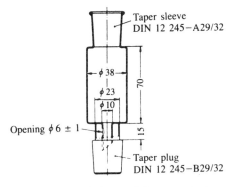

Fig. 4.91   Extractor for determining asphaltene content as per DIN 51 551

Measuring method
The specimen is in an open crucible inside a covered crucible, which
stands on sand in another covered crucible, and is heated. The residue
is weighed.

Measuring equipment
The equipment used to determine coke residue as per DIN 51 551 is
shown in Fig. 4.92.

Conclusion
Carbon residue as a measure of the coking tendency of lubricating oils
gives an indication of the tendency of lubricating oils to form coke-like
residues at similar operating temperatures. In the case of doped lubricating oils, the carbon residue is increased due to the residues from
additives.

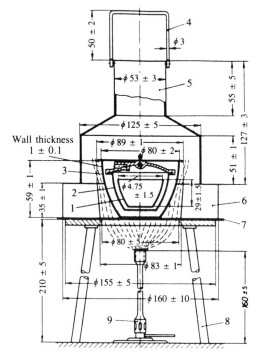

Fig. 4.92   Test equipment for determining the carbon residue as per DIN 51 551

4.8.2.2.5   Neutralization value as per DIN 51 558

Definition

Neutralization value (sour) – NV(s)
The quantity in mg KOH which is necessary to neutralize the acid constituents in 1 g of mineral oil.
(Total acid number – TAN.)

Neutralization value (alkaline) – NV(a)
That quantity of acid in equivalent mg KOH which is necessary to neutralize the alkaline constituents of 1 g of mineral oil.
(Strong base number – SBN.)

Neutralization value (water-soluble acids) – NV(wsa)
That quantity of lye which is necessary to neutralize the water-soluble acid constituents in 1 g of mineral oil.
(Strong acid number – SAN.)

Unit of measurement
mg KOH/g.

Measuring method
To determine the neutralization value (acid or alkaline) the specimen is dissolved in a mixture of toluene and propanol-(2) (isopropyl alcohol), which contains set quantities of water.

Titrate the solution at 18 to 28°C with 0.1 M alcoholic potassium hydroxide solution or hydrochloric acid to the end point (p-naphthol-benzein solution as indicator solution). To detemine the neutralization value (water-soluble acids), part of the specimen is extracted with hot water and the aqueous extract is titrated with potassium hydroxide solution and methyl orange as an indicator.

Measuring equipment
Erlenmeyer flask, chemicals.

Conclusion
The neutralization value is a measure of the amount of alkaline and acid constituents in mineral oil. In new and used oils the acid constituents include organic and inorganic acids, esters, phenol compounds, lactones, resins, heavy metal salts and additives such as inhibitors and detergents.

The alkaline constituents include organic and inorganic bases, amino compounds, salts of weak acids (soaps), basic salts, heavy metal salts and additives such as inhibitors and detergents.

It gives indications of relative changes in the oil during use under oxidative conditions, if the corresponding values of the new oil are known.

4.8.2.2.6   Saponification value as per DIN 51 559

Definition

Amount of saponifiable constituents in a mineral oil. This indicates the quantity of potassium hydroxide necessary to neutralize the free acids contained in the oil and to saponify the esters present.

Unit of measurement

mg KOH/g.

Measuring method

The specimen is dissolved in a mixture of toluene and propanol-(2) (isopropyl alcohol), heated with an alcoholic KOH solution, and the amount of unused KOH is determined by titration with hydrochloric acid against phenolphthalein as indicator.

Measuring equipment

Erlenmeyer flask, cooler, chemicals.

Conclusion

The saponification value gives an indication of the change in a lubricating oil during service. It is determined for new and used oils, and for those with saponifiable additives. Additives which contain substances which consume alkali or acid, such as sulphur, chlorine or phosphorus can affect the test result.

4.8.2.2.7   Total acid number (TAN), Strong acid number (SAN) and Strong base number (SBN)

These are the English names for the various definitions and specifications of the neutralization value. For definitions, units of measurement, measuring method, measuring equipment and conclusion for TAN, SAN, and SBN see Section 4.8.2.2.5.

4.8.2.2.8   Total base number as per EN 55 (draft)

Methods for determining the neutralization value are not suitable for determining the alkaline constituents of lubricating oils with basic additives. This applies especially to doped engine oils. They can be determined in accordance with DIN EN 55 (draft), 'Determination of the total base number in mineral oil products by potentiometric perchloric acid titration'.

Definition

The quantity of perchloric acid, as an equivalent quantity of potassium hydroxide, which is necessary to neutralize all basic constituents in 1 g of oil.

Unit of measurement

mg KOH/g.

Measuring method

The specimen is dissolved in a water-free mixture of chlorobenzene and glacial acetic acid and titrated with a solution of perchloric acid in glacial acetic acid by means of a potentiometer. A glass indicator electrode and a reference electrode are used. The readings of potential are recorded against the volumes of titrant; the end point of titration is the last point of inflection in the titration curve.

Measuring equipment

Potential measuring instrument with glass and reference electrodes, titration equipment, and chemicals.

Conclusion

The total base number is a measure of the basic constituents in lubricating oils. These include organic and inorganic bases, amino compounds, salts of weak acids, basic salts of polyacid bases, and salts of heavy metals.

In the case of fresh oils, the total base number is a measure of the type and extent of doping, and with used oils it indicates the consumption of additives and of additives remaining.

4.8.2.2.9   Ageing behaviour

By ageing of a lubricating oil we mean the changes, especially in chemical characteristics, which occur during service, primarily through oxidation.

(a)   Ageing resistance as per DIN 51 554

Definition

By Baader ageing resistance we mean the change in mineral oils, in particular of additive-free insulation oils, of additive-free and additive-containing lubricating oils, and of additive-free hydraulic oils.

DIN 51 554, Part 1: Preferentially insulation oils.

DIN 51 554, Part 2: Lubricating oils C, C–L and C–LP, hydraulic oils H

Unit of measurement
DIN 51 554, Part 1
- Saponification value in mg KOH/g
- Dielectric loss factor tan δ_{90}
- Sludge content in wt%
DIN 51 554, Part 2
- Saponification value in mg KOH/g

Measuring method
The specimens are aged for a set period of time at set temperatures while air is introduced and a copper wire coil is immersed periodically.
DIN 51 554, Part 1
- Test temperature 110°C
- Ageing time 140 h
DIN 51 554, Part 2
- Test temperature 95°C
- Ageing time three days

Measuring equipment
Baader test device as in Fig. 4.93.

Conclusion
Measure of changes which have occurred in lubricating oils due to oxidation during service. Comparison with the values for the new oil reinforces the conclusion.

(b) *Ageing performance to DIN 51 352*

Definition
Change in undoped and additive-containing lubricating oils (oxidation inhibitors, ash-containing detergents), expressed by the Conradson increase in carbon residue (Part 1).
Change in lubricating oils VD–L (Part 2).

Unit of measurement
Carbon residue in wt%.

Measuring method
DIN 51 352, Part 1
- Ageing of the specimen while air is introduced at 200°C for two periods of 6 h. Determine carbon residue and compare with value for new oil.
DIN 51 352, Part 2
- After iron (III) oxide is added as a catalyst, the specimen is aged while air is introduced at 200°C over a period of 24 h. The carbon residue is determined and compared with the value for new oil.

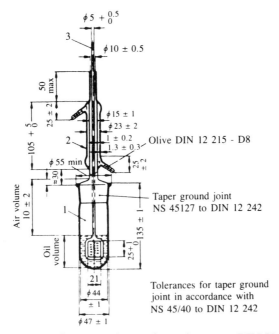

Fig. 4.93   Baader test equipment for ageing test to DIN 51 554

Measuring equipment
Ageing vessel as in Fig. 4.94.

Conclusion
Measure of changes occurring in lubricating oil during service due to oxidation.

(c)   Ageing characteristics to DIN 51 586

Definition
Change in lubricating oils for higher pressures, e.g., lubricating oils C–LP, expressed by a rise in viscosity and an increase in the precipitation value.

Unit of measurement
Viscosity in mm²/s.
Viscosity increase in percent.
Precipitation value in ml.

Measuring method
Measurement of viscosity at 100°C and precipitation value of new oil. Ageing of specimen while air is introduced at 95°C over a period of 312 h.

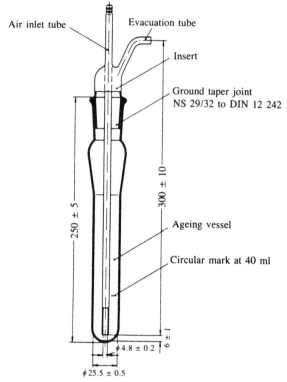

Fig. 4.94　Test equipment for ageing test to DIN 51 352

Measurement of viscosity at 100°C and precipitation value of the aged specimen.

Measuring equipment

Heating bath with agitator, centrifuge with centrifuge flask as in Fig. 4.95.

Conclusion

Measure of the changes caused by oxidation during service and indication of the suitability of a lubricating oil for the application.

(*a*)　*Ageing performance to DIN 51 587*

Definition

Change in additive-containing steam turbine oils and hydraulic oils, indicated by an increase in the neutralization value.

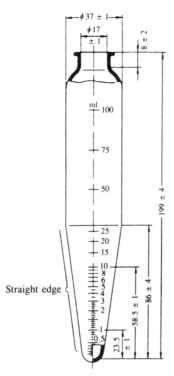

Fig. 4.95   Centrifuge flask for determining the precipitation value to DIN 51 586

Unit of measurement

Neutralization value in mg KOH/g.

Ageing time in hours.

Measuring method

Distilled water is added to the specimen and it is aged while oxygen is introduced in the presence of steel and copper wire over a period of at most 1000 h. NV is determined at intervals of about one week until a maximum NV of 2.0 mg KOH/g is obtained.

Measuring equipment

Heating bath with agitator and test equipment as in Fig. 4.96.

Conclusion

Measure of the duration of the effectiveness of the additives mixed with steam turbine and hydraulic oils to delay oxidation. The result of the test cannot be used to indicate the ageing characteristics of the base oil when the additives are exhausted.

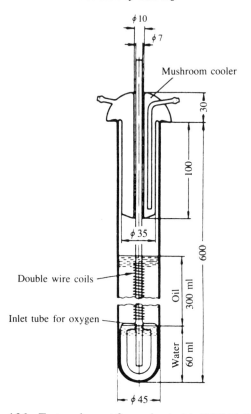

Fig. 4.96　Test equipment for ageing test to DIN 51 587

4.8.2.3　Technological characteristics

By technological characteristics we mean those characteristics which cannot be described by definite physical or chemical analytical results. The following technological characteristics and how they are determined by standard methods will be described briefly:

－ water separation ability;
－ air separation ability;
－ anti-corrosion behaviour;
－ shear stability;
－ anti-scoring and anti-wear behaviour (various test procedures);
－ cold flow behaviour, channel point.

4.8.2.3.1 Water separation ability after vapour treatment to DIN 51 589, Part 1

Definition

By water separation ability (WSA) we mean the capacity of a lubricating oil to separate itself from the condensate after treatment with water vapour.

Unit of measurement

Seconds (s).

Measuring method

The lubricating oil is emulsified by means of a constant stream of water vapour. The separation time of the aqueous phase is then determined.

Measuring equipment

Emulsifier with immersion heater as in Fig. 4.97 (DIN 51 589, Part 1).

Conclusion

In the case of lubricating oils which come into contact with water vapour and hot condensation water, it is a measure of the speed with which they separate from the water again.

4.8.2.3.2 Air separation ability (impinger method) to DIN 51 381

Definition

By air separation ability (ASA) we mean the time in which air dispersed in oil separates out down to 0.2 percent.

Unit of measurement

Minutes (min).

Measuring method

At various test temperatures, e.g., 25, 50, or 75°C, air is blown into the oil under pressure. After the air-in-oil dispersion has formed, the time taken for the air to leave the oil is determined by obtaining the density. The amount of dispersed air is recorded against time on a graph.

Measuring equipment

Test device for determining the air separation ability by impinger. Figure 4.98 shows the schematic layout and test vessel (DIN 51 381).

Conclusion

Primarily with steam turbine oils and hydraulic oils, which are also used as gear oils, it is a measure of the separation time of the dispersed air, which can have a deleterious effect on operating performance.

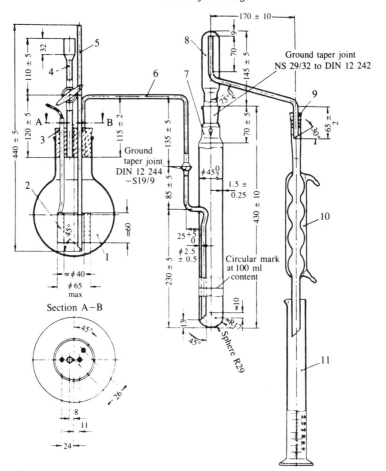

Fig. 4.97 Emulsifier with immersion heater for determining water separation ability to DIN 51 589

4.8.2.3.3 Corrosion-prevention characteristics with regard to steel in the presence of water (agitation method) to DIN 51 355

Definition

By corrosion we mean the reaction of a metallic material with its environment, which produces a measurable change in the material. By corrosion-preventing characteristics we mean the extent of the reduction of corrosion-related surface change of a steel sheet, which is exposed to the action of a mixture of gear oil and water.

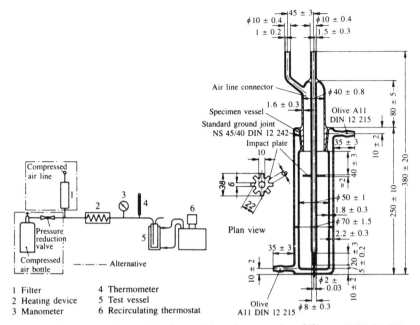

Fig. 4.98 Test equipment for determining air separation ability to DIN 51 381

Unit of measurement
Corrosion level as in Table 4.47.

Measuring method
A rotating steel sheet is exposed to the action of a mixture of 200 g gear oil and 5 ml distilled water at 80°C for 4 h or at 130°C for 8 h. The corrosion level is then determined.

Table 4.47 Corrosion levels to DIN 51 355

Corrosion level	Significance	Description
0	No corrosion	Unchanged
1	Traces of corrosion	At most three corrosion areas, of which each has a maximum diameter of 1 mm
2	Slight corrosion	Up to 5 percent of the surface corroded
3	Moderate corrosion	5–20 percent of the surface corroded
4	Severe corrosion	Over 20 percent of the surface corroded

Measuring equipment
Agitator and heating bath.

Conclusion
A measure of the protection against corrosion afforded by gear oil to steel gear components in the presence of water.

4.8.2.3.4   Corrosion-preventing characteristics, humidity chamber test to DIN 51 359

Definition
See Section 4.8.2.3.3.

Unit of measurement
Corrosion level as in Table 4.47 (see Section 4.8.2.3.3).

Measuring method
Steel sheets of various compositions, external shapes and surface finishes are immersed in the test oil and then hung in the humidity chamber. Conditions: 50°C, 875 l/h, and relative air humidity 100 percent. The level of corrosion is then determined.

Measuring equipment
Humidity chamber as in Fig. 4.99 (DIN 51 359).

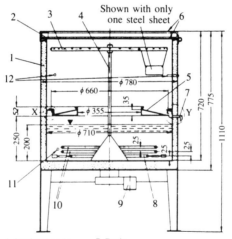

1 Insulated casing	7 Drain
2 Removable lid	8 Air inlet ring
3 Rotating rack	9 Geared motor
4 Drive shaft	10 Heating element
5 Drip tray	11 Temperature measuring point for water
6 Cotton fabric	12 Temperature measuring points for air

Fig. 4.99   Humidity chamber for corrosion test to DIN 51 359

Conclusion

A measure of the corrosion protection afforded by lubricating oils and anti-corrosion oils to structural steel components.

4.8.2.3.5   Anti-corrosion characteristics of additive-containing steam turbine and hydraulic oils to DIN 51 585

Definition

See Section 4.8.2.3.3.

Unit of measurement

Level of corrosion as in Table 4.48.

Measuring method

A mixture of 300 ml of oil and 30 ml of distilled water or artificial sea water is agitated for 24 h at 60°C, with a steel bar immersed in it. The corrosion level is then determined.

Measuring equipment

Test device as in Fig. 4.100 (DIN 51 585).

Conclusion

Indicates the characteristics of the oil which protect steel components from corrosion, if they come into contact with water or water vapour during service.

4.8.2.3.6   Anti-oxidation and anti-corrosion characteristics of low viscosity lubricating oils to DIN 51 394

Definition

Protection of various metals against oxidation and corrosion.

Unit of measurement

mg/cm^2 for weight change of the metal test pieces.

mg KOH/g for change in the neutralization value.

% for change in kinematic viscosity at 40°C.

Table 4.48   Corrosion levels of steel bar to DIN 51 585

Corrosion level	Significance	Description
0	No corrosion	Unchanged
1	Slight corrosion	Maximum of six corrosion areas, of which none has a diameter greater than 1 mm
2	Moderate corrosion	Not more than 5 percent of the surface corroded, but more than corrosion level 1
3	Severe corrosion	Over 5 percent of the surface corroded

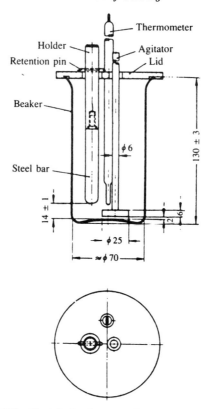

Fig. 4.100   Test device for corrosion test to DIN 51 585

Measuring method

In a test tube 100 ml of oil is aged in the presence of five test blocks of different metals at 121°C for 168 h while 5 l of air is introduced. Then the corrosion-related change in weight of the test blocks and the change in neutralization value and viscosity of the oil are determined.

Measuring equipment

Glass apparatus with heating bath and agitator and test blocks as in Fig. 4.101 (DIN 51 394).

Conclusion

Indicates the ageing stability and anti-corrosion characteristics of low-viscosity lubricating oils in the presence of various structural metals.

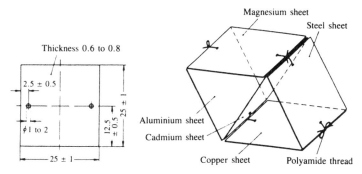

Fig. 4.101 Dimensions and arrangement of test sheets for corrosion test to DIN 51 382

4.8.2.3.7 Shear stability of lubricating oils with polymer additives to DIN 51 382

Definition

The shear stability of polymer-containing lubricating oils is defined as the relative fall in viscosity VA

$$VA = \frac{v_0 - v_1}{v_0} \times 100 \; F$$

v_0, v_1 = kinematic viscosity of unsheared or sheared oil
F = correction factor

Unit of measurement

%.

Measuring method

The oil specimen is subjected to shear stress in an injection pump for a certain number of cycles. The kinematic viscosity is measured before and afterwards.

Conditions

Engine oils: 30 cycles, viscosity measurement at 100°C.
Hydraulic oils: 250 cycles, viscosity measurement at 40°C.

Measuring equipment

Test equipment as in Fig. 4.102 (DIN 51 382).

Conclusion

In the case of lubricating oils with polymer additives, e.g., multigrade oils, gives an indication of the irreversible fall in viscosity due to mechanical stress to be expected during service.

1 Injection nozzle
2 Atomisation chamber
3 Outflow from atomisation chamber
4 Distributor plate
5 Cool reservoir
6 Three-way valve on cool reservoir
7 Storage reservoir
8 Three-way valve on storage reservoir
9 Stand
10 Condensation hose
11 Twin piston injection pump
12 Pump regulating screw
13 Electric motor
14 Pump vent screw
15 Pump stroke counter
16 Injector nozzle pressure pipe
17 Overflow return pipe

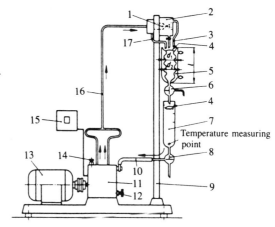

Fig. 4.102   Equipment for testing shear stability to DIN 51 382

4.8.2.3.8   *Welding load of liquid lubricants per DIN 51 350, Part 1*
Definition

Welding load means the test force under which welding occurs in a four-ball system in a test device. By OK-force we mean that force at which welding does not occur. The test runs for OK-force and welding force are taken as the FBA welding load.

Unit of measurement

Newton (N).

Measuring method

In a system consisting of a rotating ball which is pressed against three stationary balls, the lubricant is subjected to sliding stress under increasing load until welding of the four-ball system occurs.

Measuring equipment

Four-ball apparatus and test piece system as in Fig. 4.103 (DIN 51 350, Part 1).

Conclusion

Indicates the presence of anti-scoring additives in lubricating oils. It is not possible to make a direct assessment of the performance of the lubricants in machine components.

4.8.2.3.9   *Wear parameters of liquid lubricants to DIN 51 350, Part 2*
Definition

Domes produced on the three stationary balls in a four-ball system by wear during mixed friction.

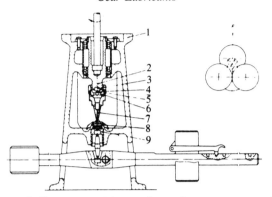

Fig. 4.103   Four-ball apparatus (FBA) and test piece system as per DIN 51 350

Unit of measurement
mm.

Measuring method
In a system consisting of a rotating ball which is pressed against three stationary balls, the wear which has occurred after one hour's operation at constant load is determined as the diameter of the dome.

Measuring equipment
Four-ball apparatus and test piece system as in Fig. 4.103 (DIN 51 530).

Conclusion
Indicates the presence of wear-reducing additives in lubricating oils. It is not possible to make a direct evaluation of the performance of the lubricants in machine components.

4.8.2.3.10   Performance of hydraulic oils in a vane pump as per DIN 51 389, Parts 1 and 2

Definition
Wear of the vanes and the thrust ring in a Vickers V–104–C10 and V–105–C10 vane pump.

Unit of measurement
Mass loss in mg.

Measuring method
The hydraulic oil is circulated by a vane pump at a set temperature and pressure on a test rig. The wear of vanes and ring in the pump is then determined.

Conditions:
Quantity of oil 70 l
Delivery 25 l/min
Speed 140 min^{-1}
Pressure 140 bar
Test duration 250 h
Temperature selected so that the oil viscosity is 13 mm^2/s
Measuring equipment
Test rig with vane pump (Fig. 4.104) with wear insert (Fig. 4.105) (DIN 51 389).
Conclusion
Indicates the wear-reducing characteristics/additives of hydraulic fluids when used in vane pumps.

4.8.2.3.11 Performance of lubricants in the FZG gear test machine to DIN 51 354, Parts 1 and 2
Definition
Stress limit of lubricants, especially gear lubricants, indicated by the wear of gearwheels and scoring of tooth surfaces.
Unit of measurement
mg for wear-related change in mass.
mg/kWh for work-related change in mass.
m for surface roughness.
Damage force level.

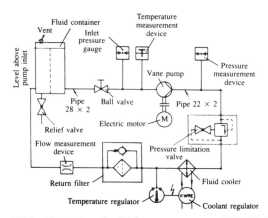

Fig. 4.104   Test set-up for Vickers pump test to DIN 51 389

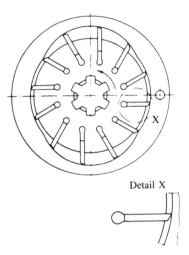

Detail X

Fig. 4.105   Test piece arrangement for Vickers pump test to DIN 51 389

Measuring method
The lubricant is subjected to a load, increasing by stages, at constant velocity and constant temperature in an FZG machine until scores occur on the surfaces or the wear rises to a high level (damage force level). See also Fig. 4.106.

Measuring equipment
FZG gear test machine as in Fig. 4.107 (DIN 51 354).

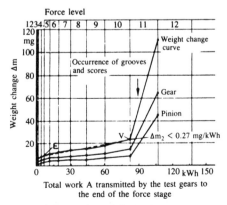

Fig. 4.106   Curve of change in weight in the FZG test to DIN 51 354

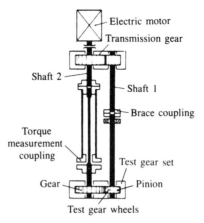

Fig. 4.107 FZG test machine as per DIN 51 354

Conclusion

Indicates the maximum loading of gear oils determined by anti-scoring and anti-wear additives.

4.8.2.3.12   Cloud point to FTM 791a/3456

Definition

The tendency of an oil to form channels at low temperatures and to delay flow back into a channel.

Unit of measurement

'Non-channelling' or 'channelling' at test temperature.

Measuring method

After a period of cooling (18 hours) at the test temperature in the test device, a steel strip immersed in oil is drawn carefully through the oil and then removed, creating a 'channel', and flow back into the channel is observed. If oil flows back in within 10 seconds, it is rated as 'non-channelling'.

Measuring equipment

Test device as in Fig. 4.108.

Conclusion

Indicates flow and flow-back performance at low temperatures.

4.8.3   Characteristics of (fluid) gear greases

The characteristics of (fluid) gear greases are divided into physical, chemical, and technological characteristics.

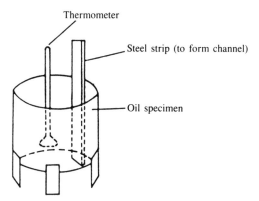

Thermometer

Steel strip (to form channel)

Oil specimen

Fig. 4.108   Test device for determining channel point to FTM 791a/3456

4.8.3.1   Physical characteristics

The following physical characteristics and standard methods of measuring them will be described briefly:

- apparent viscosity;
- solid foreign matter content;
- drop point;
- consistency and penetration;
- solids content (graphite, MoS_2).

4.8.3.1.1   Apparent viscosity and flow characteristics

(a) *Principles.* Lubricating greases belong to the non-Newtonian substances, and more precisely to the Bingham bodies. For Newtonian fluids, the flow characteristic is indicated by

$$\tau = \eta \, \frac{du}{dy}$$

Plastic substances, which are also Bingham bodies, have a pronounced flow limit, τ_0, before a flow process starts. Their flow behaviour can, therefore, be described by the following expression

$$\tau = \tau_0 + \eta^* \, \frac{du}{dy}$$

with η^* = viscosity with grease lubrication.

The yield point τ_0 and viscosity η^* of a given lubricating grease depend on the temperature (Fig. 4.109) and must be determined for each grease in a rotary viscometer.

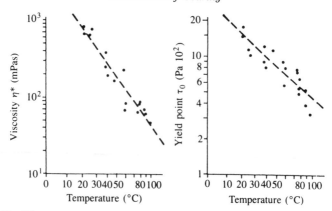

Fig. 4.109   Viscosity and yield point of a lithium-saponified lubricating grease with EP additives (Borchert)

To determine the apparent viscosity η_s, the above expression is modified as follows

$$\eta_s = \frac{\tau}{du/dy} = \frac{\tau_0}{du/dy} + \eta^*$$

For the shear rate

$$du/dy = u/h$$

in general. By modification we obtain

$$du/dy = \frac{\omega}{\psi}\frac{1}{1 + \varepsilon\cos\rho}$$

and

$$\eta_s = \frac{\tau_0\,\psi}{\omega\left(\dfrac{1}{1 + \varepsilon\cos\rho}\right)} + \eta^*$$

Figure 4.110 shows, as an example, the flow curve $\tau = f(du/dy)$ of a lithium-saponified lubricating grease with EP additives (**88**).

(b) *Measurement.* The flow behaviour of lubricating greases is studied in rotary viscometers. There are designs with cylindrical or conical stator/rotor systems (see also Fig. 4.77). Figure 4.111 shows the Haake rotary viscometer in schematic form. The measuring system consists of

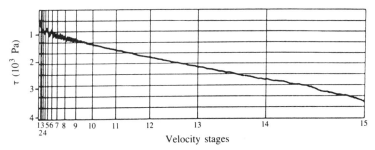

Fig. 4.110 Flow curve for a lithium-saponified lubricant with EP additives at 49.5°C (Borchert)

two coaxial circular cylinders, of which one is driven by an electric motor and the other remains still. In most designs the inner cylinder is driven. Between the cylinder walls is the substance to be examined. The rotation of the inner cylinder creates a Couette flow in the viscous medium with corresponding friction moments. The drive moment at the rotating cylinder is measured. The shear stress τ is determined from the drive torque M, the cylinder radius r and the wetted cylinder area A.

$$\tau = \frac{m}{rA} \; (N/m^2)$$

this is indicated at the control unit. With an auxiliary assembly the flow behaviour can be automatically recorded in the form

shear stress $\tau = f$ (shear rate du/dy)

Rotary viscometers can be modified so that shear rates of up to 14000 s^{-1} can be achieved.

4.8.3.1.2 *Solid foreign matter content to DIN 51 813*
Definition
By solid foreign substances we mean solid impurities of a foreign nature.
Unit of measurements
mg/kg (ppm).
Measuring method
A grease specimen is pressed through a test filter mesh. Then a small quantity of grease is mixed with the solid foreign matter contained in the filtered specimen. This specimen is then dissolved and filtered. The

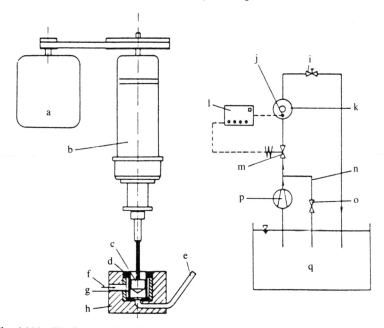

Fig. 4.111 Haake rotovisco for low temperatures:

(a)	drive motor	(i)	flow regulator valve
(b)	dynamometer	(j)	platinum resistance sensor
(c)	rotating element	(k)	temperature equalization
(d)	measuring dish		dish
(e)	delivery tube	(l)	temperature display
(f)	temperature	(m)	magnetic valve
	equalization fluid	(n)	short circuit wire
(g)	temperature	(o)	fixed throttle
	equalization chamber	(p)	pump
(h)	insulation jacket	(q)	low-temperature reservoir

difference in weight is used to determine the solid foreign substance content.

Measuring equipment

Press equipment, test filter mesh, and filtration equipment.

Conclusion

In the case of soap-based lubricating greases without solid lubricant additives it is a measure of the cleanliness of the lubricating grease.

4.8.3.1.3 Drop point to DIN ISO 2176

Definition

That temperature at which a lubricating grease reaches a certain flow capacity under specified conditions.

Unit of measurement

°C.

Measuring method

Lubricating grease in a test nipple is heated in an oil bath according to set heating methods until the grease begins to drip from the nipple.

Measuring equipment

Test equipment as in Fig. 4.112 (DIN ISO 2176).

Conclusion

The service temperature of the lubricating grease must lie below the drop point. A more precise conclusion is not possible.

4.8.3.1.4 Penetration to DIN ISO 2137

Definition

Penetration depth of a cone into the grease under set conditions.

Resting penetration:

penetration without previous mechanical load.

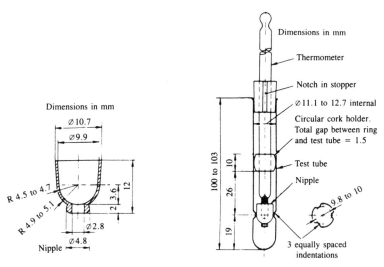

Fig. 4.112   Test equipment for drop point measurement (DIN ISO 2176)

Worked penetration:
> penetration after mechanical working (60 double strokes in the kneading machine)

Unit of measurement
1/10 mm.

Measuring method
After the grease is temperature equalized to 25°C, the cone is allowed to sink into the grease for 5 s.

Measuring equipment
Penetrometer with cone as in Fig. 4.113 and kneader as in Fig. 4.114 (DIN ISO 2137).

Conclusion
The penetration is a measure of the consistency, that is, the mechanical strength of a lubricating grease. It is the basis of the classification of lubricating greases according to consistency into the NLGI classes in Table 4.49 (DIN 51 818).

4.8.3.1.5   Solids and solid lubricant content to DIN 51 831

Definition
By solids we mean the total quantity of substances which are insoluble in solvents. By solid lubricant we mean the total quantity of graphite and molybdenum disulphide.

Unit of measurement
% (mass proportion).

Table 4.49   Classification of lubricating greases according to consistency as per DIN 51 818

NLGI class	*Worked penetration (mm^{-1}) (to DIN 51 804, Sheet 1)*
000	445–475
00	400–430
0	355–385
1	310–340
2	265–295
3	220–250
4	175–205
5	130–160
6	85–115

Gear Lubricants 295

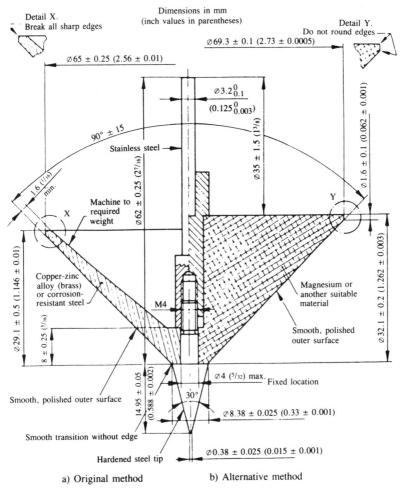

Fig. 4.113 Penetrometer with cone (DIN ISO 2137)

Measuring method
The total solids are flocculated with a mixture of solvents and their quantity is determined. The extraction residue is treated with a sulphuric acid/nitric acid mixture, which causes molybdenum (IV) sulphide to be precipitated. The graphite is determined from the mass loss occurring on roasting.

Dimensions in mm
(inch values in parentheses)

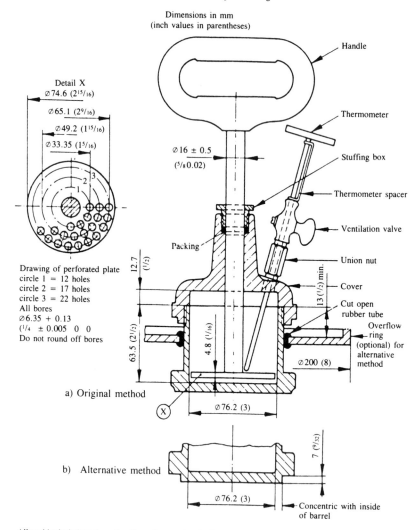

Drawing of perforated plate
circle 1 = 12 holes
circle 2 = 17 holes
circle 3 = 22 holes
All bores
⌀6.35 + 0.13
(¹/₄ ± 0.005 0 0
Do not round off bores

a) Original method

b) Alternative method

Allowable deviations from the dimensions, unless indicated otherwise: ± 0.25 mm (0.01 in)

Fig. 4.114   Lubricating grease kneader (to DIN ISO 2137)

Measuring equipment

Extraction and filtration equipment and chemicals.

Conclusion

Indicates the solid lubricant content of new metal soap lubricating greases.

4.8.3.2 Chemical characteristics

The following chemical characteristics and standard methods of measuring them will be described briefly:

- ash;
- mineral oil and soap content;
- water resistance;
- oxidation resistance.

4.8.3.2.1 Ash to DIN 51 803

Definition

Roasting residue after oxidation of a lubricating grease. It is determined as sulphate ash or oxide ash.

Unit of measurement

% mass.

Measuring method

About 2–5 g of the specimen is heated in a crucible which has been roasted and weighed previously, until the vapours leaving it can be ignited.

Measuring equipment

Kjeldahl flask, crucible, evaporating dish, desiccator, and muffle furnace.

Conclusion

In the case of non-graphitized lubricating greases, it provides an indication of the quantity and type of the metal soaps, inorganic thickeners, inorganic fillers and ash-producing additives.

4.8.3.2.2 Mineral oil and soap content to DIN 51 814

Definition

Mass fraction of mineral oil and soap in mineral oil lubricating greases containing alkaline soaps.

Unit of measurement

% mass.

Measuring method

A specimen of lubricating grease is separated into the components soap and mineral oil in a dialysis device by means of a semi-permeable dialysis membrane and petroleum ether. After the petroleum ether is distilled off, the components are determined gravimetrically.

Measuring equipment

Analytical equipment as in Fig. 4.115 (DIN 51 814).

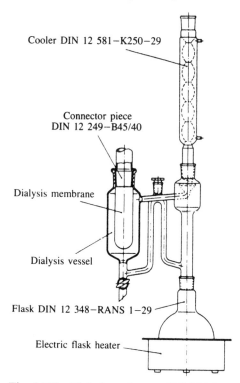

Cooler DIN 12 581−K250−29

Connector piece
DIN 12 249−B45/40

Dialysis membrane

Dialysis vessel

Flask DIN 12 348−RANS 1−29

Electric flask heater

Fig. 4.115   Dialysis equipment (DIN 51 814)

Conclusion

Assists assessment of the structure of lubricating greases. The separated components can be investigated further.

4.8.3.2.3   Response to water
(a)   *Static test to DIN 51 807, Part 1*
Definition

Effect of still water of specific origin at various temperatures on a lubricating grease which is not subjected to mechanical load.

Unit of measurement

Assessment levels as in Table 4.50 (DIN 51 807).

Measuring method

A specimen of lubricating grease on a glass plate is exposed to the action of still water for three hours. Any changes are then determined.

Table 4.50   Response to water (static test). Assessment levels as per DIN 51 807

Assessment level	Significance	Description
0	No change	None of the features mentioned for the following assessment levels
1	Slight change	Colour change (paler) of the surface of the grease, due to slight water absorption by the surface layer
2	Moderate change	Lubricating grease starts to dissolve, detectable by the formation of a white/yellow mucilaginous surface layer and moderate to severe clouding of the water
3	Severe change	Partial or complete dissolution of the grease, mainly with oil precipitation and formation of milky-white oil-in-water emulsion

Measuring equipment

Glass plate, test tube and heating device.

Conclusion

Measure of the water resistance of lubricating greases. Extrapolation to the practical situation is difficult as it is a static test.

(b) *Dynamic test to DIN 51 807, Part 2*

Definition

Effect of water spray of specified origin at various temperatures on a lubricating grease subjected to mechanical/dynamic loading.

Unit of measurement

Assessment levels as in Table 4.51 (DIN 51 807).

Table 4.51   Response to water (dynamic test). Assessment levels to DIN 51 807

Assessment level	Mass loss due to grease being washed out (%)
0	0–10
1	10–30
2	30–50
3	> 50

Measuring methods

The grease is tested in an annular ball bearing, the casing of which is sprayed with water for one hour. Water penetrates the bearing through a test orifice and acts on the grease. Its loss of mass is determined.

Measuring equipment

Test apparatus as in Fig. 4.116 (DIN 51 807).

Conclusion

A measure of the water resistance of K, KT, and KP lubricating greases under the test conditions.

4.8.3.2.4   Oxidation resistance to DIN 51 508

Definition

Resistance of a lubricating grease to the absorption of oxygen.

Unit of measurement

Pressure drop in bar.

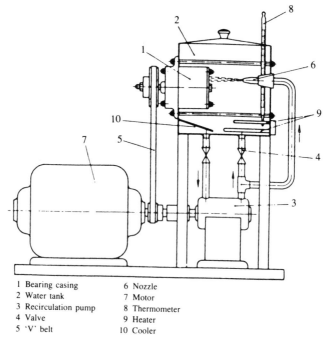

1 Bearing casing	6 Nozzle
2 Water tank	7 Motor
3 Recirculation pump	8 Thermometer
4 Valve	9 Heater
5 'V' belt	10 Cooler

Fig. 4.116   Test apparatus for dynamic tests of water resistance (DIN 51 807, Part 2)

Measuring method
The specimen of lubricating grease is placed in a test device, which is filled with oxygen at a pressure of 7 bar, and oxidized in a heating bath. The pressure drop after a specified period of time is a measure of the degree of oxidation.

Measuring equipment
Test equipment as in Fig. 4.117 (DIN 51 808).

Conclusion
Indicates the behaviour of lubricating greases which are exposed to the action of the atmosphere for a long time under static conditions. No indication of behaviour in storage or under dynamic conditions.

4.8.3.3 Technological characteristics
The following technological characteristics and standard methods for measuring them will be described briefly:

– oil precipitation;
– flow pressure;
– delivery behaviour;
– anti-corrosion behaviour (various methods);
– anti-wear behaviour;
– mechanical/dynamic behaviour.

4.8.3.3.1 Oil precipitation to DIN 51 847
Definition
That quantity of oil which precipitates out of the lubricating grease under the test conditions.

Unit of measurement
g/100 g, % mass.

Measuring method
The specimen is in a vessel with a conical wire mesh base and is loaded with a weight. The normal test takes place at 40°C over seven days, and the short test at the same temperature over eighteen hours.

Measuring equipment
Test equipment as in Fig. 4.118 (DIN 51 847).

Conclusion
Indicates the tendency of a lubricating grease to oil precipitation during storage. However, no direct conclusions can be drawn regarding the practical behaviour of lubricating grease.

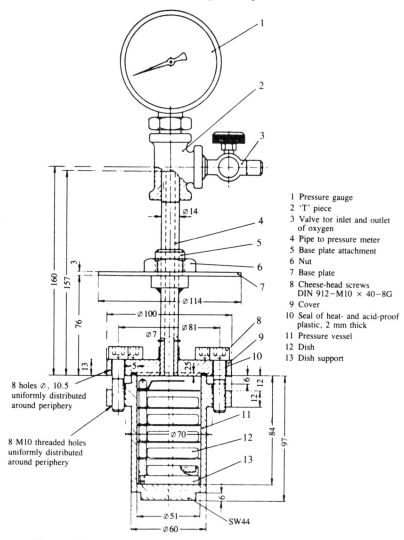

1 Pressure gauge
2 'T' piece
3 Valve for inlet and outlet
  of oxygen
4 Pipe to pressure meter
5 Base plate attachment
6 Nut
7 Base plate
8 Cheese-head screws
  DIN 912−M10 × 40−8G
9 Cover
10 Seal of heat- and acid-proof
   plastic, 2 mm thick
11 Pressure vessel
12 Dish
13 Dish support

8 holes ∅, 10.5
uniformly distributed
around periphery

8 M10 threaded holes
uniformly distributed
around periphery

Fig. 4.117   Equipment for testing oxidation resistance (DIN 51 808)

4.8.3.3.2   Flow pressure to DIN 51 805
Definition
That pressure which is necessary to press a thread of lubricating
grease out of a nozzle under specified conditions.

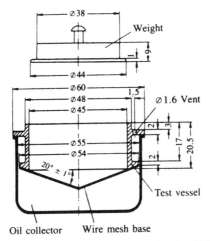

Fig. 4.118   Equipment for testing oil precipitation (DIN 51 847)

Unit of measurement
mbar, °C.
Measuring method
The nozzle filled with lubricating grease is connected to a pressure source. At the specified test temperature (+25; 0, −20 or 35°C) the pressure is increased at intervals of 30 s until the grease strand breaks, i.e., until the compressed gas flows through the nozzle.
Measuring equipment
Test equipment as in Fig. 4.119 (DIN 51 805).
Conclusion
A measure of the consistency and flow behaviour of a lubricating grease.

4.8.3.3.3   Delivery behaviour to DIN 51 816, Part 2
Definition
The delivery or pressure release behaviour of a lubricating grease indicates the drop in dynamic pressure of a lubricating grease over time after a sudden reduction of pressure to the normal level.
Unit of measurement
Minutes (min), bar, °C, mm²/s.
Measuring methods
The lubricating grease is pumped, free of air bubbles, into a pipe which can be sealed off, and brought to the test temperature (−10, 0,

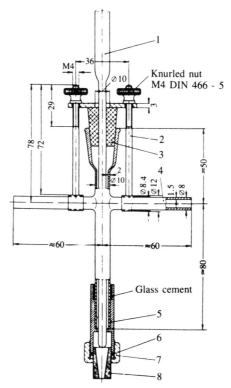

Fig. 4.119   Equipment for testing the flow pressure (DIN 51 805)

+ 20°C). The test pressure is then applied. After the valve at the end of the pipe is suddenly opened, the residual pressure at the start of the pipe is measured at various time intervals.

Measuring equipment

Test equipment as in Fig. 4.120 (DIN 51 816).

Conclusion

Indicates the performance in central lubrication systems.

4.8.3.3.4   Anti-corrosion characteristics to DIN 51 802

Definition

Corrosive change in the races of the roller bearing outer rings used in the test, caused by the influence of a mixture of lubricating grease and water.

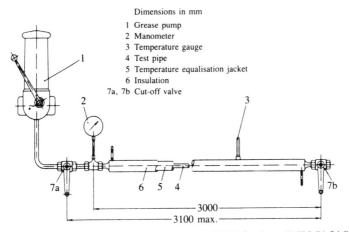

Dimensions in mm
1 Grease pump
2 Manometer
3 Temperature gauge
4 Test pipe
5 Temperature equalisation jacket
6 Insulation
7a, 7b Cut-off valve

Fig. 4.120   Equipment for testing pressure relief behaviour (DIN 51 816)

Unit of measurement

Corrosion levels as in Fig. 4.121 and Table 4.52 (DIN 51 802).

Measuring method

The lubricating grease is tested in spherical roller bearings with the addition of water. After a specified cycle of a period of operation (80 min^{-1}) without heat and load and a stationary period, the races of the outer rings are examined for corrosion.

Table 4.52   Corrosion levels of ball races to DIN 51 802

Corrosion level	Significance	Description
0	No corrosion	Unchanged
1	Traces of corrosion	At most three corrosion areas, of which none has a diameter of more than 1 mm
2	Slight corrosion	Not more than 1 percent of the surface corroded, but more or larger corrosion areas than for level 1
3	Moderate corrosion	More than 1 percent, but not more than 5 percent of the surface corroded
4	Severe corrosion	More than 5 percent, but not more than 10 percent of the surface corroded
5	Very severe corrosion	More than 10 percent of the surface corroded

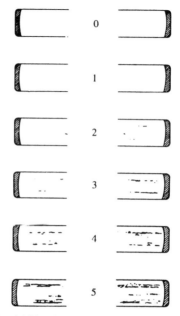

Fig. 4.121   Corrosion levels (DIN 51 802)

Measuring equipment
SKF Emcor machine with eight test bearing units.
Conclusion
Indicates the corrosion protection afforded by lubricating greases to roller bearings, plain bearings, and contact surfaces in the presence of water.

4.8.3.3.5   Corrosive effect on copper to DIN 51 811
Definition
By corrosion we mean discoloration produced on a copper strip or a coating produced on it, which can be rubbed or flaked off.
Unit of measurement
Corrosion levels as in Table 4.53 (DIN 51 811).
Measuring method
A ground copper strip is left for 24 h in the lubricating grease specimen at test temperature (e.g., 50°C, 100°C). The level of corrosion is determined from the discoloration.

Table 4.53   Corrosion levels for copper strip to DIN 51 811

Corrosion level	Significance	Description
0	No tarnishing and no corrosion	Unchanged
1	Slight tarnishing	Slightly orange, barely different from a freshly ground copper strip Dark orange
2	Moderate tarnishing	Burgundy Lavender Multi-coloured with lavender blue and/or silver haze on burgundy Silvery Brass-coloured or golden
3	Severe tarnishing	Magenta (aniline-red) coating on brass-coloured strip Multi-coloured with a red and green sheen (peacock-like), but not grey
4	Corrosion	Transparent black, dark grey or brown with peacock-like, slightly green sheen Graphite black or dull black Shiny black or pitch black

Conclusion

Indicates the extent of the corrosive action of a lubricating grease on copper.

4.8.3.3.6   Wear characteristics of consistent lubricants in the four-ball test apparatus to DIN 51 350, Part 5

Definition

Domes produced on the three stationary balls in a four-ball system by wear in mixed friction.

Unit of measurement

mm.

Measuring method

In a system consisting of a rotating ball pressed against three stationary balls, the wear present after 1 min or 1 h at constant load is measured, determined as the average wear dome diameter.

Measuring equipment

Four-ball apparatus and test piece system as in Fig. 4.103 (see Section 4.8.2.3.9).

Conclusion

Indicates the presence of anti-wear additives in lubricating greases. It is not possible to make a direct assessment of the performance of the lubricating greases in machine components.

4.8.3.3.7   Mechanical/dynamic behaviour in roller bearings to DIN 51 806

Definition

By mechanical/dynamic behaviour we mean the behaviour of lubricating greases in roller bearings under specified conditions.

Unit of measurement

Lubricating grease rating as in Table 4.54 and absolute wear by weight for roller sets and cages of the bearing as in Table 4.55 (DIN 51 806).

Measuring method

The lubricating grease is subjected to specified operating tests in two similar spherical roller bearings under radial loading. For test conditions see Table 4.56 (DIN 51 806).

Measuring equipment

Test machine for roller bearing grease as in Fig. 4.122 (DIN 51 806).

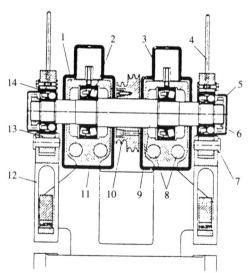

Fig. 4.122   Test machine for roller bearing grease (DIN 51 806)

Table 4.54 Abbreviations and ratings for lubricating grease to DIN 51 806

No.	*Lubricating grease rating in test bearing*			*Lubricating grease rating in test bearing casing*			
	T	*OC*	*CC*	*S*	*OP*	*G*	*EC*
	Wetting (resistance to dry contact)*	*Oil carbon†*	*Consistency change*	*Sealing‡ Residual grease*	*Oil precipitation*	*Gelling§*	*Entrained circulation¶*
1	Adequate	None	Slight	More than $\frac{2}{3}$ of grease	Slight	None	None
2	Moderate	Moderate	Moderate	More than $\frac{1}{2}$ of grease	Moderate	—	—
3	Inadequate	Severe	Severe	At most $\frac{1}{2}$ of grease	Severe	Occurs	Occurs

* Wetting means lubricant film on rollers and contact surfaces.

† Oil carbon is solid, black lubricating grease residue on the bearing surfaces which cannot be washed off with petrol. The measure of consistency is the worked and static penetration to DIN 51 804 Sheet 2 or the flow pressure to DIN 51 805 (partly still in draft form).

‡ Sealing is the self-sealing property of a lubricating grease by which it forms a stable, sealing bead at the junction with the shaft.

§ Gelling is the conversion of the lubricating grease into a flexible state which is no longer capable of lubrication.

¶ Entrained circulation means that the lubricating grease is circulated by rotating parts.

A lubricating grease in which gelling and/or entrained circulation occurs, inevitably causes bearing failure and does not complete the test. A distinction can therefore be made only between G1 and G3 or between EC1 and EC3.

Table 4.55   Wear assessment to DIN 51 806

Number	Absolute wear by weight (mg) of roller sets	of cage
1	<25	<100
2	25–100	100–200
3	>100	>200

Table 4.56   Test conditions to DIN 51 806

Test	Quantity of grease (g)	Rotational speed (min⁻¹)	Test time (days)	Test temperature (°C)	Remarks
A	200 + 30	2500 ± 125	20	Steady-state temperature*	Test run without heating
			1	Steady-state temperature*	
B	150	1500 ± 75	19	Constant t, where t means 70, 80, or larger whole multiple of 10. The constant test temperature should be set as in section 6.5.1, e.g., 100 ± 2.5°C	Test to determine maximum allowable operating temperature

* Steady-state temperature is the temperature of the test bearing which occurs in constant test conditions without heating.

Conclusion

Indicates the behaviour of lubricating greases in roller bearings under service-like conditions at various temperatures and rotational speeds.

4.9   STANDARDS FOR GEAR LUBRICANTS

4.9.1   Lubricant Standards

By a standard we mean a technical description which is accessible to all and has been produced by the cooperation and agreement of or with the general approval of all interested parties. It is based on the coordinated results of science, technology, and practice (DIN 820 Standardization procedures).

For a series of lubricants there are so-called requirement standards. The characteristics contained in them are then tested to the test standards given in Section 4.8.

Table 4.57 contains a list of the currently existing requirement standards for lubricants.

4.9.2   Designation of lubricants to DIN 51 502

In DIN 51 502 the designation of mineral oil-based and synthetic lubricating oils and lubricating greases is standardized. Each type of lubricant is assigned an identification letter and symbol. Table 4.58 shows this for mineral oils, low-flammability hydraulic fluids and synthetic fluids. The identification letters for certain liquid lubricants can be found in Table 4.59. The corresponding comparison for lubricating greases is contained in Table 4.60, while Table 4.61 contains additional letters for lubricating greases which indicate the response to water as well as the service temperature range.

4.9.3   Standards for selecting gear lubricants

4.9.3.1   Selecting lubricants for toothed gear lubricating oils as per DIN 51 509, Part 1

Part 1 of DIN 51 509 gives information on the selection of liquid lubricants for toothed gears. Table 4.62 shows that various standard lubricating oils with and without anti-wear additives are suggested for gear lubrication.

4.9.3.2   Selection of lubricants for toothed gears – plastic lubricants as per DIN 51 509, Part 2

Part 2 of DIN 51 509 (draft) gives information on the selection of plastic lubricants for toothed gear lubrication. The following lubricants were suggested:

(a)   G lubricating greases (fluid gear greases) as per DIN 51 826 (draft);
(b)   Bituminous adhesive lubricants. These include B lubricating oils as per DIN 51 513;
(c)   Bitumen-free adhesive lubricants. There is no requirement standard for these products which are designated as spray-on adhesive lubricants.

Table 4.57   Specifications for lubricants

	Lubricant Description	Standard
L–AN lubricating oils	Pure minerals for relatively undemanding lubrication tasks	DIN 51 501
Refrigeration machine oils	Mineral oils or carbons used for refrigeration machines with ammonia or halogenated hydrocarbons as the refrigerant	DIN 51 503
VB and VC lubricating oils with and without additives and VD–L oils	Pure mineral oils or mineral oils with L additives (oxidation and corrosion inhibitors) for air compressors and air vacuum pumps	DIN 51 506
Z oils	Pure mineral oils for steam engines	DIN 51 510
B oils	Dark, bituminous mineral oils for gear sets, open gears, guideways, and wire ropes	DIN 51 513
L–TD lubricating and control oils	Mineral oils with oxidation and corrosion inhibitors (no anti-wear additives) for steam turbines, stationary gas turbines, and machines by them, such as generators, compressors, pumps, gear sets	DIN 51 515 Part 1
C oils	Ageing-resistant mineral oils (without inhibitors) for circulation lubrication. Meet higher demands than L–AN lubricating oils	DIN 51 517 Part 1
CL oils	Mineral oils with oxidation and corrosion inhibitors. Meet higher demands than C lubricating oils	DIN 51 517 Part 2

CLP oils	Mineral oils with oxidation and corrosion inhibitors as well as anti-wear additives. Compared with CL oils, meet additional demands in mixed friction	DIN 51 517 Part 3
H and HL oils	H Hydraulic oils are pure mineral oils for hydrostatic drives with no requirements. HL Hydraulic oils are mineral oils with oxidation and/or corrosion inhibitors for hydrostatic drives with high thermal requirements	DIN 51 524
HLP hydraulic oils	Mineral oils with oxidation and corrosion inhibitors as well as anti-wear additives for hydrostatic and hydrodynamic drives with high requirements	DIN 51 525
K lubricating greases	Lubricating greases based on mineral oils and synthetic oils for roller and plain bearings between −20 and +140°C	DIN 51 825 Part 1
KT greases	Lubricating greases based on mineral oils or synthetic oils, which may contain additives and/or solid lubricants, for roller bearings and plain bearings between below −50 and +120°C. They have better low-temperature properties than K greases	DIN 51 825 Part 2
KP greases	Lubricating greases based on mineral oils or synthetic oils, which may contain additives and/or solid lubricants, for roller bearings and plain bearings between −20 and +140°C. They contain additives to reduce friction and wear in mixed friction or to increase load capacity	DIN 51 825 Part 3
G greases	Lubricating greases in NLGI classes 000 to 1 with an increased mineral and synthetic oil content for gears at temperatures between −20 and +120°C. They may contain additives for reducing friction and wear and/or for increasing load capacity.	DIN 51 826

Table 4.58 Identification letters and symbols for lubricants and hydraulic fluids to DIN 51 502

1	2	3	4	5		6
				Specified		
No.	Name of substance group	Type of substance (example)	Identification letters	in	for	Symbol
1	Mineral oils	N lubricating oil (normal lubricating oils)	N	DIN 51 501	L–AN[1]	
		B lubricating oil (e.g., bituminous)	B	DIN 51 513	BA, BB, BC	
		C lubricating oils (circulation lubricating oils)	C	DIN 51 517 Part 1 to 3	C, CL, CP[2]	
		CG lubricating oils (guideway oils)	CG		[2]	
		D lubricating oils (compressed air oils)	D			
		F oils (air filter oils)	F			
		FS oils (mould separation oils)	FS			
		H oils (hydraulic oils)	H	DIN 51 524 Part 1* and Part 2*	HL, HLP[2]	
		J oils (electrical insulating oils)	J	DIN 57 370 Part 1/ VDE 0370 Part 1		
		K lubricating oil (refrigeration machine oil)	K	DIN 51 503	KA, KC	□
		L oils (hardening and quenching oils)	L	†		
		Q oils (heat transfer oils)	Q	†		
		R oils (anti-corrosion oils)	R			

		†	
S oils (cooling lubricants)	S		
T lubricating oils (steam turbine lubricating and control oils)	T	DIN 51 515 Part 1	L–TD[1]
V lubricating oils (anti-compressor oils)	V	DIN 51 506	VB, VC, VBL[4] VCL[4], VDL[4]
W oils (rolling oils)	W		
Z lubricating oils (steam cylinder oils)	Z	DIN 51 510	ZS, ZA, ZB, ZC, ZD
2 Low-flammability hydraulic fluids			
Oil-in-water emulsions	HFA[3]		
Water-in-oil emulsions	HFB[3]		
Aqueous polymer solutions	HFC[3]		
Water-free fluids	HFD[3]		
3 Synthetic or partly synthetic fluids			
Diester oil	E		
Fluorocarbon oils	FK		
Polyglycol oils	PG		
Silicon oils	SI		

* Still in draft form.
† Standard in preparation.
[1] The international class letter L (lubricants) can be omitted, as no abbreviations are currently envisaged for other mineral oil products.
[2] ISO Standard 3498 uses the following identification letters: for CL – CB, for CLP – CC, for CG – C, for HL – HL, for HLP – HM.
[3] This classification corresponds to ISO/DIS 6071 and ISO/DP 6743 and is inserted in the sixth Luxemburg Report (standard committee for works safety in coal mines).
[4] In the current version of DIN 51 506, (1977), the identification letters still have a hyphen. When it is revised, the hyphen will be omitted.

315

Table 4.59 Additional identification letters as per DIN 51 502

1	*2*
Additional identification letter	*Type of lubricant*
E	For lubricating oils which are used mixed with water, e.g., water-miscible cooling lubricants
F	For lubricants with solid lubricant added, e.g., graphite, molybdenum disulphide
L	For lubricants with additives to increase corrosion protection and/or ageing resistance, e.g., CL lubricating oil to DIN 51 517 Part 2
P	For lubricants with additives for reducing friction and wear in the mixed friction area and/or to increase load capacity, e.g., CLP lubricating oil to DIN 51 517 Part 3
V*	For lubricants which are diluted with solvents, e.g., lubricating oil DIN 51 513 – BBV†

* The additional identification letter V gives rise, in certain circumstances, to designation according to the Working Substance Decree.
† In the 1977 version of DIN 51 513, the identification letters still have a hyphen. When this edition is revised, the hyphen will be omitted.

4.9.3.3 Selection of lubricants for construction machinery to DIN 51 516
In this Standard, gear oils and gear greases are also suggested, mainly for use in construction machinery. Table 4.63 contains information on gear oils. The API-GL classifications given there are dealt with in Section 4.10. Table 4.64 also contains information on gear greases.

4.9.4 Suitable, standard lubricants for gear lubrication
Standards DIN 51 509 and DIN 51 516 list a series of lubricants for selecting suitable gear lubricants. The following standard lubricants will, therefore, be described briefly

B lubricating oils;
L–AN lubricating oils;
L–TD lubricating oils;
C, Cl, and CLP lubricating oils;
HL and HLP lubricating oils;
Automotive gear oils;

Table 4.60 Identification letters and symbols for lubricating greases as per DIN 51 502

Type of lubricating grease	*Identification letter(s)*	*Symbol*
	1	*2*	*3*
Lubricating greases for roller bearings, plain bearings and contact surfaces as per DIN 51 825 Part 1, in the service temperature range from −20 to +140°C	K[1]	For lubricating greases based on mineral oils
Lubricating greases for high pressures in the service temperature range from −20 to 140°C	KP	
Lubricating greases for service temperatures over +140°C	KH	
Lubricating greases suitable for low temperatures as per DIN 51 825 Part 2 — from −30°C to +120°C; from −40°C to +120°C; from −55°C to +120°C	KTA KTB KTC	△
Lubricating greases for enclosed gears	G	
Lubricating greases for open gears, cog systems (adhesive lubricants without bitumen)	OG	
Lubricating greases for plain bearings and slabs[2]	M	
Lubricating greases based on synthetic oils are identified by their basic characteristics like those above based on mineral oils	Addition of identification letters as in Table 1, substance group 3	For lubricating greases based on synthetic oils ◇

[1] ISO Standard 3498 uses the letters XM for the identification letter K.
[2] Lower requirements than for K lubricating greases.

Table 4.61   Additional letter for lubricating greases as per DIN 51 502

1	2	3
Additional letter	Response to water[1] as in DIN 51 807 Part 1 assessment level	Service temperature range[2] °C
B	0 or 1	−20 to +50
C	0 or 1	−20 to +60
D	2 or 3	−20 to +60
E	0 or 1	−20 to +80
F	2 or 3	−20 to +80
G	0 or 1	−20 to +100
H	2 or 3	−20 to +100
K	0 or 1	−20 to +120
M	2 or 3	−20 to +120
N	0 or 1	−20 to +140
R	0 or 1	over +140

[1] 0 means no change.
1 means slight change.
2 means moderate change.
3 mean great change.
[2] The upper service temperature is obtained up to N as in DIN 51 806 Part 2 (currently in draft form), test B (see Table 2) in DIN 51 825 Parts 2 and 3 (currently in draft form).

K, KT, and KP lubricating greases;
G lubricating greases.

B lubricating oils to DIN 51 513
These are highly viscous, dark, bituminous mineral oils which are used mainly for open toothed gears, guideways, and wire cables by means of manual, oil bath and splash lubrication. They may contain a solvent to aid application.

As Table 4.65 shows, B lubricating oils are available in three viscosities, indicated at 100°C.

Table 4.62   Comparison of viscosities of various standard types of lubricating oil as per DIN 51 509

ISO viscosity classes to DIN 51 519	Identification number	SAE viscosity classes to DIN 51 511	SAE viscosity classes to DIN 51 512	C and CL lubricating oils to DIN 51 517	L–AN lubricating oils to DIN 51 501	L–TD lubricating oils to DIN 51 515	C–LP lubricating oils to DIN 51 515	Vehicle gear oils
				without anti-wear additives			*with anti-wear additives*	
22	16			×	×	×	×	
32								
		10 W						
32	25		75	×	×	×	×	×
46								
46	36	20 W		×	×	×	×	
68								
68	49	20		×	×	×	×	
			80					×
100	68	30		×	×		×	
150	92			×	×		×	
		40						
220	114			×	×		×	
			90					×
220	144			×	×		×	
		50						
320	169			×			×	
460	225			×	×		×	
			140					×
680	324			×				

L–AN lubricating oils to DIN 51 501
These are pure mineral oils for lubrication tasks without special requirements. These oils are only suitable for gear lubrication if constant temperatures are below 50°C, extended oil-change intervals are not required and no special demands are made on load capacity.

Table 4.66 contains the minimum requirements for L–AN lubricating oils.

L–TD lubrication and control oils to DIN 51 515
These are mineral oils with additives to improve anti-corrosion properties and ageing resistance. They contain no additives to improve anti-wear or anti-scoring characteristics. L–TD lubricating oils are suitable

Table 4.63 Selection of lubricants for construction machinery to DIN 51 516

No.	Lubricant group	Type of lubricating oil	Abbreviation	Designation of lubricant container, lubrication equipment, and lubrication point to DIN 51 502				Examples of application
				Symbol	Letter(s)	Number	Colour	
1	Engine oils	API–CC engine oils	HD-CC 10 W		CC	10 W		Diesel and SI engines, mobile air compressors, hydraulic systems, mechanical linkages, compressed air equipment
2	containing additives		HD-CC 20 W-20		CC	20 W-20		
3			HD-CC 30		CC	30		
4		API–CD engine oils	HD-CD 10 W		CD	10 W		
5			HD-CD 30		CD	30		
6			HD-CD 15 W-40		CD	15 W-40		
7	Gear oils	Gear oils API–GL–4	HYP-GL-4-80	☐	GL-4	80		Spur and bevel gears, range speed gears, hypoid drives, worm gears
8	containing additives	Gear oils API–GL–4	HYP-GL-4-90		GL-4	90	White	
9		Gear oils API–GL–5	HYP-GL-5-90		GL-5	90		
10	Power transmission oils	Fluid gear oils	ATF		ATF	–		Hydrodynamic gears, torque converters, clutches, power shift transmission
11	Hydraulic oils	H-LP hydraulic oils to DIN 51 524 Part 2	H-LP 32		H-LP	32		All types of hydraulic systems and hydrostatic drives
12	containing additives		H-LP 46		H-LP	46		
13	Adhesive lubricants	e.g., B lubricating oils to DIN 51 513	–		BB	–		Open gears, cables, guide rails, wheel flanges, chains

Table 4.64 Lubricating greases for construction machinery to DIN 51 516

No.	Lubricant group	Type of lubricating grease	Abbreviation	Consistency to DIN 51 818 NLGI class	Designation for lubricant container, equipment, and points to DIN 51 502	Colour	Examples of applications
1		Multi-purpose greases to DIN 51 825 Part 1 or Part 3	K 2 K	2	△ K / 2 K		Roller and plain bearings to 140°C operating temperature, articulated shafts
2	Lubricating greases		KP 2 K	2	△ KP / 2 K		
3		High-temperature lubricating greases	KH 2 R	2	△ KH / 2 R	White	Roller and plain bearings to 180°C operating temperature e.g., drying drums
4		Gear greases	GP 00 D	00	▽ GP / 00 D		Gear lubrication with inadequate sealing or in accordance with requirements of gear manufacturer

Table 4.65   Minimum requirements for B lubricating oils to DIN 51 513

Type of lubricating oil	Requirements[1]			Test to
	BA	BB	BC	
Kinematic viscosity at 100°C mm²/s	16 to 36	49 to 114	225 to 500	[2]
Asphaltene content min. g/100 g oil		2		DIN 51 595
Flash point in open crucible Cleveland min. °C	145	165	165	DIN 51 376
Pour point equal to or lower than °C	−6	+6	+30	DIN 51 597 DIN prEN 6
Water content max. g/100 g oil		0.2		DIN 51 582
Ash (oxide ash) max. g/100 g oil	0.1	0.2	0.3	DIN EN 7
Water soluble acids Reaction		Neutral		DIN 51 558 Part 1
Solid foreign substances max. g/100 g oil	0.2	0.3	0.4	DIN 51 592

[1] These requirements relate to B lubricating oils without solvents and to those from which the solvent has evaporated.
[2] Test to DIN 51 550 in conjunction with DIN 51 561, DIN 51 562, or DIN 51 015. The test may also be performed with a Cannon Fenske Opaque Viscometer or a BS/IP U-Tube Reserve Flow Viscometer. (Test standards for measuring viscosity with these two viscometers are in preparation.)

for gears where there are higher demands on corrosion protection and ageing resistance, but not on load capacity.

Table 4.67 contains the minimum requirements for L–TD lubricating oils.

C lubricating oils to DIN 51 517, Part 1

These are pure mineral oils which have better ageing stability without additives than L–AN lubricating oils. They can be used in gears if no special demands are made on wear protection in the mixed friction region.

Table 4.68 contains the minimum requirements for C lubricating oils.

CL lubricating oils to DIN 51 517, Part 2

These are mineral oils with additives for improving corrosion protection and ageing resistance. They can be used in gears where higher demands are made on the characteristics mentioned than can be met by C lubricating oils. However, no special demands may be made on anti-scoring and anti-wear capacity.

Table 4.69 contains the minimum requirements for CL lubricating oils, and one can recognize the requirements regarding corrosion protection and ageing stability compared with the minimum requirements for C lubricating oils.

Table 4.66 Minimum requirements for L–AN lubricating oils to DIN 51 501

Type of lubricating oil[1]		AN 5	AN 7	AN 10	AN 22	AN 46	AN 68	AN 100	AN 150	AN 220	AN 320	AN 680	Test to	Comparable ISO*/ASTM†/IP‡ standards
ISO viscosity class		ISO VG 5	ISO VG 7	ISO VG 10	ISO VG 22	ISO VG 46	ISO VG 68	ISO VG 100	ISO VG 150	ISO VG 220	ISO VG 320	ISO VG 680		ISO 3348
Kinematic[2] Viscosity at 40°C	mm²/s min	4.14	6.12	9.00	19.8	41.4	61.2	90.0	135	198	288	612	DIN 51 550 in conjunction with DIN 51 562 Part 1	ISO 3104 ASTM D 445 IP 71
	mm²/s max	5.06	7.48	11.0	24.2	50.6	74.8	110.0	165	242	352	748		
Kinematic viscosity at 50°C approximately[3]	mm²/s	3 4	4 5	6 9	14 17	25 30	36 44	53 68	75 95	105 130	150 180	300 360		
Density at 15°C	g/ml				To be indicated by supplier								DIN 51 757	ISO 3675 ASTM D 1298 IP 160
Cleveland flash point in open crucible	°C min	80[5]	100[5]	120		145		170		200		250	DIN 51 376	ISO 2592 ASTM D 92 IP 36
Pour point (flow limit) equal to or lower than	°C		−12	−18	−15		−12		−9		−6	−3	DIN 51 597 DIN EN 6[7]	ISO 3016 ASTM D 97 IP 15
Neutralization number (water-soluble acids)	mg KOH/g						0						DIN 51 558 Part 1	ASTM D 974 IP 139
Neutralization number (acid)	mg KOH/g max						0.15							
Saponification number	mg KOH/g max						0.3						DIN 51 559	ASTM D 94 IP 136 A
Ash (oxide ash)	g/100 g max	0.01				0.02					0.05		DIN EN 7	ASTM D 482 IP 4
Asphaltene content	g/100 g max				0.05						0.2		DIN 51 595	IP 143
Water content	g/100 g max				0.2						0.5		DIN ISO 3733[7]	ISO 3733 ASTM D 95 IP 74
Content of undissolved substances	g/100 g					Below the limit of quantitative detectability[6]							DIN 51 592[7]	

* International Organisation for Standardization (ISO).
† American Society for Testing and Materials (ASTM).
‡ Institute of Petroleum (IP).
1 The numbers correspond to the rounded values of mid-point viscosity in mm²/s at 40°C. They are derived from the new ISO viscosity classes in DIN 51 519.
2 The SI unit of kinematic viscosity is m^2/s. 1 mm²/s = 1×10^{-6} m²/s. The SI unit of dynamic viscosity is the Pascal second. 1 m Pa s = 1×10^{-3} N s/m². Conversion in accordance with DIN 51 550 can be performed with an average value for density of 0.900.
3 The values serve to classify lubricants used in the past. They are based on relevant viscosity columns mm²/s at 50°C from DIN 51 519 (Table 2).
4 Density is not a qualitative feature. It serves for converting weight to volume. For analysis of plain bearings, the density and viscosity at the same temperature (e.g. 40°C) should be inserted in the calculation. If density and viscosity are indicated with other reference temperatures, the two values must be converted to the same temperature (e.g. 40°C) for the bearing calculation.
5 If the Abel Pensky flash point in an enclosed crucible as per DIN 51 755 is between 55 and 100°C. Dangerous Good Class A III applies to transport and storage.
6 Due to testing error in the test procedure, reliable figures are not possible below 0.03 g/100 g.
7 Currently in draft form.

Table 4.67   Minimum requirement for TD lubricating oils to DIN 51 515

Type of lubricating oil		Requirements				Test² to	Comparable ISO* standards
		TD 32	TD 46	TD 68	TD 100		
ISO viscosity class¹		ISO VG 32	ISO VG 46	ISO VG 68	ISO VG 100	DIN 51 519	ISO 3448
Kinematic viscosity³ at 40°C	mm²/s (cSt) min	28.8	41.4	61.2	90.0	DIN 51 550 in conjunction with DIN 51 561 or DIN 51 562	ISO 3104
	max	35.2	50.6	74.8	110		
Flash point in open crucible (Cleveland)	°C min	160	185	205	215	DIN 51 376	ISO 2592
Density at 15°C	g/ml max	0.900	0.900	0.900	0.900	DIN 51 757	ISO 3675
Pour point equal to or lower than	°C	−6	−6	−6	−6	DIN 51 597 DIN prEN 6†	ISO 3016
Neutralization number	mg KOH/g	To be indicated by supplier				DIN 51 558 Part 1	–
Ash (oxide ash)	g/100 g	To be indicated by supplier				DIN EN 7	–
Water content	g/100 g	Below the limit of quantitative detectability⁴				DIN 51 582	ISO 3733

Property	Unit					DIN	ISO[*]
Solid foreign substance content	g/100 g	Below the limit of quantitative detectability[5]				DIN 51 592	–
Water separation capacity after steam treatment	seconds max	300	300	300	300	DIN 51 589 Part 1	–
Air separation capacity[6] at 50°C	minutes max	5	5	6	No stipulations	DIN 51 381	–
Corrosive effect on copper	Corrosion level max			2-100 A 3		DIN 51 759	ISO 2160
Anti-corrosion properties with regard to steel	Corrosion level			0-A[7]		DIN 51 585	–
Ageing behaviour[8] Increase in neutralization number after 1000 h	mg KOH/g oil max	2.0[9]	2.0[9]	2.0[9]	2.0[9]	DIN 51 587	–

* International Organisation for Standardization (ISO).

† Currently in draft form.

1 The figures represent the mid-point viscosity in mm²/s (cSt) at 40°C. They are derived from the new ISO viscosity classes in DIN 51 519.

2 Specimens should be stored without exposure to light until the test is carried out.

3 The SI unit of kinematic viscosity is m^2/s. $1 \ mm^2/s = 1 \times 10^{-6} \ m^2/s \ (= 1 \ cSt)$.

4 Due to test error in the test procedure, it is not possible to obtain reliable figures below 0.1 g/100 g.

5 Due to test error in the test procedure, it is not possible to obtain reliable figures below 0.5 g/100 g.

6 In view of starting difficulties, it may be advisable to test air separation capacity to DIN 51 381 at a temperature of 25°C. In such cases, these values must be requested from the supplier. A test procedure for determining foaming characteristics on the basis of ASTM D 892 is in preparation.

7 If there is a danger of sea water penetrating the lubricating oil system, it may be agreed that the test be carried out to DIN 51 585 with artificial sea water.

8 Due to long test time, ageing behaviour should not be determined on acceptance, but at type testing.

9 With an increase in the neutralization number compared with the neutralization number of the new oil of up to 2.0 mg KOH/g oil, it is not possible to make a differential assessment of the L–TD lubricating oils.

Table 4.68 Minimum requirements for C lubricating oils to DIN 51 517

Type of lubricating oil Designation to DIN 51 552		Requirements											Test to
		C 7	C 10	C 22	C 46	C 68	C 100	C 150	C 220	C 320	C 460	C 680	
ISO viscosity class to DIN 51 519		ISO VG 7	ISO VG 10	ISO VG 22	ISO VG 46	ISO VG 68	ISO VG 100	ISO VG 150	ISO VG 220	ISO VG 320	ISO VG 460	ISO VG 680	DIN 51 550 in conjunction with DIN 51 562 Part 1
Kinematic viscosity at 40 °C[2] mm²/s min		6.12	9.0	19.8	41.4	61.2	90.0	135	198	288	414	612	
max		7.48	11.0	24.2	50.6	74.8	110	165	242	352	506	748	
Density at 15 °C[3] g/ml							To be agreed if necessary						DIN 51 757
Flash point in open crucible (Cleveland) °C min		105	125	165	175	185	200	210	220	230	240	250	DIN 51 376
Pour point equal to or lower than °C		−21			−15		−12	−9		−6		−3	DIN 51 597 prEN 6*
Neutralization number (acid) mg KOH/g max							0.15						DIN 51 558 Part 1
Ash (oxide ash) wt% max							0.02						DIN EN 7
Water content wt%						Below the limit of quantitative detectability[4]							DIN ISO 3733*
Undissolved solids content g/100 g						Below the limit of quantitative detectability[5]							DIN 51 592
Demulsification capacity		Testing not possible					To be indicated if necessary						DIN 51 599
Baader ageing behaviour Increase in saponification number mg KOH/g max							No testing envisaged						DIN 51 554 Part 1 DIN 51 554 Part 3
Increase in coke residue wt% max			1.2			1.5		2.0			2.5	No requirement	DIN 51 352 Part 1
Response to NBR SRE 1 seal material to DIN 53 538 Part 1* ±2 h at (100 ± 1)°C								To be agreed if necessary					
Relative change in volume vol%													DIN 53 538 Part 1* DIN 53 521
Change in Shore A hardness							To be agreed if necessary						DIN 53 538 Part 1* DIN 53 505

Table 4.69 Minimum requirements for CL lubricating oils to DIN 51 517

Type of lubricating oil[1] / Designation to DIN 51 502		CL 5	CL 10	CL 22	CL 32	CL 46	CL 68	CL 100	CL 150	CL 220	CL 460	Test to
Designation to ISO 3498 or DIN 8659 Part 2*		ISO VG 5	FC 10	FC 22	CB 32		CB 68		CB 150			
ISO viscosity class to DIN 51 519		ISO VG 5	ISO VG 10	ISO VG 22	ISO VG 32	ISO VG 46	ISO VG 68	ISO VG 100	ISO VG 150	ISO VG 220	ISO VG 460	
Kinematic viscosity at 40°C² mm^2/s min		4.14	9.0	19.8	28.8	41.4	61.2	90.0	135	198	414	DIN 51 550 in conjunction with DIN 51 562 Part 1
Kinematic viscosity at 40°C² mm^2/s max		5.06	11.0	24.2	35.2	50.8	74.8	110	165	242	506	
Density at 15°C g/ml		To be agreed if necessary										DIN 51 757
Flash point in open crucible (Cleveland) °C min		105	125	165	170	175	185	200	210	220	240	DIN 51 376
Pour point equal to or lower than °C		-21				-15		-12	-9		-6	DIN 51 597 prEN 6*
Neutralization number⁴ acid or alkaline mg KOH/g		To be indicated by supplier										DIN 51 558 Part 1
Ash (oxide ash)⁴ wt%		To be indicated by supplier										DIN EN 7
Water content wt%		Below the limit of quantitative detectability⁵										DIN ISO 3733*
Undissolved solids content g/100 g		Below the limit of quantitative detectability⁵										DIN 51 592
Demulsification capacity		To be indicated if necessary										DIN 51 599
Corrosive effect on copper max		Corrosion level 2 100 A3										DIN 51 759
Anti-corrosive properties with regard to steel method A max		Corrosion level 0 A										DIN 51 585
Ageing behaviour⁷ Increase in neutralization number after 1000 h mg KOH/g oil max		2.0⁸										DIN 51 587
Response to NBR SRE 1 seal material to DIN 53 538 Part 1* after 7 days ±2 h at (100 ± 1)°C		To be agreed if necessary										DIN 53 538 Part 1* DIN 53 521
Relative change in volume vol%												DIN 53 538 Part 1* DIN 53 505
Change in Shore A hardness		To be agreed if necessary										

CLP lubricating oils to DIN 51 517, Part 3
These are mineral oils with additives for improving corrosion protection, ageing resistance, and protection against wear and scoring. They are used in gears where there is a requirement for increased wear-protection in mixed friction due to the load or for protection against scoring where load is excessive. Table 4.70 contains the minimum requirements for CLP lubricating oils. It can be seen that to identify the anti-scoring and anti-wear capacity in the FZG test, a damage force level of at least twelve is required with wear of at most 0.3 mg/kWh.

HL hydraulic oils to DIN 51 524, Part 1
These are mineral oils with additives for improving corrosion protection and ageing resistance. They can be used in gears in the same conditions as CL lubricating oils.

Table 4.71 contains the minimum requirements for HL hydraulic oils.

HLP hydraulic oils to DIN 51 524, Part 2
These are mineral oils with additives for improving corrosion protection, ageing resistance, and protection against scoring and wear. The load capacity in gears is lower than that of CLP lubricating oils, however.

Table 4.72 contains the minimum requirements for HLP hydraulic oils, and it can be seen that a FZG damage force level of at least ten with a specific wear of at most 0.3 mg kWh is required as a criterion for the scoring and wear protection characteristics. Furthermore, wear in a vane pump test must not exceed certain limits.

Automotive gear oils
The requirements for automotive gear oils are not laid down in DIN Standards, but in classifications and specifications (see Section 4.10).

K lubricating greases to DIN 51 825, Part 1
These are lubricating greases for the service temperature range from $-20°C$ to $+140°C$, which contain a mineral oil-based and/or synthetic base oil in addition to a thickener. They may contain additives for improving corrosion protection and/or ageing behaviour (KL lubricating greases).

They are manufactured in NLGI classes 0, 1, 2, 3, or 4 for various service temperature ranges and levels of water resistance (Table 4.73). Table 4.74 contains the minimum requirements for K lubricating greases.

KT lubricating greases to DIN 51 825, Part 2
These are lubricating greases with better low temperatures than K lubricating greases for the service temperature range from below $-20°C$ to

Table 4.70 Minimum requirements for CLP lubricating oils to DIN 51 517

Type of lubricating oil[1] Designation to DIN 51 502		Requirements									Test to
			CLP 46	CLP 68	CLP 100	CLP 150	CLP 220	CLP 320	CLP 460	CLP 680	
ISO viscosity class to DIN 51 519			ISO VG 46	ISO VG 68	ISO VG 100	ISO VG 150	ISO VG 220	ISO VG 320	ISO VG 460	ISO VG 680	
Designation to ISO 3498 or DIN 8659 Part 2*						CC 150		CC 320	CC 460		DIN 51 550 in conjunction with DIN 51 562 Part 1
Kinematic viscosity at 40°C[2]	mm²/s	min	41.4	61.2	90.9	135	198	288	414	612	DIN 51 757
		max	50.6	74.8	110	165	242	352	506	748	
Density at 15°C[3]	g/ml										DIN 51 376
Flash point in open crucible (Cleveland)	°C	min	175	185	200	200	200	200	200	200	DIN 51 597 prEN 6*
Pour point equal to or lower than	°C			−15	−12	−9		−6		−3	DIN 51 558 Part 1
Neutralization number[4] acid or alkaline	mg KOH/g					To be agreed if necessary					DIN 51 575
Sulphate ash[4]	g/100 g					To be agreed if necessary					DIN ISO 3733*
Water content	wt%					Below the limit of quantitative detectability[5]					DIN 51 592
Undissolved solids content	g/100 g					Below the limit of quantitative detectability[6]					DIN 51 579 without bomb 3 h at 100°C
Corrosive effect on copper						To be agreed if necessary					DIN 51 355
Anti-corrosion properties with regard to steel agitation method A						Corrosion level 0–A					DIN 51 586
Ageing behaviour[7] Increase in velocity	%					To be agreed if necessary[8]					
Precipitation number	ml					To be agreed if necessary[8]					
Mechanical test in FZG machine machine[7] Load limit: Damage force level		min					12				DIN 51 354 Part 2
Work-related weight change	mg/kWh	max					0.3				
Response to NBR SRE 1 seal material to DIN 53 538 Part 1* after 7 days ± 2 h at (100 ± 1)°C											DIN 53 538 Part 1* DIN 53 521
Relative change in volume	vol%					To be agreed if necessary					
Change in Shore A hardness						To be agreed if necessary					DIN 51 538 Part 1* DIN 53 505

Table 4.71 Minimum requirements for HL hydraulic oils to DIN 51 524

Type of hydraulic oil[3]			Requirements							Test to
			HL 10	HL 22	HL 32	HL 46	HL 68x	HL 100		
ISO viscosity class to DIN 51 519			ISO VG 10	ISO VG 22	ISO VG 32	ISO VG 46	ISO VG 68	ISO VG 100		—
Kinematic viscosity[4]	at −20°C	max	600	–	–	–	–	–		DIN 51 550 in conjunction with DIN 51 561, DIN 51 562 Part 1 or DIN 51 569
	at 0°C	max	90	230	420	780	1400	2560		
	at 40°C	max	11.0	24.2	35.2	50.6	74.8	110		
	at 40°C	min	9.0	19.8	28.8	41.4	61.2	90.0		
	at 100°C	min	2.4	4.1	5.0	6.1	7.8	9.9		
Pour point equal to or lower than		°C	−30	−21	−18	−15	−12	−12		DIN 51 597 DIN ISO 3016*
Flash point higher than		°C	125	165	175	185	195	205		DIN 51 376
Content of undissolved substances		g/100 g	Below the limit of quantitative detectability[5]							DIN 51 592*
Water content		g/100 g	Below the limit of quantitative detectability[6]							DIN ISO 3733
Anti-corrosion characteristics with regard to steel (Method A)		max				Corrosion level 0-A				DIN 51 585
Corrosive effect on copper		max			Corrosion level 2-100 A 3					DIN 51 759
Ageing behaviour[7] Increase in neutralization number after 1000 h		mg KOH/g max				2.0[8]				DIN 51 587
Response to SRE-NB R1 seal material to DIN 53 538 Part 1 after 7 days ± 2 h at (100 ± 1)°C[7]		relative vol% change	To be indicated by supplier[9]							DIN 53 538 Part 1 in conjunction with DIN 53 521

Change in Shore A hardness		To be indicated by supplier[9]			DIN 53 538 Part 1 in conjunction with DIN 53 505
Air separation capacity at 50°C[11]	min max	5	10	9	DIN 51 381
Foaming characteristics ml max (process B)	S1 S2 S3	To be indicated by supplier[9]	To be indicated by supplier[9]		DIN 51 566*
Demulsification capacity[10] at 54°C	min max	30	40	60	DIN 51 599
Density at 15°C[11]	g/ml	To be indicated by supplier			DIN 51 757
Ash (oxide ash)	g/100 g	To be indicated by supplier			DIN EN 7
Neutralization number[11] acid or alkaline	mg KOH/g	To be indicated by supplier			DIN 51 558 Part 1

* Currently in draft form.

1 The figures correspond to rounded values of mid-point viscosity in mm^2/s at 40°C. They are derived from the new viscosity classes in ISO 3448, see DIN 51 519. To prevent the only possible point of confusion with the earlier viscosity series, the lubricating oil ISO VG 68 is temporarily assigned the additional letter x, making it HL 68x.

4 The legal unit of kinematic viscosity is m^2/s, 1 mm^2/s = 1 × 10^{-6} m^2/s (= 1 cSt).

5 Due to the degree of accuracy, it is not possible to give accurate values below a mass content of 0.05 percent.

6 Due to the degree of accuracy, it is not possible to give accurate values below a mass content of 0.1 percent.

7 Due to the length of the test, these values should not be determined on acceptance, but on type approval.

8 With an increase in the neutralization number compared with the neutralization number of the new oil of up to 2.0 mg KOH/g oil, it is not possible to make a differential assessment of HL hydraulic oils.

9 The limits are set after the completion of round-robin tests.

10 These limits do not apply to hydraulic oil with detergent properties.

11 These test processes are necessary as proof of identity for monitoring goods supplied. As they depend on the type of the base oils and additives used, it is not possible to indicate general requirement limits.

Table 4.72   Minimum requirements for HLP hydraulic oils to DIN 51 524

Type of hydraulic oil³			Requirements						Test to
			HLP 10	HLP 22	HLP 32	HLP 46	HLP 68x	HLP 100	
ISO viscosity class to DIN 51 519			ISO VG 10	ISO VG 22	ISO VG 32	ISO VG 46	ISO VG 68	ISO VG 100	—
Kinematic viscosity⁴ mm²/s	at −20°C	max	600						DIN 51 550 in conjunction with DIN 51 561 DIN 51 562 Part 1 or DIN 51 569
	at 0°C	max	90	300	420	780	1400	2560	
	at 40°C	max	11.0	24.2	35.2	50.6	74.8	110	
	at 40°C	min	9.0	19.8	28.8	41.4	61.2	90.0	
	at 100°C	min	2.4	4.1	5.0	6.1	7.8	9.9	
Pour point equal to or lower than		°C	−30	−21	−18	−15	−12	−12	DIN 51 597 DIN ISO 3016*
Flame point higher than		°C	125	165	175	185	195	205	DIN 51 376
Content of undissolved substances		g/100 g	Below the limits of quantitative detectability⁵						DIN 51 592*
Water content		g/100 g	Below the limits of quantitative detectability⁶						DIN ISO 3733*
Anti-corrosion characteristics with regard to steel (method A)		max	Corrosion level 0–A						DIN 51 585
Corrosive effect on copper		max	Corrosion level 2–100 A 3						DIN 51 579
Ageing behaviour			To be indicated by supplier						Standard in preparation
Response to SRE-NBR 1 seal material to DIN 53 538 Part 1 after 7 days ± 2 h at (100 ± 1)°C⁷	Relative vol% change		To be indicated by supplier⁷						DIN 53 538 Part 1 in conjunction with DIN 53 521
	Change in Shore A hardness		To be indicated by supplier⁷						DIN 53 538 Part 1 in conjunction with DIN 53 505

Property	Limit					Test method
Air separation capacity at 50°C[8]	min max		5		[7]	DIN 51 381
Foaming characteristics (process B)	S1 S2 S3	To be indicated by supplier[7]	10			DIN 51 566*
Demulsification capacity at 54°C[8]	min max	30	40	60		DIN 51 599
Mechanical test in FZG gear test machine[7]	Damage force level min		10			DIN 51 354 Part 2
	Work-related weight change mg/kWh max		0.27[10]		[9]	
Mechanical test in vane pump (mg abrasion)[7]	Ring max Vane max		120[10] 30[10]		[9]	DIN 51 389 Part 2*
Density at 15°C[11]		To be indicated by supplier				DIN 51 757
Ash (oxide ash)[11]	g/100 g	To be indicated by supplier				DIN EN 7
Neutralization number[11] acid or alkaline	mg/KOH g	To be indicated by supplier				DIN 51 558 Part 1

* Currently in draft form.

3 The figures correspond to rounded values of mid-point viscosity in mm^2/s at 40°C. They are derived from the new viscosity classes in ISO 3448, see DIN 51 519. To prevent the only possible point of confusion with the earlier viscosity series, the lubricating oil ISO VG 68 is temporarily assigned the additional letter x, making it HL 68x.

4 The legal unit of kinematic viscosity is m^2/s. $1\ mm^2/s = 1 \times 10^{-6}\ m^2/s$.

5 Due to the degree of accuracy of the test procedure, it is not possible to give accurate values below a mass content of 0.05 percent.

6 Due to the degree of accuracy of the test procedure, it is not possible to give accurate values below a mass content of 0.1 percent.

7 Limits are set after completion.

8 These limits do not apply to hydraulic oil with detergent properties.

9 Process cannot currently be used for the viscosity class. Suitable additives such as for viscosity classes ISO VG 32–ISO VG 100 should be used.

10 Differential assessment is not possible below the limit given.

11 These test processes are necessary as proof of identity for monitoring goods supplied. As they depend on the type of the base oils and additives used, it is not possible to indicate general requirement limits.

Table 4.73 Classification of *K* lubricating greases as per DIN 51 825

NLGI class to DIN 51 818	Auxiliary letter to DIN 51 502	Service temperature range to (°C)		Response to water to DIN 51 807 Pt. 1 assessment level
0, 1, 2, 3, or 4	B	−20	+50	0 or 1
	E	−20	+80	0 or 1
	F	−20	+80	2 or 3
	G	−20	+100	0 or 1
	H	−20	+100	2 or 3
	K	−20	+120	0 or 1
	M	−20	+120	2 or 3
	N	−20	+150	By agreement

+120°C, which contain cold-resistant base oils, mineral oils and/or synthetic oil in addition to a thickener. They may also contain additives and/or solid lubricants.

They are made in NLGI classes 0, 1, 2, or 3 for various service temperature ranges and degrees of water resistance (Table 4.75). Table 4.76 contains the minimum requirements for KT lubricating greases.

KP lubricating greases to DIN 51 825, Part 3
These are lubricating greases with a higher load capacity, protection against wear and scoring than K and KT lubricating greases for the service temperature range from below −20°C to +140°C, which contain mineral oil and/or synthetic oil as the base oil, in addition to a thickener. They may contain additives to improve corrosion protection and ageing resistance (KLP lubricating greases). They are manufactured in NLGI classes 00, 0, 1, 2, or 3 for various service temperature ranges and degrees of water resistance (Table 4.77). Table 4.78 contains the minimum requirements for KP lubricating greases.

G lubricating greases to DIN 51 826
These are lubricating greases in the NLGI Classes 000, 00, 0, or 1, which are also designated as liquid greases, for the service temperature range between −20°C and +100°C. In addition to a thickener they contain a higher proportion of base oil, mineral oil, and/or synthetic oil as well as, in some cases, additives to improve the friction and anti-wear behaviour. G lubricating greases are used in enclosed gears with splash lubrication.

Table 4.79 contains the minimum requirements for G lubricating greases.

Table 4.74 Minimum requirements for K lubrication greases as per DIN 51 825

Characteristic		Requirements for auxiliary letter								Test to
		B	E	F	G	H	K	M	N	
Response to water Assessment level at DIN 51 807 test temperature		0-40 1-40	0-40 1-40	2-40 3-40	0-90 1-90	2-90 3-90	0-90 1-90	2-90 3-90	By agreement	DIN 51 807 Part 1
Test in SKF roller bearing grease test machine (applies to NLGI classes 1, 2, 3, and 4) Test A		No requirement	Test bearing finding: No. 1 Grease finding: No. ≤2							DIN 51 806 Part 1* and Part 2*
Test B At test temperature	°C	Steady-state temperature	80		100		120		140	
Flow pressure at −20°C (applies only to NLGI classes 0, 1, 2, 3)	mbar					≤1400				DIN 51 805
Oxidation resistance pressure drop after 100 h	bar	No requirement					≤0.8			DIN 51 808
Anti-corrosion characteristics from SKF Emcor process	Corrosion level					0 and 0				DIN 51 802
Corrosive effect on copper	Corrosion level	0 at 50	0 at 80		0 at 100		1 at 120		1 at 140	DIN 51 811
Content of solid foreign substances over 25 μm indicated as mass concentration	mg/kg						≤20			DIN 51 813 Part 1
Water content, indicated as mass percentage	%	≤3.0					≤0.4			DIN ISO 3733*
Oil separation in normal test indicated as mass percentage	%	To be agreed with supplier								DIN 51 817
Thickener		Type and quantity to be indicated by supplier on request								DIN 51 814 DIN 51 820 Part 1*
ASTM drop point		To be indicated by supplier on request								DIN 51 801 Part 1
Base oil		Type and viscosity to be indicated by supplier on request								DIN 51 820 Part 1* DIN 51 550 in conjunction with DIN 51 561 DIN 51 562 Part 1
Response to SRE seal material as per DIN 53 538 Part 1* after 7 days ± 2 h at (100 ± 1)°C relative volume change		To be indicated by supplier on request								DIN 53 521 DIN 53 538 Part 1* DIN 53 538 Part 2*

* Currently still in draft form.

Table 4.75 Classification of KT lubricating greases as per DIN 51 825

NLGI class to DIN 51 818	Identification letters (DIN 51 502)	Auxilliary letter	Service temperature range from (°C)	to	Response to water to DIN 51 807 Part 1 assessment level to test temperature given
0, 1, 2, or 3	KTA	C	−30	+60	0 or 1
		D	−30	+60	2 or 3
		E	−30	+80	0 or 1
		F	−30	+80	2 or 3
		G	−30	+100	0 or 1
		H	−30	+100	2 or 3
		K	−30	+120	0 or 1
		M	−30	+120	2 or 3
	KTB	C	−40	+60	0 or 1
		D	−40	+60	2 or 3
		E	−40	+80	0 or 1
		F	−40	+80	2 or 3
		G	−40	+100	0 or 1
		H	−40	+100	2 or 3
		K	−40	+120	0 or 1
		M	−40	+120	2 or 3
	KTC	C	−50	+60	0 or 1
		D	−50	+60	2 or 3
		E	−50	+80	0 or 1
		F	−50	+80	2 or 3
		G	−50	+100	0 or 1
		H	−50	+100	2 or 3
		K	−50	+120	0 or 1
		M	−50	+120	2 or 3

4.10 SPECIFICATIONS AND CLASSIFICATIONS

4.10.1 General

By specifications, especially technical specifications, we mean documents in which features, especially characteristics, are specified, such as, quality levels, suitability for use, safety, dimensions etc.

By classification, we mean, as a rule, the ordering of features into specific ranges or classes. There is no quality classification as such, but in practice the differentiation between specifications and classifications is not always clearly made.

Table 4.76 Minimum requirements for K lubricating greases as per DIN 51 825

Characteristic		C	D	E	F	G	H	K	M	Test to
						Requirements				
Response to water	Assessment level at test temperature	0–40 or 1–40	2–40 or 3–40	0–40 or 1–40	2–40 or 3–40	0–90 or 1–90	2–90 or 3–90	0–90 or 1–90	2–90 or 3–90	DIN 51 807 Part 1
Test in SKF roller bearing grease test machine	Test A / Test B / At test temperature °C	No requirement	Without heating	80		Test bearing finding: No. 1	Grease finding: No. ≤2 / 100	120		DIN 51 806 Part 1* and Part 2*
Flow pressure for KTA at −30°C / for KTB at −40°C / for KTC at −50°C	mbar				≤1400					DIN 51 805
Oxidation resistance Pressure drop after 100 h	bar	No requirement					≤0.8			DIN 51 808
Anti-corrosion characteristics from SKF Emcor process	Corrosion level				0 and 0					DIN 51 802
Corrosive effect on copper	Corrosion level		0 at 60		0 at 80			0 to 120		DIN 51 811
Content of solid foreign substances over 25 µm	mg/kg				≤20		0 or 100			DIN 51 813 Part 1
Water content	wt%		≤3.0							DIN ISO 3733*
Oil separation in normal test	wt%						≤0.4			DIN 51 817
Thickener		To be agreed with supplier / Type and quantity to be indicated by supplier on request									DIN 51 814 / DIN 51 820 Part 1*
ASTM drop point		To be indicated by supplier on request									DIN 51 801 Part 1
Base oil		Type and viscosity to be indicated by supplier on request									DIN 51 820 Part 1* / DIN 51 550 in conjunction with DIN 51 561 / DIN 51 562 Part 1
Solid lubricant		Type and quantity as well as particulate size to be indicated by supplier on request									DIN 51 318 Part 1* / DIN 51 832
Response to SRE NBR 1 seal material as per DIN 53 538 Part 1 after 7 days ± 2 h at (100 ± 1)°C relative volume change											DIN 53 521 / DIN 53 538 Part 1*

* Currently still in draft form.

Table 4.77   Classification of KP lubricating greases to DIN 51 825

NLGI class to DIN 51 818	Auxilliary letter to DIN 51 502	Service temperature range from (°C)	to	Response to water to DIN 51 807 Part 1 assessment level
00, 0, 1, 2, or 3	A	< −20		By agreement
	B	−20	+50	0 or 1
	E	−20	+80	0 or 1
	F	−20	+80	2 or 3
	G	−20	+100	0 or 1
	H	−20	+100	2 or 3
1, 2, or 3	K	−20	+120	0 or 1
	M	−20	+120	2 or 3
	N	−20	+150	By agreement

Characteristics which are measured by means of standard and non-standard methods are used, to test and assess the features specified in specifications and classifications.

4.10.2   Industrial gear lubricants
Gear oils can be specified and classified according to viscosity and quality or service characteristics. The ISO viscosity classification for liquid industrial lubricants as per DIN 51 519 also applies to gear oils. Table 4.80 contains these viscosity classes.

Apart from the stipulation of lubricating oils which are basically suitable for gear lubrication with the comparison of viscosity ranges in DIN 51 509, there are no generally applicable and recognized specifications and classifications regarding quality in the former Federal Republic of Germany, if one disregards the in-house requirements of larger users.

In the USA, two specifications passed by the American Gear Manufacturing Association (AGMA) have achieved significance. These specifications are:

- AGMA 250.02: Standard Specification for Lubrication of Industrial Enclosed Gearing
- AGMA 252.01: Standard Specification for Mild Extreme-Pressure Lubricants for Industrial Enclosed Gearing.

Table 4.81 lists the AGMA oils in AGMA 250.02 according to their viscosity, while Table 4.82 contains details of the viscosity and the oil type. Table 4.83 contains the AGMA–EP oils in AGMA 252.01. Details of gear type and oil type can be found in Table 4.84.

The gear lubricants listed in Table 4.85 have been specified by US Steel. Table 4.86 contains, as examples, the minimum requirements for high-performance gear oils as per US Steel 222 and Table 4.87 contains those for gear lubricants for open gears as per US Steel 226/236.

The following Thyssen Standards for gear lubricants are examples of German company specifications:

TH–N 256 140
C/CL lubricating oils (based on DIN 51 513, Parts 1 and 2)
TH–N 256 142
CLP lubricating oils (based on DIN 51 517, Part 3)
TH–N 256 540
G lubricating greases (based on DIN 51 826)

Table 4.88 contains the most important requirements for TH–N 256 140, Table 4.89 those for TH–N 256 142, and Table 4.90 requirements for TH–N 256 540.

4.10.3 Lubricants for vehicle gears

4.10.3.1 Viscosity classification
The viscosities of motor vehicle gear oils are classified in SAE J 306 (**89**); this classification has been adopted by DIN. Table 4.91 shows this classification. Figure 4.123 contains Brookfield viscosities as a function of temperature, and gives the viscosity/temperature curves for some multigrade gear oils.

4.10.3.2 Quality classification
In the matter of quality classification, one has to differentiate between civil and military applications.

4.10.3.2.1 Civil specifications/classifications
The American Petroleum Institute (API) has adopted some classifications which are based on test methods for evaluating load capacity which have been devised by the Coordinating Research Council (CRC). The reference oils (Reference Gear Oil (RGO)) necessary for the comparison have been defined by the American Society for Testing of Materials (ASTM).

Table 4.78 Minimum requirements for KP lubricating greases to DIN 51 825

Characteristic		A	B	E	F	G	H	K	M	N¹	Test to
						Requirements					
Response to water Assessment level DIN 51 807 at test temperature		By agreement	0–40 or 1–40	0–40 or 1–40	2–40 or 3–40	0–90 or ?–90	2–90 or 3–90	0–90 or 1–90	2–90 or 3–90	By agreement	DIN 51 807 Part 1
Test in SKF roller bearing grease test machine											DIN 51 806 Part 1* and Part 2*
Test A		2	No requirement			Test bearing finding: No. 1					
Test B		2				Grease finding: No. ≤2					
At test temperature	°C		Steady-state temperature	80		100		120		140	
Test in Shell four-ball apparatus Dome diameter process C	mm				≤1.8						DIN 51 530 Part 5*
Flow pressure at −20°C	mbar	2		≤1400						2	DIN 51 805
Oxidation resistance pressure drop after 100 h	bar	2	No requirement			≤0.8				2	DIN 51 808
Anti-corrosion characteristics from SKF Emcor process	Corrosion level				0 and 0						DIN 51 802

		Corrosion level	0 at 50	0 at 80	0 at 100	1 at 120	By agreement	
Corrosive effect on copper		2	0 at 50	0 at 80	0 at 100	1 at 120	By agreement	DIN 51 811
Content of solid foreign substances over 25 μm, indicated as mass concentration	mg/kg			≤20				DIN 51 813 Part 1
Water content, indicated as mass fraction	%	3	≤3.0					DIN ISO 3733*
Oil separation in normal test indicated as mass fraction	%			To be agreed with supplier	≤0.4			DIN 51 817
Thickener			Type and quantity to be indicated by supplier on request					DIN 51 814 / DIN 51 820 Part 1*
ASTM drop point				To be indicated by supplier on request				DIN 51 801 Part 1
Base oil			Type and viscosity to be indicated by supplier on request					DIN 51 820 Part 1* / DIN 51 550 in conjunction with DIN 51 561 / DIN 51 562 Part 1
Solid lubricant			Type and quantity as well as particle size to be indicated by supplier on request					DIN 51 831 Part 1* / DIN 51 832
Response to SRE seal material as per DIN 53 538 Part 1* after 7 days ± 2 h relative volume change at (100 ± 1)°C				To be indicated by supplier on request				DIN 53 521 / DIN 53 538 Part 1* and Part 2*

* Currently still in draft form.
1 At higher temperatures, classification to DIN 51 502, Table 6 (1979).
2 To be agreed with supplier.
3 Dependent on thickener.

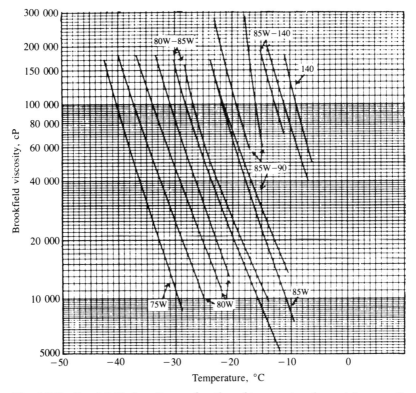

Fig. 4.123 Brookfield viscosity as a function of temperature for certain gear oils

Table 4.92 contains the API classification for gear oils. Table 4.93 lists the API classes and the associated reference oils and tests which are to be performed.

Additional specifications and classifications for vehicle gear oils are compiled by major vehicle manufacturers. These include, for example:

- Daimler–Benz AG (DB);
- Volkswagenwerk AG;
- American Motors Corporation;
- Chrysler Corporation;
- Ford Motor Company;
- General Motors Corporation (GM);
- International Harvester Company (IHC);

Table 4.79 Minimum requirements for G lubricating greases to DIN 51 826

Characteristic		*Requirements for NLGI class DIN 51 818*				*Test to*
		000	*00*	*0*	*1*	
Worked penetration	0.1 mm	445–475	400–430	355–385	310–340	DIN ISO 2137
Response to water Assessment level at test temperature		To be indicated by manufacturer				DIN 51 807 Part 1
Test in Shell four-ball apparatus[4] VKA welding force	N	2000				DIN 51 350 Part 4[5]
Flow pressure at lower application temperature	mbar	To be agreed where necessary				DIN 51 805
Oxidation resistance pressure drop after 100 h	bar	0.8				DIN 51 808
Anti-corrosion characteristics from Emcor process	Corrosion level	0 and 0				DIN 51 802
Anti-corrosion effect on copper at 80°C	Corrosion level	$\leqslant 1$				DIN 51 811
Content of solid foreign substances over 25 mm, indicated as mass fraction	mg/kg	$\leqslant 20$				DIN 51 813 Part 1[6]
Water content, indicated as mass fraction	%	$\leqslant 2.0$				DIN ISO 3733
ASTM drop point	°C	To be indicated by supplier on request				DIN ISO 2176
Base oil		Type and viscosity to be indicated by supplier on request				DIN 51 820 Part 1[5] DIN 51 550 in conjunction with DIN 51 561
Thickener		Type and quantity to be indicated by supplier on request				DIN 51 814 DIN 51 820 Part 1[5]
Solid lubricant		Type and quantity as well as particle size to be indicated by supplier on request				DIN 51 831 Part 1 DIN 51 832
Response to SRE 1 seal material as per DIN 53 538 Part 1 after 7 days ± 2 h at (100 ± 1)°C relative volume change	%	To be indicated by supplier on request				DIN 53 538 Part 1 DIN 53 521

[4] Load capacity is still being evaluated by testing DIN 51 350 Part 4[5] until the method has been tested with the FZG gear test machine to DIN 51 354 Parts 1 and 2.
[5] Currently still in draft form.
[6] Insofar as the test to this Standard is applicable.

Table 4.80   ISO viscosity classes

ISO viscosity class	Mid-point kinematic viscosity (cSt) at 40°C	Limits of kinematic viscosity (cSt) at 40°C min	max
ISO VG 2	2.2	1.98	2.42
ISO VG 3	3.2	2.88	3.52
ISO VG 5	4.6	4.14	5.06
ISO VG 7	6.8	6.12	7.48
ISO VG 10	10.0	9.00	11.00
ISO VG 15	15.0	13.50	16.50
ISO VG 22	22	19.8	24.2
ISO VG 32	32	28.8	35.2
ISO VG 46	46	41.4	50.6
ISO VG 68	68	61.2	74.8
ISO VG 100	100	90.0	110.0
ISO VG 150	150	135.0	165.0
ISO VG 220	220	198	242
ISO VG 320	320	288	352
ISO VG 460	460	414	506
ISO VG 680	680	612	748
ISO VG 1000	1000	900	1100
ISO VG 1500	1500	1350	1650

Table 4.81   Viscosity ranges for various AGMA lubricants to AGMA 250.02

AGMA number	Viscosity ranges, SSU^{1}* (mm^2/s) at 100°F (37.8°C)	at 210°F (98.9°C)
1	180–240 (39–52)	
2	280–360 (63–80)	
3	490–700 (110–155)	
4	700–1000 (155–215)	
5		80–105 (16–22)
6		105–125 (22–26)
7 comp†		125–150 (26–32)
8 comp†		150–190 (32–41)
8A comp†		190–250 (41–54)
		1000–2000 (215–430)
		3000–10000 (650–2000)

* Saybolt universal seconds.
† Mixed with 3–10 percent animal fats, e.g., acid-free tallow. Maximum pitch circle velocity 7.5 m/s.

Table 4.82 AGMA oils for enclosed industrial gears to AGMA 250.02

Oil type
- high-quality, highly refined mineral oil
- non-corrosive to gears, roller bearings and plain bearings
- neutral reaction capacity, free of abrasive impurities, good air separation capacity, good oxidation resistance
- undoped mineral oils with the following exceptions:
  - for worm gears it is advisable to add 3–10 percent animal fat, e.g., acid-free tallow
  - for unusual velocities, loads and temperatures it is permissible to incorporate additives

Viscosity
- A minimum viscosity index of 30 is acceptable for normal operation. At operating temperatures higher than 27°C, a VI of at least 60 is desirable. Oils in AGMA 7 comp, 8 comp and 8A comp must have a VI of at least 90.
- The viscosity of the AGMA must be within the ranges in Table 4.81.

- Rockwell International;
- Chark Equipment Company;
- John Deere;
- Mack Trucks, Inc.

Table 4.94 contains a list of the specified viscosity levels of gear oils and hypoid gear oils for vehicle gears. The penetration of multi-grade oils can be seen clearly. Table 4.95, based on DB operational materials

Table 4.83 Viscosity ranges for undoped AGMA–EP oils to AGMA 252.01

	Viscosity ranges, SSU* (mm²/s)	
AGMA oils	*at 100°F (37.8°C)*	*at 210°F (98.0°C)*
2EP	280–400 (63–89)	
3EP	400–700 (89–155)	
4EP	700–1000 (155–215)	
5EP		80–105 (16–22)
6EP		105–125 (22–26)
7EP		125–150 (26–32)
8EP		150–190 (32–41)
		500–575† (112–125)
		800–1000† (185–215)

* Saybolt Universal seconds.
† Maximum pitch circle velocity 7.5 m/s.

Table 4.84   Low-additive AGMA–EP oils for enclosed industrial gears to AGMA 252.01

Application range
For use in the following gear sets
 – spur gears
 – helical spur gears
 – double helical gears
 – straight bevel gears
 – spiral bevel gears
For use in worm gears

Oil type
No additive separation
Little foaming tendency in normal operating conditions
Non-corrosive to metals, even in the presence of water
Minimum Timken OK load of 30 lb
No abrasive impurities
Adequate stability at operating temperatures up to 70°C

Viscosity
Minimum viscosity index of 60
Viscosity ranges as in Table 4.83

regulations, Sheet 234 (**92**), shows how complex the relationship between viscosity and gear set can be.

Chemical, physical, and technological characteristics are stipulated in the individual specifications. Daimler-Benz, for example, specifies the following performance for hypoid gear oils:

 – compliance with the requirements of MIL–L–2105 B/C;
 – classification in API GL–5;
 – successful completion of company tests.

Table 4.85   US Steel specification for gear lubricants

Requirement No.	Lubricant
220	Extreme pressure (EP) oil
221	Hypoid gear oil
222	High-performance gear oil
223	EP gear oil for oil mist lubrication
225	Special gear oil
226, 236	Gear lubricants for open gears
227	Gear lubricant for construction machinery
343	Special gear lubricant

Table 4.86   Requirements for high-performance gear oils in US Steel 222

	Lubricant requirements
General requirements	Particularly high values for load capacity, oxidation stability and water separation
Additives	Sufficient for extreme-pressure properties for high-performance gear oils. Additives must not separate out by centrifuging
Viscosity	ASTM classes 50, 70, 90, 110, 130, 150, 280, and 500 SSU at 210°F (98.9°C)
Viscosity index	At least 85
Copper strip corrosion	Assessment 1 a
Flash point	At least 400°F (205°C)
Pour point	Maximum 20°F (-7°C) for oils up to 130 SSU at 210°F. Lower values may be required for special purposes
Dynamic negation behaviour at 180°F (82°C)	Upper maximum 7.5% water Lower minimum 75% water
Oxidation performance (312 h, 203°F (95°C), 10 l dry air/h)	Viscosity increase maximum 7%
Timken test	Load minimum 60 lb
VhA test	Weld load minimum 250 kp, Load Wear Index mininum 40
FZG test	Damage force minimum 9
Field tests	Suitable for intended purpose

Application
Gear pairs and bearings in high-performance gear oils in operating conditions characterized by much mixed friction and entry of water and dust, as well as external heating. Splash and pressure feed lubrication. Oil maintenance and conditioning by precipitation, centrifuging and filtering.

As additional examples we can present some specifications for hypoid gear oils of major manufacturers of passenger cars and trucks. Table 4.96 contains the most important stipulations of the VW/AUDI Standard and the TL–W 726 Technical Delivery Conditions for SAE 75, SAE 80 and SAE 90 mono-grade gear oils corresponding to API GL–4, and

Table 4.87   Requirements for lubricants for open gears in US Steel 226 and 236

	Lubricant requirements	
	226	236
General requirements	Adhesive, EP properties, applied by spray and brush	Adhesive, no EP properties applied by spray and brush
Viscosity, as-supplied state	Suitable for handling and distribution for types with 500, 750, 1000, 1500, 2000 and 3000 SSU at 210°F (98.9°C)	
Vaporization loss	To be indicated	
US Steel Rotation test	1800 seconds	Not applicable
Copper strip	Assessment maximum 2b	Assessment maximum 10
VhA–EP test	Weld load minimum 200 lnp Wad Wear Index minimum 30	Not applicable
VhA wear test	Dome diameter maximum 0.8 mm at 20 lnp, 1 h, 1800 min^{-1}, 55°C	Dome diameter maximum 0.6 mm at 7.5 lnp, 1 h, 1800 min^{-1}, 55°C

Application
Lubrication of gears where splash and pressure feed lubrication are not possible. The lubricant film must adhere to the tooth sides without becoming detached or thrown off. The as-supplied viscosity is of subsidiary importance, as the solvents added to easier handling evaporate. What is important is the permanent residual viscosity which must be matched to the specific application.

Table 4.97 contains those in TL–VW 727 for SAE 90 gear oils corresponding to GL–5. Table 4.98 contains an extract from the VW Standard 501 50 for an SAE 75W–90 synthetic multigrade gear oil. Table 4.99 contains extracts from the Chrysler Specification MS–3725 for SAE 140 truck rear axle gear oils. Performance must be proved by, among other things, successful completion of tests CRC L–37 and L–42. The most important requirements for SAE 90 hypoid gear oils contained in Ford's Specification ESW–M2C105–A can be found in Table 4.100. It can be seen that sulphur and phosphorus contents are specified, and that adequate performance must be demonstrated in tests on laboratory test rigs and in axle tests. From Table 4.101 it can be seen that General Motors specifies minimum values for sulphur and phosphorus content for a SAE 80W hypoid gear oil in GMC 99850044, and forbids the addition of

Table 4.88 Requirements for C lubricating oils in TH–N 256 140 company standard (Thyssen)

Characteristic	Test	Units	C 10	C 22	C 32	C 68	C 100	C 150	C 220	C 320	C 460	C 680
Viscosity												
at 40°C	DIN 51 502	mm²/s	10	22	32	68	100	150	220	320	460	680
at 20°C			min 21									
at 100°C				*	*	*	10.2	13.3	17	21.6	27.3	35.2
Viscosity index	DIN ISO 2909	—					80	85	85	85	85	85
Boiling point	DIN 51 350	°C	250	280	300	330	340	350	360	380	400	400
Vaporization loss, ≤	DIN 51 581	g/100 g				*	*	5	5	3	3	3
Flash point, minimum	DIN ISO 2592	°C	150	170	190	210	230	240	240	260	280	290
Pour point, ≤	DIN ISO 3016	°C	−27	−24	−21	−15	−15	−12	−9	−6	−6	−6
Ash, maximum	DIN EN 7	%	0.01	0.01	0.01	0.01	0.01	0.01	0.01	0.01	0.01	0.01
Carbon residue, maximum	DIN 51 551	%	0.5	0.5	0.5	0.5	0.5	0.5	0.8	0.8	1	1
Ageing due to rise in:												
carbon residue, maximum	DIN 51 352	%	1	1	1	1	1	1	1	2	2	2
neutralization number	DIN 51 587											
after 500 h	DIN 51 558	mg KOH/g				*	*	1	1	1	1	1
after 1000 h	DIN 51 558	mg KOH/g				*	*	2	2	2	2	2
viscosity at 40°C after 1000 h	DIN 51 562	%				*	*	15	15	15	15	15
Demulsification capacity at 54°C	DIN 51 599	min				10	20	20	20	30	30	30
Air separation capacity at 50°C, ≤	DIN 51 381	min	5	10	10	10	15	*	*	*	*	*
Anti-corrosion effect on steel	DIN 51 585	Corrosion level	0–A	0–A	0–A	0–A	0–A	0–A	0–A	0–A	0–A	0–A
Corrosive effect on copper	DIN 51 759	Corrosion level					2-100 A3					
Solid foreign matter, maximum	DIN 51 592	mg/kg	300	300	300	300	300	300	300	300	300	300

Neutralization number, mg KOH/g, DIN 51 558.
* To be indicated.
Foaming behaviour, ml, DIN 51 566.
Response to seal materials, % volume/hardness change, DIN 53 521.
Density, g/ml, DIN 51 757.
Colour value, DIN ISO 2049.
Water content, %, DIN ISO 3733 not detectable.
Asphaltene content, %, DIN 51 595.
IR diagram: to be provided.

Table 4.89 Extreme pressure requirements for CLP lubricating oils in TH–N 256 142 company standard (Thyssen)

Characteristic	Test to	Units	CLP 68	CLP 107	CLP 220	CLP 320	CLP 460	CLP 680
Chemical, physical and tribological characteristics			See requirements for C lubricating oils (partly increased)					
EP additives: type		–						
fraction		%	*	*	*	*	*	*
Timken Test load	SEB 181 302	lbs	25	40	40	40	40	40
abrasion, maximum			6	6	6	6	6	6
VKA Test weld force	DIN 51 350	N	*	*	*	*	*	*
wear		mm	*	*	*	*	*	*
FZG Test, A/8.3/90 Damage force level, >	DIN 51 354	Force level	12	12	12	12	12	12
Specific weight change, maximum		mg/KOH	0.2	0.2	0.2	0.2	0.2	0.2

* to be indicated.

**Table 4.90   Requirements for G lubricating greases in company standard TH–N
256 540 (Thyssen)**

Characteristic	Test	Units	G0F	G1F	G2F
Sealant:					
type			Na	*	Na
fraction	DIN 51 844	%	*	*	*
Base oil:					
type				Mineral oil	
fraction	DIN 51 814	%	*	*	*
viscosity at 40°C, min	DIN 51 502	mm²/s	100	680	220
viscosity at 100°C, min	DIN 51 502	mm²/s	*	30	*
Water content, maximum	DIN ISO 3733	%	0.2	0.2	0.2
Penetration:	DIN ISO 2137				
worked penetration 60 DH					
at 25°C			400	340	260
difference from 60 DH to					
5000 DH			*	*	*40
Solid foreign matter	DIN 51 813	mg/g	<10	<10	<10
Drop point, minimum	DIN ISO 2176	°C	*	*	170
Corrosive effect on Cu	DIN 51 811	(°)	1	1	1
at 80°C					
Oil separation	DIN 51 842	%	*	*	1–5
FZG Test, A/2.8/8	DIN 51 354				
damage force level		Force level	>12	>12	
specific weight change		mg/kWh	*	*	
Timken Test					
abrasion at 40 lb, maximum	SEB 181 302	mg	5	5	5
EMCOR Test	DIN 51 502			0 and 0	

EP additives, type and fraction.
Solid lubricants, Yes/No, type and fraction.
Worked penetration at 50°C:
  difference 50°C to 75°C.
Thickness, DIN 51 803, %.
Neutralization number, DIN 51 809, mg KOH/g.
Oxidation resistance after 100 h, DIN 51 808, bar.
VKA test, weld force and wear, N or mm, DIN 51 350.
Response to seal materials, % volume/hardness change.
IR diagram of grease is to be provided.

substances containing chlorine, lead, and zinc. Furthermore, this oil must
meet the requirements of MIL–L–2105 B. In accordance with Mack's
Specification GO–G, gear oils must meet the higher requirements of
MIL–L–2105 C (Table 4.102), which are specified for various viscosity
levels. Table 4.103 contains extracts from IHC Specification B–22 for
hypoid gear oils (135 HEP), which are specified for various viscosity
levels. Adequate performance must be demonstrated by, among other
things, the successful completion of tests CRC L–37 and L–42.

Table 4.91　SAE viscosity classes for motor vehicle gear oils to DIN 51 512

SAE viscosity class	Maximum temperature for an apparent viscosity of 150 000 mPas† to DIN 51 398* (°C)	Kinematic viscosity at 100°C as per DIN 51 550 (10⁻⁶ m²/s‡)	
		Min	Max
75W	−40	4.1	−
80W	−26	7.0	−
85W	−12	11.0	−
90	−	13.5	<24.0
140	−	24.0	<41.0
250	−	41.0	−

* Currently in draft form.
† SI unit of apparent viscosity: Pa s
　1 mPa s = 1 × 10⁻⁶ Pa s (= 1 cP).
‡ SI unit of kinematic viscosity: m²/s
　1 mm²/s = 1 × 10⁻⁶ m²/s (= 1 cSt).

4.10.3.2.2　Military specifications
The most important military specification for gear oils is undoubtedly MIL–L–2105C, which replaced the older MIL–L–2105B and specifies gear oils SAE 75W, SAE 80W–90 and SAE 85W–140. The class SAE 75W replaces the earlier specification MIL–L–10324A for gear oils for low-temperature use.

Table 4.104 contains the requirements for viscosity, channel point and flash point. Table 4.105 contains the physical and chemical data to be measured and the tests for demonstrating performance and service characteristics.

4.10.4　Special tests for gear oils
A series of special tests is carried out to study the performance of standardized vehicle gear oils. These include road tests with certain vehicles based on specific programmes as well as test-rig studies of certain axle gears to examine scoring and wear reduction properties. In addition, there are tests with certain gears for studying oxidation and corrosion performance as well as special tests to study flow behaviour at low temperatures, foaming performance and stability. Table 4.106 contains a brief description of these tests.

Table 4.92   API classification for motor vehicle gear oils

API–GL–1
Designates the operating conditions in vehicle rear axle drives with spiral-tooth bevel gears and with worm gears as well as in some manual transmission gears which operate under such mild conditions, characterized by low specific surface loads and low sliding velocities, that undoped mineral oils can be used satisfactorily. To improve the characteristics of the lubricants for these operating conditions, rust or oxidation improvers can be used. Friction improvers and high-pressure additives are not used.
Applicable to a small number of manual transmissions.

API–GL–2
Designates service conditions in vehicle axle drives with worm gears, which operate under such conditions of load, temperature and sliding velocities that lubricants which can be used satisfactorily in the conditions in API–GL–1 are not suitable.
Designed for worm gears and industrial gears.

API–GL–3
Designates service conditions in vehicle manual transmissions and in final drives with spiral-tooth bevel gears which operate in moderately severe load and velocity conditions. These service conditions require lubricants with a load capacity greater than that of lubricants which can be used satisfactorily in the conditions in API–GL–1, but below the requirements of lubricants which must meet the service conditions of API–GL–4.
Designed for manual transmissions and moderately loaded spiral-tooth bevel gear final drives.

API–GL–4
Designates the service conditions in gears, particularly hypoid gears, for vehicles as well as in other equipment for vehicles which operates in conditions of high velocities/low torque or low velocities/high torque.
GL–4 gear oils must possess at least the anti-scoring characteristics of CRC Reference Gear Oil RGO–105 (**91**).
Designed for manual transmissions, spiral-tooth bevel gear final drives and lightly loaded offset bevel gears.

API–GL–5
Designates the service conditions in gears, particularly hypoid gears, for vehicles as well as in other equipment for vehicles which operates under service conditions of high velocities/shock loading, high velocities/low torque, or low velocities/high torque.
Designed for offset bevel gears; corresponds to MIL–L–2105C.

API–GL–6
Designates service conditions in gears, particularly hypoid gears with a large offset (offset greater than 50.8 mm (2″), which equals 25 percent of the crown gear diameter), for passenger cars, and in other vehicle components which operate in high-velocity and high-output conditions.
Designed for bevel gears with a large offset; corresponds to Ford M2C105A. This class is no longer valid as the equipment required for tests is no longer available.

353

Table 4.93　Comparison of API class, reference oil and tests

API class	Reference oil	Tests
GL–1	RGO–100	–
GL–2	>RGO–100	–
GL–3	<RGO–104	–
GL–4	RGO–105	L–19, L–20, L–21
GL–5	⩾RGO–110	L–33, L–37, L–42, L–60
GL–6	–	L–33, L–42, L–60, D vag Start Test

4.11　COMPOSITION AND STRUCTURE OF GEAR OILS

Gear lubricants require a balanced formulation of base oil/base oil mixture and additive combination in order to meet the many demands made of them. The overall formulation depends on the gear lubricant type, which has to meet various requirements. Table 4.107 gives a rough classification of the various gear lubricants.

When formulating a lubricant, especially when providing for the main requisites, i.e., wear and scoring prevention, one must take into account that the response of the additives depends largely on the operating conditions. The activity of the relevant additives must, therefore, be matched to the velocity, temperature, and load conditions. If the formulation is less than optimal, there may be disadvantages, e.g., additive-related deposits in the gear set and the oil circuit. It is particularly difficult to satisfy the contrasting operating conditions of high velocity/low loads and low velocity/high loads. This is shown in Table 4.108 (**93**).

In general, a gear oil formulation is as follows:

- 50–95 percent base oil;
- 0–35 percent viscosity index improvers;
- 0–2 percent pour point improvers;
- 5–12 percent performance additives, e.g., extreme pressure additives.

Vehicle gear oils have the general formulation shown in Table 4.109 (**94**).

Today one can assume the formulation shown in Table 4.110 for various industrial and vehicle gear oils. The high total additive content of up to 37 percent necessary for wide-ranging multigrade oils is evident, as is the fact that modern gear oils mostly contain sulphur/phosphorus compounds as EP additives (**95**). This gives gear oil data as contained in

Table 4.94 Specified viscosity levels for oils and hypoid oils for vehicle gears

	DB		VW		AMC		Chrysler		Ford		GM		IHC		Rodwell		Clark		Deere		Mack	
	1	2	1	2	1	2	1	2	1	2	1	2	1	2	1	2	1	2	1	2	1	2
SAE 70W	*																					
SAE 75W						*										*			*	*		
SAE 80W	*			*	*		*									*			*	*		
SAE 85W												*	*									
SAE 90		*						*		*				*		*	*		*	*		*
SAE 140								*						*					*	*		*
SAE 250																						
SAE 80W/85W	*																					
SAE 80W/90								*				*		*		*			*	*		*
SAE 80W/140						*										*						*
SAE 85W/90	*			*		*								*		*			*			
SAE 85W/140												*		*		*			*	*		*

1 Gear oil.
2 Hypoid gear oil.

Table 4.95 Gear oil for one manufacturer's assemblies (DB operating materials regulations)

Assembly	SAE class	(1)*	(2)	(3)	(4)	(5)	(6)	(7)	(8)
Passenger cars:									
rear axle	90, 85W/90	×							
limited-slip differential	90								
mechanical steering	90, 85W/90	×							
L 075 Z mechanical steering				×					
servo steering						×			
mechanical transmission					×	×			
automatic MB transmission							×		
Cross-country vehicles:									
axles, mechanical steering	90, 85W/90	×							
servo steering									(×)
mechanical MB transmission					×				
transfer box	80, 80W, 80W/84W†		×		×				
automatic MB transmission							×		
Commercial vehicles:									
hypoid axle	90, 85W/90	×							
pinion axle, MB and ZF transfer box	80, 80W, 80W/85W†		×						
planetary hub reduction axle	80, 80W, 80W/85W†		×						
	90, 88W/90	(×)							
	90, 85W/90	×							
MB transfer box‡, mechanical steering: G1/15, 1/17, 1/18, 2/24, 3/36, 3/40, 3/50, 3/60, 3/70					×				(×)

	Lubricant	(1)	(2)	(3)	(4)	(5)	(6)	(7)	(8)
MB transmission: G3/65, 3/90, GO3/60, 3/80, G4/65–6, G4/65–7, 4/95–6, 4/95–7, 4/110–6, GV4/65–6, 4/95–6, 4/110–6, GO4/105–130	80, 80W, 80W/85W†								×
ZF transmission, Fuller transmission, Voith retarder	80, 80W, 80W/85W†								×
Allison transmission								×	
automatic MB transmission							×		
servo steering						×	(×)		
UNIMOG, MG trac: axles	90, 85W/90	×							
manual transmission§, mechanical steering	80, 80W, 80W/85W					×			×
power take-off gear, power take-off intermediate gear	80, 80W, 80W/85W†								×
DB front cable winch, fan	80, 80W, 80W/85W†								×
hydraulic diff. lock, Hydrostat (406)					×				

* (1) Hypoid gear oil; (2) Gear oil; (3) Hypoid gear oil (limited slip gears); (4) Liquid gear oils; (5) Steering gear oils; (6) ATFs for DB–ATFs; (7) ATFs for Allison transmissions; (8) Gearbox oils.
† Hot zones SAE 90, 85W/90.
‡ On vehicles with hypoid axles only in temperate areas.
§ Engine oil type SAE 10W.
× Should be used.
(×) Can be used.

Table 4.96   Requirements for API GL–4 gear oils, SAE 75, SAE 80, and SAE 90 to TL–VW 726 (AUDI/VW)

| (*1*) Oil type | Gear oil with EP additives, qualified to MIL–L–2105 |
| | Additive bases: sulphur/phosphorus |

(*2*) *Composition, characteristics, density, viscosity*

	TL–VW 726 *Type Y* *SAE 75*	*TL–VW 726* *SAE 80*	*TL–VW 726* *Type X* *SAE 90*	*Proof to*
IR spectrum		By sample		E DIN 51 820
EP additive content (%)		By sample		P–VW 1424
Sulphate ash (%)		By sample		DIN 51 575
Density at +15°C (g/cm³)		By sample		DIN 51 757
Viscosity at −18°C (mm²/s)	≤3000	≤22000	≤76000	DIN 51 757
				P–VW 1408
at +100°C (mm²/s)	≥5	≥8	≥16	DIN 51 562
				P–VW 1425
VW corrosion test:				
steel:				
rust deposit		Unacceptable		
copper:				
tarnishing		Acceptable		
coating		Acceptable		
weight loss (%)		≤0.5		
Pour point (°C)	≤−35	–	–	DIN 51 597

(*3*) *Functional requirements (for samples only)*

Load limit (FZG normal test A/8.3/90)				DIN 51 534
Damage force level	≥12			(CEC–L–07–T–81)

(*4*) *Compatibility with elastomers (determination of modulus −1/3)*

Test temperature (°C)	110	120	140	P–VW 3334
Test duration (h)		96		
Change in tensile stress (%)				
Acrylate elastomers	±25	±25	0–50	
Nitrile elastomers	±25	±25	–	
Fluorine elastomers	±25	–	±25	

Table 4.111 for various types (**96**). Table 4.112 (**96**) gives the correlation between gear oil type, specification, and gear type on the one hand, and EP additive content on the other.

4.12 CHANGES IN GEAR LUBRICANTS DURING SERVICE – USED OIL ANALYSIS

4.12.1 General

During the generally very long service periods of gear oils, considerable changes must be expected in the oil, which are the result of stress due to temperature, pressure, shearing, atmosphere, and humidity. The extent of the changes can be influenced catalytically, i.e., reinforced, by the structural materials (**97**).

Table 4.97   Requirements for API GL–5 SAE 90 gear oils in TL–VW 727 (AUDI/VW)

(1) Oil type	Gear oil with EP additives, qualified to MIL–L–2105B

(2) *Composition, characteristics, density, viscosity*

	TL–VW 727	*TL–VW 727*	*Test to*
IR spectrum	By sample		E DIN 51 820
EP additive type	Anglamol 95	Anglamol LZ 6043	
EP additive content (%)	⩾ 6.5	⩾ 7	P–VW 1424
Sulphate ash (%)	By sample		DIN 51 575
Density at +15°C (g/cm^3)	By sample		DIN 51 757
Viscosity at −18°C (mm^2/s)	⩽ 76000		DIN 53 015 (P–VW 1408)
at = 100°C (mm^2/s)	⩾ 16		DIN 51 562 P–VW 1425
VW corrosion test: steel:			
rust deposit	Unacceptable		
copper:			
tarnishing	Acceptable		
coating	Acceptable		
weight loss (%)	⩽ 0.5		

(3) *Functional requirements (only for testing samples)*

			DIN 51 354
Strain limit (FZG test A/16.6/90)			
damage force level		⩾ 12	(CEC–L–07–T–81)

(4) *Compatibility with elastomers (determination of modulus −1/3)*

Test temperature (°C)		140	P–VW 3334
Test duration (h)		96	
Change in tensile stress (%)			
Acrylate elastomers		0–50	
Fluorine elastomers		±25	

The following effects are particularly important for gear oils:
– interaction with air (oxygen), especially at higher temperatures;
– interaction with water;
– reduction of additive content;
– contamination by solid foreign matter;
– effect of shear stresses.

4.12.2   Interaction with air

4.12.2.1   General

The gear oil can contain air in three forms:
– dissolved in the oil;

Table 4.98   Requirements for multigrade gear oils SAE 75W–90, API GL–4/GL–5 to VW 50 150 (AUDI/VW)

(1) Oil type	Synthetic multigrade gear oil SAE 75W–90 with EP additives, quantified to API GL–4/GL–5		
(2) Composition, characteristics, density, viscosity			
IR spectrum	E DIN 51 820		By sample
EP additive type			To be indicated
EP additive content	P–VW 1424	(%)	By sample
Sulphate ash	DIN 51 575	(%)	By sample
Density at +15°C	DIN 51 757	(g/cm³)	By sample
Viscosity at −18°C	DIN 51 015	(mPas)	⩽ 2000
	(P–VW 1408)		
at +100°C	DIN 51 562	(mm²/s)	⩾ 13.5
at +150°C	P–VW 1430	(mPas)	⩾ 4.5
Pour point	DIN 51 597	(°C)	⩽ −4.5
VW corrosion test:	P–VW 1425		
steel:			
rust deposit			Unacceptable
copper:			
tarnishing			Acceptable
coating			Acceptable
weight loss		(%)	⩽ 0.5
(3) Functional requirements			
Gear oils corresponding to this VW Standard must meet the most recent requirements in API GL–4/GL–5. Test evidence must be provided to VW Central Laboratories.			
Strain limit	DIN 51 354		
(FZG test A/16.6/90)	(CEC–L–07–T–81)		
damage force level			⩾ 12
Shear stability:			
viscosity after 20 h			
operation (100°C)	DIN 51 562	(mm²/s)	⩾ 12.0
shear loss		(%)	
Compatibility with elastomers	P–VW 3334		
(Determination of modulus −1/3):			
Test temperature		(°C)	130
Test duration		(h)	96
Change in tensile stress:			
Acrylate elastomers		(%)	±25
Nitrile elastomers		(%)	0–50
Fluorine elastomers		(%)	±25

– dispersed in the oil, i.e., finely distributed;
– as surface foam.

In addition to the physical and mechanical effects of air in dispersed form or as surface foam, the intimate contact of the air with the oil, especially at high temperatures, can be expected to cause ageing of the oil, that is, oxidation processes.

4.12.2.2   Forms of air in oil

(*a*)   *Air dissolved in oil.*   The amount of air dissolved in the oil depends to a certain extent on the oil type, but mainly on the pressure and temperature. Under normal conditions, 20°C and 1013 mbar, the quantities

Table 4.99   Requirements for Extreme Pressure gear oils in MS 3725 (Chrysler Corporation)

Characteristics	
SAE class	140
Viscosity at 98.9°C (mm²/s)	25, 7–34, 3
Viscosity index (min)	85
Flash point (°C, min)	190
Channelling (FTMS 791a-3456)	Passed at −7
Copper strip corrosion, 3 h at 121°C (max)	2c
Oxidation resistance at 135°C:	
hours to viscosity rise of 50 percent (min)	375
VKA test:	
scoring load (kg, min)	120
welding load (kg, min)	260
Foaming tendency:	
after 5 min introduction of air at 93°C (ml, max)	50
Pentane insoluble (wt%, max)	0.01
Water content (vol%, max)	0.1
Effect on seal materials:	
durometer hardness change (units, max)	±3
change in tensile stress (%, max)	−15
change in length (%, max)	−20
change in volume (%)	0–3
Performance	
CRC L–37 test	Passed
CRC L–42 test	Passed
400 cycle drive train durability test	Passed
differential gear scoring test	Passed
Applications	
Works filling of truck rear axles	

of air shown in Table 4.113 dissolve in the various liquids (97). It can be seen that 7–9% vol air, containing 1 to 2% vol oxygen, can form a bright solution in mineral oil.

(b) *Air dispersed in oil.* In service, lubricating oils can absorb finely dispersed air and thus form air-in-oil dispersions. Dispersed air in the oil leads to an increase in viscosity. Ten % vol air in oil increases its viscosity by about 15 percent (97). The operational disadvantages include, among other things, the fact that the oil is more greatly compressed if dispersed air is present.

Table 4.100   Requirements for hypoid gear oils in ESW–M8C105A (Ford Motor Co.)

Characteristics

Viscosity at 98.9°C (mm^2/s)	16.8–19.2
Viscosity at −18°C (mPas, max)	150 000
Viscosity index (min)	90
Pour point (°C, max)	−23
Channel point (°C, max)	−32
Flash point (°C, min)	183
Copper corrosion, 3 h at 121°C (max)	3A
Oxidation stability (FTMS 79lb–2504):	
viscosity increase (mm^2/s) (%, max)	100
normal pentane insoluble (%, max)	3
benzene insoluble (%, max)	2
deposits	Acceptable values

	Foaming	Foam
Foaming behaviour:	tendency	stability
sequence 1 (ml, max)	25	0
sequence 2 (ml, max)	25	0
sequence 3 (ml, max)	25	0

Effect of water (FLTM BJ10–3):	
strip assessment (min)	8
cup assessment (min)	8
Total sulphur (wt%)	3.0–40
Phosphorus (wt%, min)	0.15
Moisture (%, max)	0.10

Performance

Tinken test (FLTM BJ1–2):	
weight loss (mg, max)	10
high-speed load (lb, min)	12.5
Differential gear test (E engine)	
stability	No abrasive deposits
corrosion	None
seal life	Acceptable values
viscosity change at 98.9°C (%, max)	±5
Vehicle rear axle test	No scoring on flanks

Applications
Conventional hypoid gears

(c) *Surface oils.* If small air bubbles dispersed in oil rise and collect on the surface, without bursting immediately, surface foam forms. The individual air bubbles are surrounded by thin oil skins. If the foam collapses before the oil returns to the circuit, or if it can be ensured that the oil

Table 4.101 Requirements for SAE 80W hypoid gear oils in GMC 9985044 (General Motors Corporation)

Characteristics

SAE class	80W
Viscosity at 100°C (mm^2/s, min)	8.6
Viscosity at −18°C (mPas, max)	8100
Channelling (FTMS 791a–3456)	Passed at −34°C
Sulphur due to additives (wt%, min)	2.0
Sulphur in base oil (wt%, min)	To be indicated
Phosphorus (wt%, min)	0.10
Chlorine	None
Zinc	None
Lead	None

Performance
Must meet the requirements in MIL–L–2105B

Applications
Factory fill for vehicles with three and four-speed transmissions and conventional rear axles

Table 4.102 Requirements for GO–G gear oils (Mack Trucks, Inc.)

Characteristics

SAE class	90, 140, 80W/90, 80W/140, and 85W/140
Seal compatibility 100 h at 93°C (Company test)	Passed
Corrosion performance (Company test)	Passed
Oxidation stability (Company test):	
vaporization loss (wt%, max)	10
viscosity rise at 98.9°C (%, max)	15
precipitation coefficient (max)	0.65

Performance

Mack Power Divider Snap Test (Company test)	Passed
Mack Transmission/Carrier Field Test (Company test)	Passed
Requirements of MIL–L–2105C	Met

Application
Factory fill and service fill for transmission and rear axle gears of road and construction vehicles

Table 4.103 Requirements for 135 HEP gear oils in IHC B–22 (International Harvester Co.)

Characteristics			
SAE class	90	85W/140	80W/90
Viscosity at 100°C (mm^2/s)	19.2–23.9	24.0–40.9	13.5–23.9
Viscosity at −12°C (mPas, max)	150 000	150 000	–
Viscosity at −26°C (mPas, max)	–	–	150 000
Viscosity index	85–120	85–120	85–120
Pour point (°C, max)	−23	−18	−26
Flash point (°C, min)	190	204	190
Sulphur (additive) (wt%)	1.7	1.7	1.7
Humidity (°C, max)	0.1	0.1	0.1
Channel point (FTMS 3456.1) (°C)	−18	−21	−35
Trace deposits (°C, max)	0.01	0.01	0.01
Foaming performance:			
sequence 1: tendency	10	20	20
stability	0	0	0
sequence 2: tendency	50	50	50
stability	0	0	0
sequence 3: tendency	20	20	20
stability	0	0	0
Humidity corrosion (Company test) (h, min)	40	40	40
Thermal stability (Company test):			
viscosity change at 100°C (%)	±10	±10	±10
vaporization loss (%, max)	10	10	10
pentane-insoluble (%, max)	1	1	1
Copper strip corrosion (max)	2	2	2
Humidity corrosion (FTMS 5326.1)			
seven days (%, max)	5	5	5
Thermal oxidation stability (50 h at 163°C)			
(FTMS 791a–2504):			
viscosity rise (%, max)	60	60	60
pentane-insoluble (%, max)	2	2	2
benzene-insoluble (%, max)	1	1	1
Performance			
CRC L–37 test	Passed	Passed	Passed
CRC L–42 test	Passed	Passed	Passed
VKA–EP test			
scoring load (kg, min)	120	120	120
welding load (kg, min)	250	250	250

Application
Factory and service fill of spur, bevel and worm gears as well as gearboxes of construction machinery, and also hypoid gears in trucks

Table 4.104   Physical characteristics of gear oils in MIL–L–2105C

	MIL–L–2105C Specification		
	75W	80W/90	85W/140
Viscosity at 100°C (mm²/s)			
min	4.1	13.5	24.0
max	–	<24.0	<41.0
max temperature for viscosity			
of 150 000 mPas (°C)	−40	−26	−12
Channel point, min (°C)	−45	−35	−20
Flash point, min (°C)	150	165	180

pump takes in only unfoamed oil, surface foam does not represent a danger. The tendency to form foam is increased by products of ageing in the oil, which influence the surface-active characteristics of the oil.

4.12.2.3   Ageing of the oil

By ageing, we mean oxidation of the lubricating oil under the influence of air, changes due to heat or light, and catalytic effects as well as to consequential processes such as polymerization and condensation. The most important effect is undoubtedly the chemical reaction between the

Table 4.105   Requirements for gear oils in MIL–L–2105C

Viscosity at 100°C (mm²/s)
Temperature for apparent viscosity of 150 000 mPas
Channel point (°C)
Flash point (°C)
Density (g/cm³)
Viscosity index
Pour point (°C)
Content of pentane-insoluble, sulphur, phosphorus, chlorine, nitrogen,
  organic metal compounds

Performance and in-service characteristics:
  compatibility
  foaming performance
  corrosion performance in humidity
  copper strip corrosion
  oxidation stability
  scoring and wear behaviour with
    low speeds/high torques (L–37)
    high speeds/low torques (L–42)

Table 4.106　Tests for evaluating the performance of specified vehicle gear oils

Test	Equipment	Conditions	Assessment
CRC L–19 (FTM 6504)	Chevrolet car on dynamometer	High speeds/shock loads	Anti-scoring properties
Ford vehicle rear axle score test (Drag start test)	Ford sedan car, 4-door, on test track	High speeds/shock loads	Anti-scoring properties
CRC–L42 (FTM 6507.1)	Spicer (Dana) truck axle	High speeds/shock loads	Anti-scoring properties
CRC L–20 (FTM 5317.1)	Chrysler or Dana truck axle	30 h, high torque/low speeds	Prevention of scoring and deformation and pitting on flanks; no deposits or corrosion in gears
CRC L–37 (FTM 6505.1)	Chrysler or Dana axle	100 min high speed/low torque followed by 23 h low speed/high torque	Prevention of scoring as well as deformation and pitting on flanks; no deposits or corrosion in gears
CRC L–21	Chevrolet axle	Humidity corrosion test over 10 days	No rust on gears or in bearings
CRC L–33 (FTM 5326.1)	Dana Jeep axle	Humidity corrosion test over 7 days	No rust on gears or in bearings
CRC L–60 (FTM 2504)	Laboratory spur gear transmission	Thermal oxidation stability test over 50 h	Viscosity rise: max 100% Pentane-insoluble: max 3% Benzene-insoluble: max 2%
BJ 15–1	Electrically driven Ford rear axle	Scoring, corrosion and stability test	No evidence of scoring, corrosion, or damage on tooth flanks; no deposits in casing
ASTM D–892	Laboratory equipment	As per standard	Foaming performance
ASTM D–130	Laboratory equipment	As per standard	Stability in the presence of copper and copper alloys
FTM 791	Laboratory equipment	As per standard	Channel point as a measure of cold flow performance

Table 4.107   Classification of gear oils according to application

Vehicle gear lubricants	Industrial gear lubricants	Special gear lubricants
Gear oils transmission oils axle oils hypoid gear oils locking differential hypoid gear oils Fluid gear oils ATF/A, Dexron, Ford Hydraulic oils Universal tractor oils (TOU, STOU) Traction fluids Lubricating greases	Gear oils low load high load Traction fluids Lubricating greases Adhesive lubricants Solid lubricants	Gear oils, fluids, Gear greases for: special gears special hydraulics nuclear power plants

Table 4.108   Effect of gear oil formulation on effectiveness (93)

	Use in hypoid gears	
Additive types	High speeds	High torque
Chlorine carrier	+	0 to (+)
Active sulphur in non-carbon compound	+	−
Relatively inactive sulphur in non-carbon compound	+	(−)
Relatively inactive sulphur in carboxylic acid ester	(−) t0 (+)	+
Relatively inactive sulphur in carboxylic acid	+	+
Carboxylic acid ester	(−) to −	+
Oxyphosphite neutral ester	0	0
Oxylphosphite acid ester	+	+
Oxyphosphate neutral ester	0	0
Thiophosphate neutral ester	+	+

+ Definitely effective.
(+) Slightly effective.
0 Ineffective, but not deleterious.
(−) Slightly deleterious.
− Definitely deleterious.

Table 4.109   General formulation of vehicle gear oils (94)

SAE class	Base oil	EP additives	VI improver	Pour point depressant
140	Heavy ⎫			
85W/140	Heavy ⎪			Yes
80W/140	Light ⎬	Same level All classes	Yes	Yes
90	Medium ⎪			
80W/90	Medium ⎭			Yes
75W/90	Light (Synthetic)	Higher level	Yes	Yes
75	Light	Higher level		Yes

oxygen in the air and the oil. Resistance to oxidation increases with the level of refining.

The ageing of hydrocarbons is a chain reaction. As energy is supplied, an organic unstable radical forms initially, which reacts with oxygen and forms hydroperoxides. By means of various mechanisms, the very labile hydroperoxides can decompose again; the type of hydrocarbon and the

Table 4.110   Formulation of industrial and vehicle gear oils (95)

	Gear oils for						
	Vehicles					Industry	
Oil type	MIL-B 85W/90	MIL-B 80W/140	MIL-C 85W/90	ATF Dexron	STOU 15W/30	DIN C-LP	US steel
Base oil	Paraffinic solvent raffinates						
VI improvers		30		5	12		
Pour point improvers	0.5	0.5	0.5	0.5	0.2	0.2	0.2
Oxidation inhibitors	1	1	1	1	1	0.5	0.5
Corrosion inhibitors	1	1	1	1	1	0.5	0.5
Passivators	0.1	0.1	0.1				0.1
Wear inhibitors							
EP additives	4.5	4.5	4.5	2	2	1	2
Friction modifiers			5	4	4		
DO additives				4	6		
Defoamers	0.01	0.01	0.01	0.01	0.01	0.01	0.01
Additives ($\approx$%)	7	37	12	17	26	2	3.5
Content of S (%)	2	2	2.5	0.5	1	0.6	0.7
P (%)	0.06	0.06	0.12	0.05	0.2	0.06	0.1
Cl (%)				0.1			
N (%)				0.1	0.1		
Zn (%)				0.12	0.06	0.07	
Ca (%)				0.2	0.6		
Ba (%)				0.15			

Table 4.111 Data for various types of gear oil (96)

	Transmission oil SAE 80W	Hypoid gear oil SAE 90	Hypoid gear oil SAE 85W/140	Industrial gear oil C–LP, ISO VG 150	Industrial gear oil C–LP, ISO VG 460
Density at 15°C (g/ml)	0.900	0.909	0.913	0.892	0.903
Viscosity at 40°C (mm²/s)	87.0	200	355	140	440
Viscosity at 100°C (mm²/s)	10.0	17.4	25.1	13	28
Viscosity index	94	93	97	96	97
Pour point (°C)	−30	−24	−18	−24	−12
Phosphorus content (% mass)	0.07	0.11	0.11	0.05	0.05
Sulphate ash (% mass)	0.1	0.1	0.1	0.03	0.04
FZG A/16.6/140 (damage force level)	>12	>12	>12	>12	>12

	ATF	Tractor gear oil	Power transmission oil	Super tractor oil (STUO)	Universal oil 15W/30
Density at 15°C (g/ml)	0.882	0.887	0.877		0.893
Viscosity at 40°C (mm²/s)	36.3	56.2	32.5		78.6
Viscosity at 100°C (mm²/s)	7.1	9.4	5.7		11.0
Viscosity index	162	150	116		128
Pour point (°C)	−42	−37	−33		−30
Phosphorus content (% mass)	0.06	0.11	0.03		0.16
Sulphate ash (% mass)	1.4	1.6	0.05		2.3
FZG A/8.3/90 (damage force level)	10	>12	11*		>12

* FZG A/16.6/140.

Table 4.112 API classification of motor vehicle gear oils (96)

API classification	Service conditions for gear oil	Gear type	Oil type or relative specifications	Typical additive content, % mass
GL–1	Light	Bevel (spiral) Worm Manual transmission	Undoped	–
GL–2	Medium	Worm (CV)	Special	Various formulations
GL–3	Medium	Bevel (spiral) Manual transmission	'Mild EP'	2.7
GL–4	Medium to severe	Hypoid with little offset Manual transmission	MIL–L–2105	4
GL–5	Severe	Hypoid etc.	MIL–L–2105 B and C	6.5
GL–6	Very severe	Hypoid with maximum offset Maximum stress	ESW–M2C 105 A (Ford)	10

369

Table 4.113   Air dissolved in lubricating oils (97)

	% Vol 20°C, 1013 mbar
Mineral oil	7–9
Silicone oil	15–25
Dicarboxylic acid ester	c. 9
Phosphate ester	c. 9
Polychlorinated biphenyls	c. 4
Water	1.87

external conditions of temperature, pressure, catalysts, and inhibitors present, as well as the oxygen available, are the most important factors influencing this. The following substances can be produced as end products:

– oil-insoluble substances, i.e., resin- and varnish-like products, which can cause deposits;
– acid, i.e., liquid products with, e.g., organic acids.

Figure 4.124 gives a schematic representation of the ageing process, which can be influenced catalytically by certain metals (Table 4.114). As Table 4.115 shows, some products of ageing attack some materials corro-

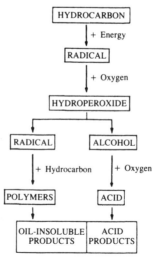

Fig. 4.124   Auto-oxidation of lubricants (schematic)

Table 4.114 Relative catalytic
effectiveness of
metals on the
oxidation of
lubricating oils
(97)

Copper	100
Lead	75
Bearing metal	60
Iron	45
Zinc	25
Tin	8
Aluminium	4

sively (97), and it can be seen that the colour coefficient can give first indications of aggresiveness.

4.12.3 Interaction with water
If water gets into the oil, there may be reversible physical processes or irreversible chemical reactions.

4.12.3.1 Physical/mechanical effects
As can be seen in Fig. 4.125, mineral oils can easily form a clear solution of 100 mg of water per kilogramme of oil. The ability to absorb water depends on the temperature and humidity of the air. When water-saturated oil is cooled, the water precipitates in the form of drops, which collect in the oil sump. The rate of precipitation depends on physical

Table 4.115 Schematic representation of the interaction of oxidation products with metals (97)

	ASTM colour	*Attacks*
Unaged lubricating oil = hydrocarbon mixture	1.5	Practically no danger
Lubricating oil and alcohols	3	Cu and Cu alloys
Lubricating oil and aldehydes/ketones	4	Practically no danger
Lubricating oil and carboxylic acids	5	Fe, Zn, Pb, Sn, and corresponding alloys
Lubricating oil and consequent products	8	Fe, Zn, Pb, Sn, Cu and corresponding alloys

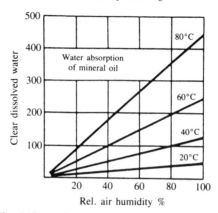

Fig. 4.125   Absorption of water by mineral oil

values, e.g., viscosity and drop size, and chemical values, e.g., interfacial tension. Small droplets reduce the rate of precipitation. The lower the interfacial tension becomes with the same energy of division, the greater is the boundary surface due to the formation of small droplets. The surface-active substances which occur with ageing, therefore, reduce the interfacial tension and thus increase the tendency of the oil to form oil–water mixtures which are difficult to separate (**97**). Free water of a few cubic centimetres in the oil can cause problems, if the temperature rises to more than 100°C and steam bubbles forming in the gear oil cause the oil to surge out through the ventilation holes.

4.12.3.2   Chemical effects
Many organic substances are split by chemical reaction with water at high temperatures. Some gear oil additives belong to this hydrolysis-sensitive group of substances. The disadvantageous effects include:

- the occurrence of oil-insoluble breakdown products or consequent products, which can form deposits;
- the occurrence of corrosive reaction products;
- exhaustion of additives, i.e., reduction in concentration.

4.12.4   Reduction of additive content
During service, the quantity of additives in the gear oil falls. The following mechanisms can be involved, separately or jointly (**97**):

- desirable chemical reactions to fulfil the functions of the additives (e.g., reaction of EP additives *with* metal surfaces);

- desirable physical processes to fulfil the functions of the additives (e.g., adhesion of anti-corrosion additives *on* metal surfaces):
- undesirable physical processes (e.g., water washing out additives, precipitation on cooling, removal with surface foam);
- decompositon due to external influences (e.g., oxidative change, thermal decomposition, hydrolytic decomposition, reaction with products of decomposition);
- other mechanisms (e.g., reaction of additives with one another).

4.12.5 Contamination by solid foreign matter
The majority of solid contaminants in gear oils are metallic abrasion particles and sand particles. For example, after vehicles have covered 40 000 km, 20–40 mg metal particles/kg oil and 5–20 mg SiO_2/kg have been found in the transmissions (**97**). If the particle size is below 3 μm and the gear oil has adequate dispersion capability, there should be no problems. If the oil has no dispersant capability or there is a greater foreign matter content, it is necessary to change the oil.

4.12.6 Effect of shear stresses
In the case of polymer-containing gear oils, that is most multigrade gear oils, it is necessary to take account of instability with respect to mechanical shear stress of the viscosity index improver. The macropolymers are decomposed mechanically at a suitably high shear gradient. This reduces the viscosity of the gear oil; in an extreme case, it may fall to that of the base oil. Between the tooth flanks of a gear, and especially in offset bevel gears which are characterized by high sliding speeds, as well as in the bearings, high shear gradients of the order of 10^7 s^{-1} occur.

4.12.7 Used oil evaluation
4.12.7.1 General
Study and evaluation of used oil permit two assessments to be made: (a) state of the gear oil; (b) state of the gear set.

Used oil analyses are useful for establishing and/or lengthening oil change intervals. However, one must always take into account the high cost of a comprehensive used oil analysis, amounting to about DM 3000–3500 (**97**). The following details must be known in order to produce a useful analysis:

- name of the product;
- period of service and quantity used;

- gear set in which it was used, age and location of gear set;
- prescribed oil type and oil change intervals;
- operating temperatures;
- noteworthy events;
- reason for used oil analysis.

Furthermore, the state of the used oil can only be assessed in comparison with the corresponding data for the fresh oil. Typical criteria for a used oil analysis are:

- appearance and odour;
- infrared spectrogram;
- hetero-elements (of the oil and/or wear particles);
- viscosity at 40°C and/or 100°C;
- TAN (Total Acid Number);
- foaming behaviour/stability;
- n-heptane-insoluble – only for used oil;
- water content – only for used oil.

In addition, care must be taken to ensure that representative oil specimens are taken and placed in clear specimen vessels. Sampling from containers and tanks should be performed in accordance with DIN 51 750, Part 2.

4.12.7.2   Used oil analysis
(*a*)   *Appearance and odour.*   First indications of great changes in the lubricating oil are obtained by comparing the appearance and odour of the used oil with those of the new oil. The following criteria can be important:

- very dark coloration, due, for example, to oxidation;
- pungent odour, due, for example, to souring or additive change;
- opacity, due, for example, to formation of an emulsion with water.

(*b*)   *Viscosity.*   In order to obtain hydrodynamically load-bearing oil films, heat supply, limited splash losses, and adequate oil supply, an optimal viscosity is necessary, which should be adequately high, but not too high. To monitor the viscosity, e.g., at 40°C and at 100°C, it is therefore necessary to compare the viscosities of the new and used oils. A fall in viscosity in polymer-containing oils due to shearing or a rise in viscosity due to ageing and foreign impurities can alter the viscosity. Viscosity changes of more than 20 percent necessitate an oil change.

(c) *Water content.* If water content exceeds 0.1 wt%, the oil should be changed.

(d) *Solid foreign matter.* If the total contamination exceeds 0.3 wt%, an oil change is recommended.

(e) *TAN, NN, or AN.* The Total Acid Number (TAN), the Neutralization Number (NN), and the Acid Number (AN) give an indication of the acidification of the oil due to ageing. If testing is carried out at intervals, a change to a steeper angle in the curve for variation with time indicates that the oil should be changed.

(f) *Foaming behaviour.* If there is a considerable deterioration in the foaming behaviour, e.g., by several hundred percent, it is necessary either to change the oil or add inhibitors to the oil in use. The latter measure should only by taken in the case of very large oil fills, if the other analytical data indicate that the oil can continue to be used.

(g) *Spectral analysis.* Spectral analysis of a lubricating oil allows evaluation of the following criteria.

- Determination of the identity of the oil by comparison with the analysis of the original oil. Contamination with products of a different nature can be identified.
- Quantitative evaluation of the remaining additive reserves and of wear particles present. Successful application of this method assumes accurate knowledge of the oil and its practical behaviour as well as of the relationship between the quantity of wear particles and the danger of damage.

4.12.7.3 Used oil analysis of synthetic oils

The information given on used oil analysis for gear oils relate primarily to mineral oil-based gear oils. There are, as yet, not standardized procedures for analysing synthetic oil.

Visual evaluation and examination of the composition by spectroscopy can be performed using the same procedure as for mineral oils. Measurement of viscosity and assessment of the results can also be carried out using the standard methods for mineral oils. For most other analytical procedures, we still need to build up experience in performing them and in forming conclusions.

5 Selection of Lubricants, DIN 51 509

5.1 GENERAL

DIN 51 509 contains useful information on selecting an optimal lubricant for a given gear set (98). After defining the applications for gear oils and gear greases, for the most important lubrication methods, and for undoped and doped gear oil, it is possible to select the viscosity with the aid of graphs. This applies to mineral oil-based gear oils. When choosing the lubricant type, one must consider not only the operating conditions, characterized by peripheral velocity and power transmitted, and the construction, but also the concept of the gear set, for instance whether it is designed for durability. In the case of multi-step gears, mean values must be established for the steps. In principle, preference should be given to a gear oil. Gear greases or adhesive lubricants should only be chosen for specific design reasons.

5.2 APPLICATIONS FOR LUBRICATION METHODS AND LUBRICANTS

Table 5.1 indicates the relationship between peripheral velocity, lubricant type, and method of lubrication for rolling gears, i.e., for spur gears and normal bevel gears. It should be borne in mind that the speed limits given for the individual lubricant types and methods of lubrication can be exceeded if appropriate design and operating measures are taken. Table 5.2 indicates the relationship between worm, lubricant type, and method of lubrication for worm gears.

Naturally, the location of the worm should be taken into account, as higher velocities can be allowed with a given lubricant if the worm is immersed. Of course, in this case, again, the limits given in Table 5.2 can be exceeded if appropriate design and operating measures are taken.

5.3 DEMARCATION LINE FOR THE USE OF DOPED AND UNDOPED GEAR OILS

Rolling gear sets with quenched and tempered gears designed for endurance fatigue strength can be lubricated with undoped oils if the limits in

Table 5.1 Recommended values for the use of various lubricant types and methods of lubrication for rolling gears

Peripheral velocity < 1 m/s – adhesive lubricants (spray lubrication)
Peripheral velocity < 4 m/s – gear greases (splash lubrication)
Peripheral velocity < 15 m/s – lubricating oils (splash lubrication)
Peripheral velocity > 15 m/s – lubricating oils (spray lubrication)

Table 5.2 Recommended values for the use of various lubricant types and methods of lubrication in worm gears

Worm immersed
Peripheral velocity of worm < 1 m/s – gear greases (splash lubrication)
Peripheral velocity of worm < 10 m/s – lubricating oils (splash lubrication)
Peripheral velocity of worm > 10 m/s – lubricating oils (spray lubrication)

Worm wheel immersed
Peripheral velocity of worm < 1 m/s – gear greases (splash lubrication)
Peripheral velocity of worm < 4 m/s – lubricating oils (splash lubrication)
Peripheral velocity of worm < 4 m/s – lubricating oils (spray lubrication)

Table 5.3 are not exceeded. In the case of hardened tooth surfaces and especially in the case of gears designed for fatigue strength for finite life, gear oils with extreme pressure and anti-wear additives must be used.

While worm gears designed for endurance fatigue strength can be lubricated with undoped gear oils in continuous operation, gears designed for finite life fatigue strength in discontinuous operation must always be lubricated with oils containing additives for improving friction behaviour, especially if the worms are hardened. It must be borne in mind that gear oils for other rolling crossed-axis gears, for instance, offset bevel gears, are not usually suitable.

Table 5.3 Conditions for the use of gear oils without extreme-pressure and anti-wear additives

(a)	Stribeck pressure	$\sigma_s < 7.5$ MPa
(b)	Pressure angle	$\alpha_n = 20$ degrees
(c)	Transverse contact ratio	$\varepsilon_\alpha = 1.3$
(d)	Addendum modification	x_1, x_2 positive
	if possible	$x_1 = x_2$
(e)	Step-down ratio	$i > 8$
(f)	Ratio of maximum sliding velocity to	$v_g/v < 0.3$
	peripheral velocity	
	Gear designed for endurance fatigue strength	

Offset bevel gears, e.g., for the hypoid gears in vehicle rear axles or for crossed-axis helical gears, always require gear oils containing extreme-pressure additives.

5.4 SELECTION OF VISCOSITY

Due to a variation in kinematic ratios, one must differentiate between rolling gears (i.e., spur gears and non-offset bevel gears), and worm gears when selecting the viscosity for a gear oil.

Figure 5.1 shows the kinematic nominal viscosity at a temperature of 40°C and 50°C as a function of a force/velocity factor for rolling gears. This curve is based on practical experience, taking account of the elasto-hydrodynamic theory of lubrication. The factor k_s/v takes account of the specific stress of the gear. The viscosities taken from this figure apply to an ambient temperature of 20°C.

Figure 5.2 shows the kinematic nominal viscosity at a temperature of 40°C and 50°C as a function of a force/velocity factor for worm gears. The same comments apply as for Fig. 5.1.

Under certain preconditions and in certain operating conditions, higher or lower viscosities must be provided than those determined from Figs 5.1 and 5.2. Tables 5.4 and 5.5 show some examples. The determination of the gear viscosity is explained below.

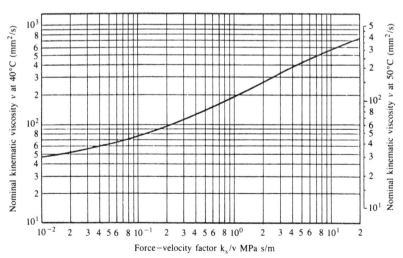

Fig. 5.1   Viscosity selection for bevel and spur gears (DIN 51 509)

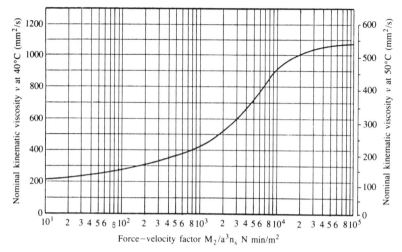

Fig. 5.2   Viscosity selection for worm gears (DIN 51 509)

Table 5.4   Conditions for higher nominal viscosity values

(a) Ambient temperature continuously over 25°C.
     Necessary viscosity rise approximately 10 percent per 10 K.

(b) Shock loading and occasional overloading.
     Shock loading and brief overloading must be taken into account when
     calculating the force/velocity factors.

(c) Similar or identical materials for both gearwheels.
     The viscosity should be increased by about 35 percent for gears of CrNi
     steels.

(d) Gear pairs sensitive to scoring, if no lubricant with wear-reducing addi-
     tives can be used.

Table 5.5   Conditions for lower nominal viscosity values

(a) Ambient temperature continuously below 10°C.
     Allowable viscosity reduction is about 10 percent per 3K.

(b) Tooth surfaces are phosphated, sulphonated and coppered.
     Allowable viscosity reduction is about 25 percent.

Table 5.6 Example of determining the necessary nominal viscosity for a spur gear

Face width	$b = 20$ mm
Pitch diameter	$d_1 = 73$ mm
Nominal peripheral force	
Peripheral velocity at generated	$F_t = 2800$ N
pitch circle	$v = 8.3$ m/s
Gear ratio	$u = 1.5$
Stribeck pressure	$k_s = \dfrac{F_t}{b \times d_1} \times \dfrac{u+1}{u} \times Z_H^2 Z_\varepsilon^2$
$Z_H^2 Z_\varepsilon^2$	about 3
Hence:	$k_s = 9.86$ N mm^{-2}
force/velocity factor	$k_s/v = 9.59/8.3 = 1.16$ MPa s m^{-1}

Table 5.6 contains design data and operating conditions for a spur gear set. It can be seen that the calculation produces a force/velocity factor k_s/v of 1.16 mPas m^{-1}. With this, one can read off a nominal viscosity of 206 mm^2 s^{-1} at 40°C from Fig. 5.1. One would, therefore, select an ISO VG 220 oil.

Table 5.7 contains the design data and operating conditions for a worm gear. For the resultant force/velocity factor M^2/a^3 one can read off a viscosity of 362 mm^2 s^{-1} at 40°C from Fig. 5.2. One would probably choose an ISO VG 320 oil.

If Stribeck contact pressure is to be converted into Hertzian stress, the relationship

$$\sigma_H = 268.4\sqrt{(k_s)}$$

applies to a steel/steel pair. It should be borne in mind that viscosities determined in this way are approximate values. One then has to choose that viscosity class range in which the nominal viscosity obtained lies.

Table 5.7 Example of determining the necessary nominal viscosity for a worm gear

Initial torque	$M_2 = 1000$ Nm
Centre distance	$a = 0.125$ m
Worm rotational speed	$n_s = 1000$ min^{-1}
Force/velocity factor	$\dfrac{k_s}{v} = \dfrac{M_2}{a^3 n_s} = 512$ N min m^{-2}

Table 5.8   Preferred lubricating oils for gears

Viscosity of lubricating oils				Lubricating oils				
				Without[4] anti-wear additives		With anti-wear additives		
VG viscosity classes to ISO/DIN 3448[1,2]	*Characteristic number*[3]	*SAE viscosity classes to DIN 51 511*[2]	*SAE viscosity classes to DIN 51 512*[2]	*C and C-T*[5] *oils to DIN 51 517 and C-L oils*	*N oils to DIN 51 501*[6]	*TD-L oils to DIN 51 515*[7]	*C-LP oils*[8]	*Vehicle gear oils*[9]
22/32	16	10W	75	×	×	×	×	×
32/46	25	10W	75	×	×	×	×	×
46/68	36	20W	75	×	×	×	×	×
68	49	20	80	×	×	×	×	×
100	68	30	80	×	×		×	×
150	92	40	90	×	×		×	×
220	114	40	90	×	×		×	×
220	144	50	90	×			×	×
320	169	50	90	×			×	×
460	225		140	×	×			×
680	324		140	×	×			×

Note. Low-additive gear oils can be assigned to group API GL–3 of the American Petroleum Institute for spiral bevel gears and manual transmissions with severe to high surface pressure and sliding velocities. High-additive gear oils may be classed in API GL–4 as normal hypoid gear oils for hypoid axles with high sliding or peripheral velocities and low load, or vice versa, or in API GL–5 for hypoid axle drives with high velocities and abruptly changing loads, in addition to the requirements in API GL–4.

[1] The ISO viscosity classes to DIN 51 519 give the mid-point viscosity (kinematic viscosity in mm²/s at 40°C).

[2] Comparison of the SAE viscosity classes and ISO viscosity classes to DIN 51 519 with the characteristic numbers can only permit an approximate comparison of viscosities.

[3] The characteristic number corresponds to the kinematic nominal viscosity in mm²/s at 50°C.

[4] These lubricating oils may contain additives for depressing pour point and setting point and for improving foaming behaviour, as well as additives for increasing corrosion protection, or ageing resistance, or both characteristics.

[5] Ageing-resistant mineral oils, which are used mainly with circulation lubrication, if greater ageing resistance is required than that of N lubricating oils in DIN 51 501.

[6] Mineral oils for lubrication applications for which there are no special requirements, e.g., regarding ageing resistance, low-temperature behaviour, etc.

[7] Mineral oils used mainly in steam turbines and in electrically driven machines or those driven by steam turbines, such as generators, turbocompressors, pumps and gear sets.

[8] A Standard on minimum requirements for C–LP lubricating oils is in preparation.

[9] In accordance with DIN 51 502, vehicle gear oils are classed as low-additive gear oils (no designation letters), high-additive-content gear oils (designation letters HYP), and fluid gear oils (designation letters ATF).

5.5 AVAILABLE MINERAL OIL-BASED GEAR OILS AS PER DIN 51 509

A great variety of lubricating oils can be used as gear oils. They only become gear oils if they achieve successful application in a gear set; they do not necessarily need to be designed exclusively for this purpose.

Table 5.8 contains various types of lubricating oil which are particularly suited to gear lubrication and, therefore, should be selected preferentially. To make selection easier, current conventional viscosity class systems are also given in this Table.

In certain cases, operating or design conditions permit, or even specify, types of lubricating oil other than those listed in Table 5.8, for the lubrication of a gear set. The most important are listed in Table 5.9.

Table 5.9 Lubricating oils suited to gear lubrication in special cases

Lubricating oils without anti-wear additives
- *Z oils as per DIN 51 510*:
  high-viscosity mineral oils primarily for steam engine lubrication.

- *Vehicle lubricating oils*:
  ageing-resistant mineral oils, which correspond approximately to C lubricating oils in DIN 51 517.

- *Engine oils*:
  ageing-resistant mineral oils, which correspond approximately to C lubricating oils in DIN 51 517.

Lubricating oils with anti-wear additives
- *Hydraulic oils as per DIN 51 525*
  HD engine oils:
  correspond to the low-additive gear oils in API GL–3.

- *C–LPF lubricating oils*:
  lubricating oils which correspond to the C–LP oils, but contain solid lubricants.

6 Gear Lubrication in Special Conditions

6.1 GENERAL SITUATION

By special operating conditions we mean low velocities, high velocities, low temperatures, high temperatures, air in the oil, and sliding velocity conditions, e.g., in worm gears, for which the use of synthetic gear oils may be considered.

These special operating conditions can be defined as follows (**99**):

- rolling velocities: 5–20 m/s
- ratio of sliding to rolling velocity: 0.2–0.3
- Hertzian stress: 600–1200 N/mm^2
- oil temperatures: 50–80°C
- centre distances: 50–1000 mm
- modules: 3–10 mm

6.2 LOW VELOCITIES

6.2.1 General situation

It is a known fact that slow-running toothed gear pairs are particularly prone to wear. In multi-step gears it is often the last stage; special gears generally have peripheral velocities below 0.5 m/s, and in difficult lubrication conditions, peripheral velocities up to 5 m/s can be low and can cause high wear. The effects of velocity, viscosity, additives, and lubricant type – among other things – have been studied in a systematic research project (**100**).

6.2.2 Effect of velocity

The basic effect of velocity can be seen in Fig. 6.1. If wear is plotted against velocity, we get a wear maximum up to v_2 for an equal operating time at each velocity. If the work transmitted is kept the same for each velocity, the result is a region of high wear for low peripheral velocities and a region of low wear for high peripheral velocities.

The variation in wear over operating time for various velocities is shown in Fig. 6.2. It is evident that wear falls considerably as velocity rises.

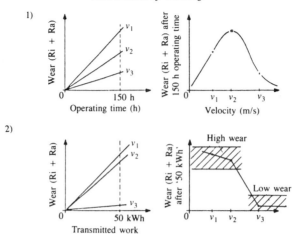

Fig. 6.1　Effect of velocity on wear (100)

6.2.3　Effect of viscosity

Figure 6.3 shows the great influence viscosity has on wear, measured at the lowest test velocity of 0.05 m/s. As viscosity rises, wear falls markedly.

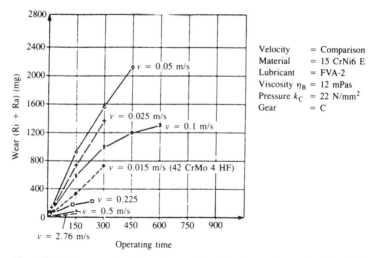

Fig. 6.2　Wear as a function of operating time for various velocities (100)

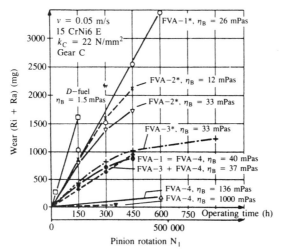

Fig. 6.3 **The effect of viscosity on wear** (100)

6.2.4 Effect of additives

Figure 6.4 summarizes the results for the effect of additives on wear.

Three different additives in the less viscous base oil gave diverging results with $v = 0.05$ m/s. The addition of a phosphorus/sulphur EP package (A99) produced an increase in wear compared with the pure base oil. At a higher velocity, however, wear declined. A definite

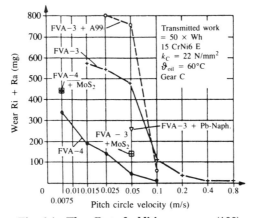

Fig. 6.4 **The effect of additives on wear** (100)

reduction in wear was evident at $v = 0.05$ m/s on addition of lead naph-thenate. This effect was more marked with the addition of MoS_2. Rising base oil viscosity obviously weakens this positive effect. Tests carried out with a more viscous base oil with the same additives even produced a slight wear increase compared with lubrication with a more viscous base oil without additives.

6.2.5 Effect of lubricant type
These investigations included some synthetic oils, used as friction gear oils, as well as some liquid gear greases. Figure 6.5 shows the results, together with the base oils as a function of viscosity.

The comparison for the friction gear fluids studied reveals behaviour dependent on the viscosity, similar to that of the mineral oils. The liquid greases studied exhibited quite different behaviour. Under EHD conditions, the soap content produces a quasi-equivalent 'effective viscosity', despite a difference in base oil viscosity. A markedly weaker velocity effect on the wear behaviour compared with mineral oil lubrication was also traced back to this altered viscosity behaviour of the liquid greases.

6.2.6 Wear analysis
The pronounced influence of velocity and viscosity indicates elasto-hydrodynamic effects. It is, therefore, possible to develop an analytical procedure which can assist in estimating the danger of wear using the calculated lubricant film thickness.

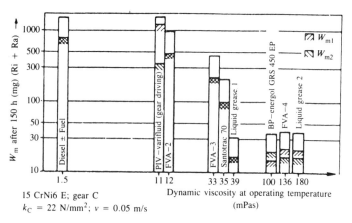

Fig. 6.5   The effect of lubricant type on wear (100)

To calculate the linear wear value in millimetres, the following equation was derived (**100**)

$$W_{\mathrm{l}} = C_{\mathrm{IT}}\left(\frac{\sigma_{\mathrm{H}}}{\sigma_{\mathrm{HT}}}\right)^{1.4}\left(\frac{\rho_{\mathrm{C}}}{\rho_{\mathrm{CT}}}\right)\left(\frac{\rho_{\mathrm{W}}}{\rho_{\mathrm{WT}}}\right)N$$

Here:

W_{l} = linear wear value
C_{IT} = wear coefficient dependent on film thickness
σ_{H} = Hertzian stress
ρ_{C} = radius of curvature
ρ_{W} = specific sliding
N = load alternation
Index T = for test

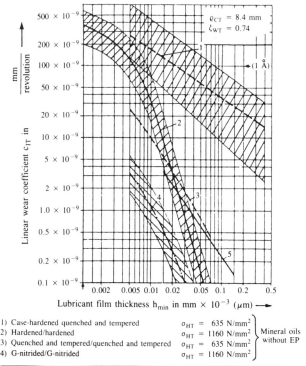

1) Case-hardened quenched and tempered	$\sigma_{\mathrm{HT}} = 635$ N/mm^2	
2) Hardened/hardened	$\sigma_{\mathrm{HT}} = 1160$ N/mm^2	Mineral oils
3) Quenched and tempered/quenched and tempered	$\sigma_{\mathrm{HT}} = 635$ N/mm^2	without EP
4) G-nitrided/G-nitrided	$\sigma_{\mathrm{HT}} = 1160$ N/mm^2	

5) Hardened/hardened	$\sigma_{\mathrm{HT}} = 1160$ N/mm^2

Lubrication: liquid without EP, NLG1:00

Fig. 6.6 Linear wear coefficient as a function of film thickness (100)

This wear value should be smaller than a so-called acceptable wear value $W_{l_{acc}}$. Therefore

$$W_l < W_{l_{acc}}$$

The linear wear coefficient C_{lT} can be taken from Fig. 6.6, as a function of the calculated film thickness (see Section 3.3.4.7). We then get the following relationship for the wear-related life L_{hw} in hours

$$L_{hw} = \frac{W_{l_{acc}}}{\left(\frac{\sigma_H}{\sigma_{HT}}\right)^{1.4}\left(\frac{\rho_C}{\rho_{CT}}\right)\left(\frac{\rho_W}{\rho_{WT}}\right)n60}$$

6.3 HIGH VELOCITIES

6.3.1 General principles

Lubrication conditions in gears at high peripheral velocities, e.g., up to 150 m/s, are characterized by high losses, especially high idling losses, as well as by difficulties regarding adequate supply of lubricant to the tooth surfaces (**99**).

6.3.2 Loss limitation

The high level of losses at high velocities produces two effects which can have negative results. On the one hand, the loss is converted into heat, which must then be removed. On the other hand, the increased heat production causes premature ageing of the gear oil.

Figure 6.7 shows the increase in idling losses with peripheral velocity (**101**). The effect of viscosity can be detected, but it is not excessively

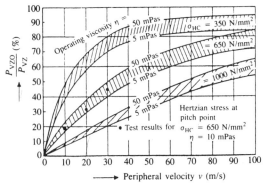

Fig. 6.7   Idling losses as a function of peripheral velocity (101)

marked. To limit these losses, as well as using low-viscosity gear oils it is necesary to ensure that not all the oil is squeezed by the meshing teeth. With spray lubrication, the oil supplied before and after the gears mesh can be controlled well. The oil sprayed on after the contact point serves to cool the gears.

The oil is exposed to considerable danger of oxidation due to the high temperatures. To ensure sufficiently long service life for the gear oils used in the normally large-fill-capacity units, they must be protected with suitable oxidation inhibitors.

As the scoring load capacity initially falls with increasing peripheral velocity, appropriate quantities of anti-scoring and anti-wear additives must be used in the oil to shift the scoring load/velocity curve upwards (Fig. 6.8).

The scoring load capacity can only be raised to a limited extent by a higher viscosity in order to maintain a thermal balance. It is at low viscosities that anti-wear and extreme-pressure additives have a particularly marked effect on scoring load capacity (Fig. 6.9).

Additional measures which may be taken to ensure gear reliability at high velocities due to adequate scoring resistance include the production of good surfaces, that is, small peak-to-valley heights, and effective

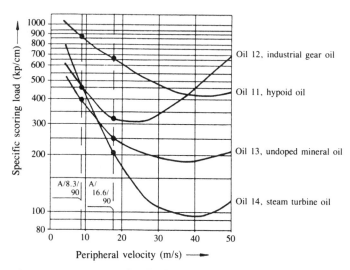

Fig. 6.8 Effect of peripheral velocity on scoring load capacity (101)

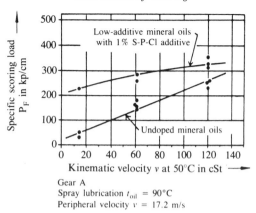

Gear A
Spray lubrication t_{oil} = 90°C
Peripheral velocity v = 17.2 m/s

Fig. 6.9   Effect of viscosity on scoring load capacity (101)

running-in of the tooth surfaces, as well as coating the tooth surfaces, e.g., with copper (99).

6.3.3   Adequate lubricant supply

At very high peripheral velocities there is a danger that, with splash lubrication, the oil on the tooth surfaces may be thrown off before it reaches the contact region. The result would be gear failure and excessive wear and scoring due to inadequate lubricant. It was mentioned in Chapter 5 that, for this reason, it is advisable to use pressure circulation lubrication or spray lubrication instead of splash lubrication above a peripheral velocity of 15 m/s. In Chapter 7, information is given on suitable structural measures to enable splash lubrication to be used at higher peripheral velocities.

6.4   HIGH TEMPERATURES

High operating temperatures in gears, which cause low viscosities, are a problem mainly if they occur alternately with low temperatures. This applies, e.g., to vehicle gearboxes, which must undergo a cold start at low temperatures, but after the warm-up phase, operate continuously at high temperatures. Due to possible difficulties in the low-temperature operating range, the low high-temperature viscosity cannot be raised with higher-viscosity gear oils.

To limit failure due to wear, scoring and pitting, and also grey spotting, it is necessary to use gear oils containing suitable AW and EP additives. The greater danger of oxidation at higher temperatures is restricted by suitable inhibiting additives. Using certain synthetic gear oils can also prove a suitable way of ensuring reliability at high temperatures.

To restrict further the tooth surface damage caused by low viscosities at high temperatures, or to restrict temperature increases by means of reduced mechanical friction, the use of solid lubricants, for example in the form of bonded lubricants, has proved successful.

6.5 LOW TEMPERATURES

Low temperatures can also prove particularly problematic, as they cause a sharp rise in viscosity and impair the flow characteristics of the gear oil. Inadequate lubrication with wear and scoring can be the consequence. These conditions must be considered, particularly in the case of vehicle transmissions.

Gear oils for low-temperature applications must exhibit adequate cold-flow performance, which is dependent on the characteristics of the base oil and the additive package. The use of multigrade gear oils, which have a low viscosity at low temperatures and a high viscosity at high temperatures, i.e. an optimal viscosity/temperature characteristic, can make the problem worse, just as in the case of particularly high operating temperatures.

6.6 REQUIREMENT FOR LOW FRICTION

Friction-reducing lubricants have become increasingly important in recent years. In developing energy-saving measures, so-called smooth-running oils have also been formulated for gear lubrication, which require additives for reducing not only wear, but primarily friction, in the mixed friction region. The following groups of products are used successfully in the so-called 'friction reducers':

- oil-soluble, long-chain carboxylic acids, phosphoric acids, and nitrogen compounds and their derivatives;
- oil-soluble molybdenum compounds;
- oil-insoluble solid lubricant suspensions.

Figure 6.10 shows how the efficiency of a vehicle final drive can be improved with such additives (oil-soluble compounds); at high temperatures (lower viscosity and lower hydrodynamic pressure) the improvement can be considerable (**102**).

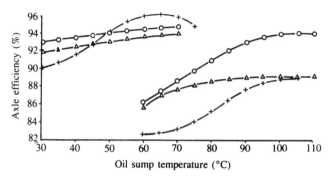

Fig. 6.10   Effect of an oil-soluble friction reducer on the efficiency of a vehicle final drive (102)

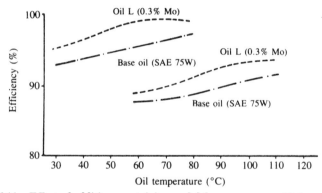

Fig. 6.11   Effect of additives containing molybdenum on gear efficiency (103)

Figure 6.11 shows that oil-soluble molybdenum compounds used as friction reducers can improve the efficiency of a vehicle final drive, in both high torque/low engine speed and high engine speed/low torque conditions (**103**).

6.7   AIR IN THE OIL

In practice it is often impossible to prevent air getting into the oil in the lubrication system of a machine installation, including gear sets. Until recently it was not clear whether the dissolved air had a negative effect on the efficiency of the gear oil, that is, on its anti-scoring and anti-wear characteristics. This aspect was studied experimentally with the FZG test machine; up to 20 percent air was dispersed in undoped and doped mineral oils (**104**).

Figure 6.12 shows that 10 percent air in the oil can raise the damage force level (scoring load) and reduce wear. Only when the air content reaches 20 percent – an impurity which is unrealistic – are the anti-scoring and anti-wear characteristics impaired.

This result can be explained by the accelerated occurrence of oxidation products due to higher temperatures (heat of compression) with air inclusions. These oxidation products have favourable tribological properties; possible disadvantageous effects, e.g., corrosion behaviour, were not studied. In fact, gear temperatures increased with the air content. With higher air contents, e.g., 20 percent, the negative effects of the higher temperature naturally predominate, so the damage force level becomes lower again and wear increases.

The favourable tribological effects of air inclusions in gear oil have also been demonstrated with other tooth shapes, in other test conditions,

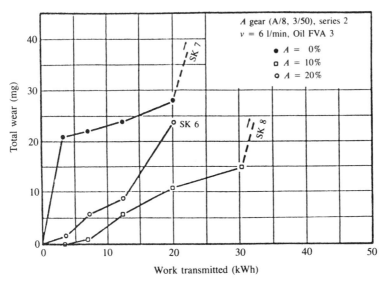

Fig. 6.12 FZG gear wear and damage force level with undoped mineral oil containing various percentages of air

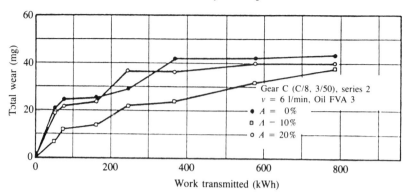

Fig. 6.13   FZG fatigue wear for undoped mineral oil with various air contents

and with doped mineral oils. Figure 6.13 shows wear as a function of operating time for undoped oils with various air contents. The lowest wear level which occurred with the oil containing 10 percent dispersed air can be seen clearly.

The damage force level 12 of a mineral oil with an extreme pressure additive is not reduced by dispersed air (Fig. 6.14). The lower wear is, however, obvious.

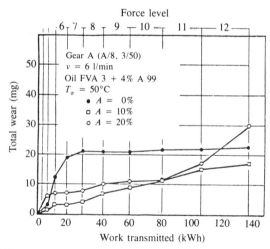

Fig. 6.14   FZG gear wear and damage force level with doped mineral oil with various air contents

6.8 LUBRICATION OF WORM GEARS WITH SYNTHETIC GEAR OILS

As already mentioned, worm gears belong to the rolling crossed-axis gears with a very high sliding component. The result is high friction with a corresponding increase in temperature, coupled with inadequate hydrodynamic pressure development due to the kinematic conditions. With the additional requirement for reduced energy consumption, the lubricant must have the following characteristics: particularly low fric-

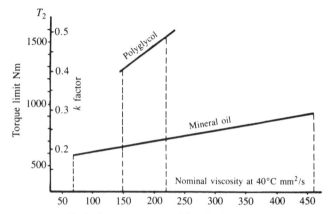

Fig. 6.15   Torque limits in worm gears with polyglycols and mineral oils (105)

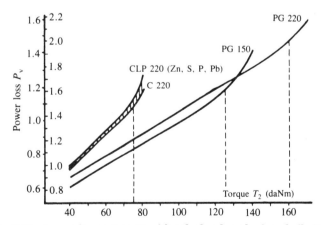

Fig. 6.16   Losses in worm gears with polyglycols and mineral oils (105)

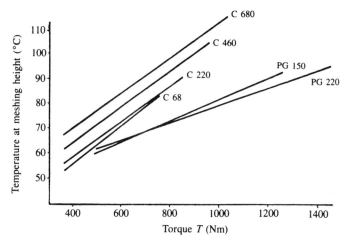

Fig. 6.17   Temperature in worm gears with polyglycols and mineral oils (105)

tion and high oxidation resistance with good wear and relatively low viscosity–temperature dependence. These requirements are met by polyglycol-based gear oils, which have proved successful in industrial worm gears (**105**).

Figure 6.15 shows the distinctly higher torque limits of polyglycols compared with mineral oils. By torque limit we mean that torque above which losses rise progressively.

Losses are considerably lower when worm gears are lubricated with polyglycols than with mineral oils (Fig. 6.16), which naturally results in lower temperatures (Fig. 6.17).

6.9   THE EFFECT OF LUBRICANTS ON THE OCCURRENCE OF GREY SPOTS

6.9.1   Description of the grey spot phenomenon and factors influencing it

Only in recent years has much attention been paid to the phenomenon of grey staining. Macroscopically, these are dull-looking areas on the surfaces of case-hardened gears. The pattern of defects and factors influencing them can be described as follows (**106**).

–   Grey spots consist of numerous eruptions of a macroscopic order of magnitude (depth c. 20 μm).

- The direction and progress of the incipient crack are determined by the sliding/rolling conditions, as in the case of pitting, i.e., the cracks start on the surface and develop at a low angle in the direction opposite to that of the friction force.
- Grey spotting occurs primarily in the region of medium peripheral velocities while using low-viscosity lubricating oils on the tooth surfaces of hardened gears.
- Operating times up to the occurrence of grey spotting can sometimes be very short, and the load can sometimes be well below the pitting fatigue strength.
- The initiation of damage resembles wear, while the development of the damage resembles fatigue.
- Lubricant viscosity and additives have a considerable influence on grey spotting.
- In the case of lubricants with a low grey spot load capacity, severe erosion occurs in the region of the grey spotting, which leads to an increase in the inner dynamic auxiliary forces and in gear noise as well as to a reduction in pitting load capacity.

6.9.2 Effect of lubricant on grey spotting

A test procedure, consisting of a stepwise test and a continuous test, and based on the FZG test machine using C gears, has been developed for relative evaluation of the grey spot load capacity of gear oils (**106**). The oils are graded into low, medium, and high grey spot classes, to which are assigned certain damage levels (Table 6.1).

Table 6.1 Damage characteristics in the grey spot test (106)

GS class	Damage force level SK in stepwise test	Damage behaviour in durability test
GSLC low	SK ⩽ 7, large grey spot area, GS sometimes over 50 percent	1×80 h operating time at level 10, GF, f_t distinctly over 20 μm
GSLC medium	SK 8–9, medium grey spot area GS c. 30–50 percent	$1–2 \times 80$ h level 10, GS, f_t 10–20 μm and/or pitting
GSLC high	SK 10, > 10 little or no grey spotting GS ⩽ 20 percent	$1–5 \times 80$ h level 10, possible GS, $(f_t < 20$ μm), pitting

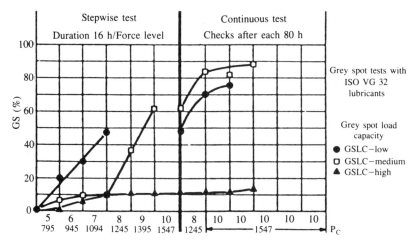

Fig. 6.18 **Development of grey spot area GS on the test pinions in the grey spot test for lubricants of different grey spot load capacities** (106)

Figure 6.18 shows, for three oils, the increase in the tooth surface area affected by grey spotting.

Figure 6.19 shows the wear behaviour of these three oils in the same test. The oils used can be described as follows.

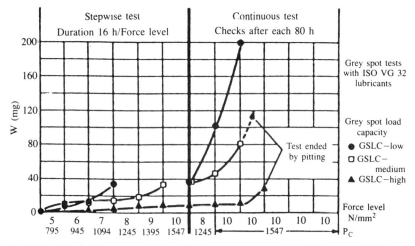

Fig. 6.19 **Wear W on test pinions in the grey spot test for lubricants of different grey spot load capacities** (106)

GSLC low: base oil A + ZnDDP additive.
GSLC medium: base oil B + ZnDDP additive.
GSLC high: base oil A + S–P additive.

Base oil A is less viscous than base oil B.

6.10 RELATIONSHIP BETWEEN GEAR OILS AND CLUTCH LININGS

6.10.1 General

Plate clutches in the oil circuit are used in many areas of propulsion engineering. Their advantage is that the friction heat produced can be removed optimally from the friction surfaces, allowing lower clutch temperatures and almost wear-free operation.

In addition, the clutches and the units in which they are installed can be compact, and high output can be produced from very small spaces.

Important criteria for smooth operation of these clutches are nevertheless the requirements determined by the application, and the correct matching of lubricant and friction material. For normal applications with mineral oils, conventional high-performance sintered friction linings have proved most successful. They have been used successfully in mass production, even at high loads, for years. So far, problems have only been encountered with those plate clutches used with EP gear oils and synthetic lubricants. New sintered friction linings had to be developed for these conditions (**107**).

6.10.2 Friction requirements to be met by clutch linings

The optimal friction characteristics of linings are characterized by the following criteria:

- high friction coefficient;
- small differences between static friction coefficient μ_{stat} and dynamic friction coefficient μ_{dyn};
- low wear;
- little change in friction due to oil ageing;
- good long-term performance;
- high load capacity;
- good compatibility with oil.

6.10.3 Effect of the lubricant on friction performance

The effect of the lubricant on the friction performance of clutch linings is examined below with the aid of selected examples (**107**).

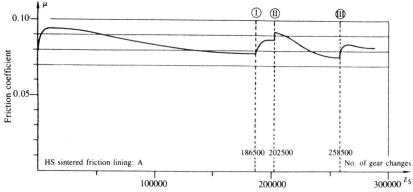

Test conditions:
$l = 0.935$ kg m^2, $q_A = 0.4$ J/mm^2, $V_{oil} = 6$ l, $n = 2500$ min^{-1}, $S_h = 225$ h^{-1}, $V = 4.6$ l/min, $A_{RC} = 77\ 400$ mm,
$O_h = 7000$ kJ, $V_A = 1$ mm^3/(mm s)
Cooling oil: ATF ESSO DEXRON B 10 696
Ⓘ Oil change with new oil ⒾⒾ Oil change with oil used for 186 000 gear changes ⓂⒾⒾ Half of cooling oil replaced
with new oil

Fig. 6.20 Effect of oil ageing on the friction performance of a plate clutch in the oil system (107)

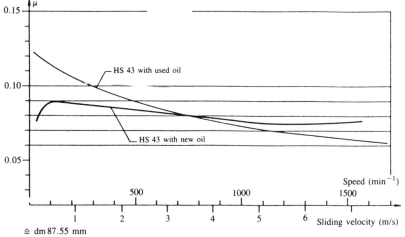

$\stackrel{\frown}{=}$ dm 87.55 mm

Test conditions:
Cooling oil: SHELL SPIRAX EP 90, Cooling oil flow rate: $V_A = 2.38$ mm^3/mm^2 s, No. of gear changes: $Z_{s_{.}} = 10\ 000$,
Gear change work $q_A = 0.94$ J/mm^2, Gear change frequency $S_h = 90$ h^{-1}, Surface pressure: $p_R = 2$ N/mm^2, Speed
max: $n = 1800$ min^{-1}
Friction lining dimensions: $D_A = 99$ mm, $D_I = 77$ mm, No. of friction surfaces: $Z_R = 4$

Fig. 6.21 Effect of oil ageing state on the friction characteristics for high-performance sintered friction lining HS 3 in an EP gear oil, Shell Spirax EP 90 (SAE 90/API–GL 4). After an oil change, the original friction characteristic is re-established (107)

402

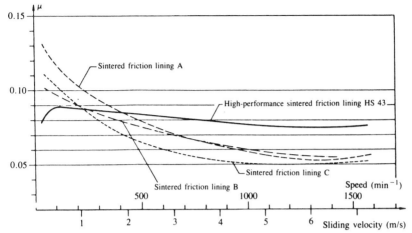

Test conditions:
Cooling oil: SHELL SPIRAX EP 90, Oil flow rate: V_A = 2.38 mm³/mm² s, No. of gear changes: Z_s = 10 000, Gear change work q_A = 0.94 J/mm², Gear change frequency S_h = 90 h⁻¹, Surface pressure: p_R = 2 N/mm², Speed max: n = 1800 min⁻¹
Friction lining dimensions: D_A = 99 mm, D_1 = 77 mm, No. of friction surfaces: Z_R = 4

Fig. 6.22 Comparison of various sintered friction materials in an EP gear oil (107)

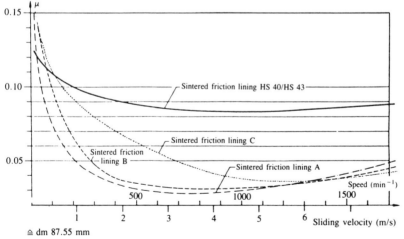

$\triangleq$ dm 87.55 mm

Test conditions:
Cooling oil: CASTROL K 390 (synthetic), Oil flow rate: V_A = 2.38 mm³/mm² s, No. of gear changes: Z_s = 10 000, Gear change work q_A = 0.94 J/mm², Gear change frequency S_h = 90 h⁻¹, Surface pressure: p_R = 2 N/mm², Speed max: n = 1800 min⁻¹
Friction lining dimensions: D_A = 99 mm, D_1 = 77 mm, No. of friction surfaces: Z_R = 4

Fig. 6.23 Comparison of various sintered friction materials in a synthetic gear oil (polyalkyl glycol) (107)

403

Figure 6.20 shows the effect of oil ageing and of an oil change on the friction of a sintered lining. As can be seen from Fig. 6.21, the friction coefficient falls markedly with rotational speed in the presence of a used oil, while, after an oil change with a new oil, a largely constant friction level is re-established. The friction performance of various sintered friction materials with EP gear oil lubrication can be seen in Fig. 6.22, and the constant friction characteristic of a newly developed friction lining material is obvious. The friction performance of various materials in the presence of a synthetic gear oil, specifically polyalkyl glycol, can be seen in Fig. 6.23. The outstanding performance of the new friction lining material is again clear.

7 Solid Lubricants for Gear Lubrication

7.1 GENERAL

Solid lubricants can be used as additives (indirect lubricant film formation) or directly as lubricants (direct lubricant film formation). Solid lubricants are available in various forms for both direct and indirect lubricant film formation. Details of this were given in Section 4.7.7.

For gear lubrication, lubricants containing solid lubricants for indirect film formation are primarily used. These are mainly the sprayed adhesive lubricants, which belong to the lubricating greases, and the gear oils containing solid lubricants.

7.2 SPRAYED ADHESIVE LUBRICANTS

7.2.1 Principles

Adhesive or sprayed adhesive lubricants are used for tooth surface lubrication of open or simply covered gears. Their use is generally restricted to peripheral velocities up to about 4 m/s. Only in special cases are higher velocities acceptable.

7.2.2 Structure and types

7.2.2.1 Structure

Sprayed adhesive lubricants are always free of bitumen and often free of solvents. They are complex, high-grade special lubricants. They usually contain a base oil, a thickener, and certain additives. Due to the thickener they are consistent lubricants, which can be classed with the lubricating greases in a wider sense.

(a) Base oils
Mineral oils, synthetic oils, or mixtures of the two are used. The base oil viscosity lies between 460 and 680 mm^2/s at 40°C with a viscosity index (VI) greater than 90 and a pour point lower than -10°C.

(b) Thickeners
Simple metal soaps or complex metal soaps (Al, complex Al, complex Ba soaps), and also inorganic non-soap thickeners are used.

(c)   Additives

Anti-wear and anti-scoring additives are used. These can be phosphorus/sulphur compounds, lead naphthenate and/or compounded oils, as well as solid lubricants. Graphite, usually natural graphite, is normally used as the solid lubricant.

Sprayed adhesive lubricants, therefore, usually have the following structure (**106**)

base oil + thickener + additives + solid lubricants

The individual constituents are used in the following proportions

base oil: 65–80 percent
thickener: about 5 percent
additives: 10–20 percent
solid lubricants: 5–10 percent

The sprayed adhesive lubricants may also contain a solvent to aid application.

The preferred consistency is NLGI class 0.

The technological ability of the lubricant to be sprayed is regarded as very important. Other important characteristics are adequate corrosion protection, good compatibility with seal materials, and, above all, outstanding anti-wear and anti-scoring properties, which are assessed, among other things, with the FZG test machine (DIN 51 354).

Sprayed adhesive lubricants are also covered by DIN 51 509, Part 2; to date there are no requirements (**108**).

7.2.2.2   Types
Depending on the requirements of open gear sets, we differentiate between various types of adhesive lubricants, which can be assigned to this family of bitumen-free special lubricants (**106**).

(a)   Priming lubricants
The priming lubricant is applied to the contact tooth surfaces of the new gear after cleaning and fills the asperities. The solid lubricant content, usually graphite, is higher than for the so-called service lubricant (about three times higher). The excess of graphite together with anti-scoring additives permits plastic deformation under local overload.

(b)   Running-in lubricants
The purpose of the running-in process for a gear is to reach full load capacity as soon as possible by optimizing the percentage contact area of the tooth surfaces while maintaining their geometrical shape.

With such a running-in process, the surface quality which can be obtained depends on the time taken to remove the asperities, and on the effective specific load.

With the running-in lubricant there is a controlled smoothing of the production asperities to increase the contact component and the load borne by the gear without damage or wear.

Due to the complex interaction of various processes and mechanisms on the surface, the question of additives for running-in lubricants is also complex. It is important to achieve optimal matching of the chemically active extreme pressure additives to the physically active solid lubricants, i.e., graphite.

(c) *Service lubricants*

The service lubricant used after running-in is applied by periodic lubrication, unlike the priming and running-in lubricants. The solid lubricant content corresponds to the detail given in Section 7.2.2.1.

(d) *Corrective lubricants*

Tooth surface damage, which occurs as a result of continuous material overload and consequent premature fatigue, cannot be prevented by the lubricant. However, it is possible to achieve a considerable improvement in the state of the surface with a new, controlled running-in process if the damage is recognized early and the causes removed.

There are corrective lubricants for restoring severely damaged working surfaces, whose constituents include, apart from graphite, other solid substances such as aluminium oxides and/or silicon dioxides. As this is an intentional correction of a damaged tooth surface, i.e., where the tooth geometry has been measurably altered, the action of classical running-in additives would not suffice.

Extremely finely distributed solid substances perform the necessary removal of material abrasively. It is obvious that such corrective measures must be applied with the greatest care and under continuous monitoring of the change in the tooth flank surface.

After treatment with the corrective lubricant, the gear set must be practically run in again, until the new contact pattern appears satisfactory. The corrective lubricant must be completely removed from the gear set beforehand.

7.2.3 Application

The choice of application method depends on the consistency, the consistency/temperature performance, the transport characteristics and the spray characteristics.

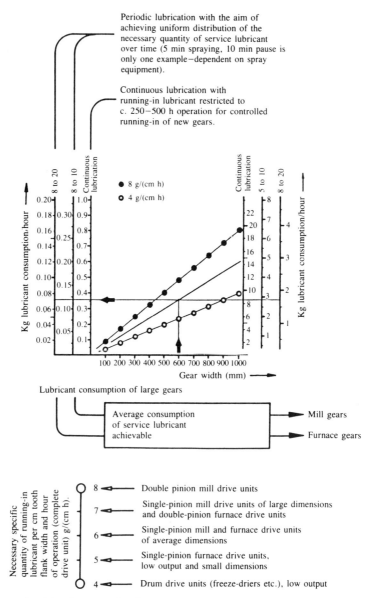

Fig. 7.1   **Consumption diagram for sprayed adhesive lubricants** (109)

The following methods are used:

- manual application with brush, spatula, etc.;
- manual spraying with aerosol can, compressed air or electric spray gun;
- transfer lubrication by paddle wheel or transfer pinion;
- automatic spray lubrication.

The most economical method of applying these special lubricants is undoubtedly automatic spraying. It produces relatively low consumption rates of between 4 and 8 g/h cm (grammes of lubricant per hour and cm pinion width). Figure 7.1 shows some examples (**109**). These values relate to continuous lubrication; with central lubrication, the consumption can fall to 2 g/h cm, and in exceptional cases to 1.2 g/h cm.

Table 7.1 gives an indication of the quantities of lubricant to be applied as a function of gear size and lubrication frequency as per AGMA Standard 251.01. It can be seen that frequent re-application is preferable. The total lubricant consumption with lubrication intervals of 1 h is considerably lower than with lubrication intervals of 4 h.

7.2.4 Wear protection performance
The anti-wear behaviour of sprayed adhesive lubricants can be tested with the FZG test machine. For this purpose, a special test was devised with test gear type A at a peripheral velocity of 2.76 m/s. A short test until the damage force level was reached and a long-term test at a load below the damage force level give an indication of the performance of different lubricants (Fig. 7.2). These tests can be performed with both splash lubrication and spray lubrication, and comparable results are obtained (Fig. 7.3) (**110**).

7.3 GEAR OILS CONTAINING SOLID LUBRICANTS

7.3.1 General
Solid lubricants in gear oils have to prevent increased wear and scoring, prevent or delay pitting, and/or influence friction. They are thus in competition with oil-soluble additives, and must be matched to these and to other additives.

7.3.2 Structure
Solid lubricants are dispersed as additional anti-wear or extreme-pressure additives in formulated gear oils or, less frequently, as the only

Table 7.1   Quantities of lubricant to be applied as a function of gear size and frequency of periodic lubrication at pitch circle velocities below 7.5 m/s to AGMA 251.01

Gear diameter (m)	Quantity of lubricant per interval (ml)*											
	$\frac{1}{4}$ h				1 h				4 h†			
3.0480	5.92	8.87	11.83	14.79	23.66	35.49	37.32	59.15	147.90	177.40	236.60	295.70
3.6576	8.87	8.87	11.83	14.79	35.49	41.40	53.23	65.06	177.40	207.00	266.20	325.30
4.2672	8.87	11.83	14.79	17.74	41.40	47.32	59.15	70.98	207.00	236.60	295.70	354.90
4.8768	11.83	14.79	17.74	20.70	47.32	59.15	70.98	82.81	236.60	295.70	354.90	414.00
5.4864	14.79	17.74	20.70	23.66	59.15	70.98	82.81	94.64	295.20	354.90	414.00	473.20
Flank width	20.32	40.64	60.96	81.28	20.32	40.64	60.96	81.28	20.32	40.64	60.96	81.28

* The spray time should extend over one or two revolutions to ensure that all surfaces are coated.
† Lubrication intervals longer than 4 h are not acceptable.

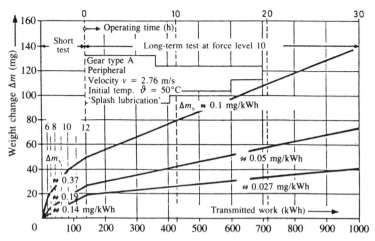

Fig. 7.2 Wear performance of adhesive lubricants in the short test (A/2.76/50) and in the long-term test (110)

additive in the base oil. To prevent precipitation, that is to create a stable dispersion or suspension, an effective dispersant has to be added. The stability of the dispersion depends on bringing the solid lubricant particles into intimate contact with the dispersant; this is achieved by grinding it.

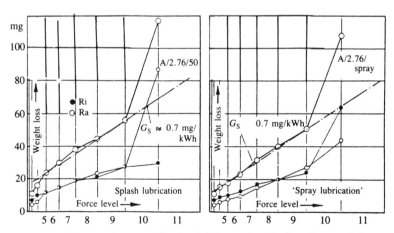

Fig. 7.3 FZG tests of spray adhesive lubricants in splash and spray lubrication (100)

In less common cases, solid lubricant concentrates in the form of suspensions are added by the gear set user to the ready-formulated gear oil. In this case it is clearly not possible to match the various additive components to one another. The concentration of solid lubricant is between 0.1 and 1.0 wt%, with concentrations of a few tenths of a percent being normal. Suspensions to be mixed into the ready-made oil generally have a solid lubricant concentration of 10 wt%.

Average particle sizes of the order of 1 μm have proved to be appropriate for the application of solid lubricant powder, especially with MoS_2 and graphite.

7.3.3 Anti-wear and anti-scoring performance

Graphite is one of the oldest solid lubricants, and is still used as a suspension or dispersed in the gear oil. In addition to natural graphite, synthetic graphite is increasingly being used (**111**).

During the course of the development of an energy-saving industrial gear oil with 0.5 percent graphite and a combination of sulphur/phosphorus-containing and organo-metallic EP additives, the EP and AW performance of this oil was studied (**112**). The higher scoring load capacity confirmed in the IAE test and the low wear values found in the Timken test can be seen in Table 7.2.

The solid lubricant molybdenum disulphide (MoS_2) is undoubtedly still the most important lubricant. Adding MoS_2 to gear oils is easily shown to increase load-bearing capacity and to reduce wear in tests on laboratory test equipment. In gear tests, however, it is quickly found that the effectiveness of MoS_2 depends on the remainder of the additive system (**113**). Table 7.3 shows results of the CRC–L axle test (high velocities with shock loading = passenger car operation), and it is clear that MoS_2 gives positive results in conjunction with a lead-containing additive. However, an unusually high MoS_2 concentration of 10 percent was necessary. Improvements were also obtained in the CRC–L–37 axle test

Table 7.2 Wear results with an industrial gear oil containing graphite (112)

	IAG gear test acceptable load (N)	Timken test (steel/bronze). Average wear mark width (mm)
ES gear oil 320	735	2.61
S–P gear oil 320	441	3.30

Table 7.3   CRC L–42 test results with various gear oils (113)

No.	Lubricant	Result	Comments
1	Base oil SAE 90 + 40 percent MoS_2	Failed	Almost passed, no noise, no wear, CRC rating 8–90
2	Base oil SAE 90 2 percent lead diamyl dithiocarbamate	Failed	Failure in phase 2; gears worn
3	Oil no. 2 + 10 percent MoS_2	Failed	Failure in phase 4, very slight wear
4	Base oil SAE 90 + 6.5 percent lead diamyl dithiophosphate	Failed	86 percent scoring; CRC rating 8–90
5	Oil no. 4 + 10 percent MoS_2	Passed	No scoring on tension flank, 3 percent scoring on thrust flank

(low velocity with high torque = truck operation) with MoS_2. The test was passed with 5 percent and 10 percent MoS_2 in conjunction with an antimony-containing additive (Table 7.4).

There need be no concern that the solid lubricant may cause damage to rolling bearings. Figures 7.4 and 7.5 show that both the friction moment and the temperature can be reduced by MoS_2 lubrication (**114**).

According to findings lead/sulphur-containing additives have an outstanding load-bearing capacity at low sliding velocities, and sulphur/phosphorus-containing additives do so at high sliding velocities. On the basis of these findings, attempts were made to find an additive combination which would be satisfactory in both ranges. This led to the development of a borate-containing gear oil, which has finely dispersed potassium triborate, i.e., a solid which is synergistically matched with an oil-soluble zinc dialkyl dithiophosphate (**115**). These oils proved to be at

Table 7.4   CRC L–37 test results with gear oils containing solid lubricants (113)

Test number	1	2	3	4	5	6
State of tooth flank	Untreated	Untreated	Untreated	Untreated	Untreated	Phosphated
Oil	A	A + 5 percent MoS_2	D	E	E + 10 percent MoS_2	E + 10 percent MoS_2
Results Running-in	Passed	Passed	Failed	Passed	Passed	Passed
Overall	Passed	Passed	Failed	Failed	Failed	Passed

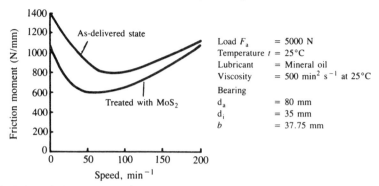

Fig. 7.4   Friction moment of a tapered roller bearing with and without MoS$_2$ (114)

least as good as conventional sulphur/phosphorus-containing vehicle gear oils. During operation in trucks at high loads this new formulation proved to be free of problems, while conventional gear oils in API GL–5 had often failed to prevent damage. The distinctly lower temperatures in the oil when the borate-containing additive is used are evident. Table 7.5 gives examples from various actual applications (**115**).

In a high-speed running-in test on a rear axle, the equilibrium temperature was distinctly lower for the borate-containing gear oil (Fig. 7.6).

Used-oil analysis after the tests, and also in the field test, revealed distinctly lower iron contents in the borate-containing lubricant than in conventional sulphur/phosphorus-containing gear oils (Fig. 7.7), an effect which indicates another mechanism without 'corrosive' removal of material. The excellent surface condition of the tooth flanks even after long periods of operation is evident from Table 7.6 and Fig. 7.8 (**116**).

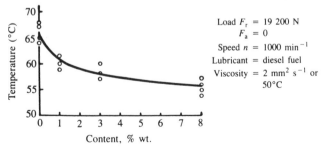

Fig. 7.5   Effect of MoS$_2$ on the operating temperature of a radially loaded deep-groove ball bearing (114)

Table 7.5   Temperature reduction by borate-containing gear oils (115)

Application	Gear type	Oil type	Conventional gear oil sump temperature (°C)	Borate-containing gear oil sump temperature (°C)
Heavy truck rear axle	Hypoid	GL–5, s–p	95	66
Heavy truck gear box	Helical spur gear	Mineral oil	110	80
Process engineering 1	Double helical gear	Mineral oil	70	52
Process engineering 2	Bevel gear	s–p gear oil	56	50
Stone crusher	Helical spur gear	GL–5, s–p	70	63
Food industry	Helical spur gear	GL–5, s–p	75	69
Timber processing machinery	Tapered roller bearing	GL–5, s–p	82	66

Another new solid lubricant has been developed as an additive for industrial gear oils, which has proved to have outstanding anti-wear and anti-pitting properties (**117**). This is graphite fluoride (CF_x), which, like MoS_2 and graphite, has a lamellar structure. Between 0.1 and 0.5 percent CF_x finely dispersed in the base oil produced better results than the classic solid lubricants and graphite and MoS_2.

Table 7.7 shows that 0.5 percent of this additive in an undoped mineral oil can raise the FZG damage force to values greater than twelve.

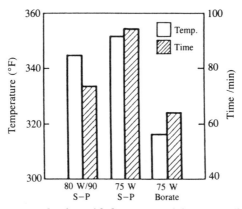

Fig. 7.6   Temperature reduction with borate-containing gear oil in a high-speed axle test (116)

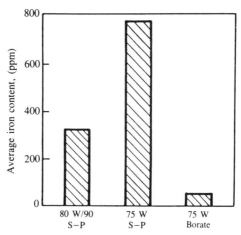

Fig. 7.7   Iron content (abrasive wear) in used borate-containing gear oil (116)

Comparable effects have been achieved by applying graphite fluoride in bonded form as a coating on tooth flanks.

Cerium fluoride (CeF_3) with an average particle size of 3 μm has also been suggested as a solid lubricant in suspensions (**118**). Tests with lubricating greases have shown that this solid lubricant is somewhat better than MoS_2 and distinctly better than graphite. No tests have yet been made with it in suspensions.

Table 7.6   Results of a long-term test with borate-containing gear oil (116)

Gear no. Oil type No. of km	*68785* *contains s–p* *297 310*	*68791* *contains borate* *302 320*
Ring gear	Slight smooth abrasion	Slight smooth abrasion
Drive pinion	Slight smooth abrasion	Slight smooth abrasion
Pinion bearing	No wear	No wear
Differential:		
bearing	Slight pitting	No wear
star	Moderate smooth abrasion	Slight smooth abrasion
axle bevel	Severe smooth abrasion,	No wear
gears	slight scratching	
gears	Severe smooth abrasion,	No wear
	severe polishing	

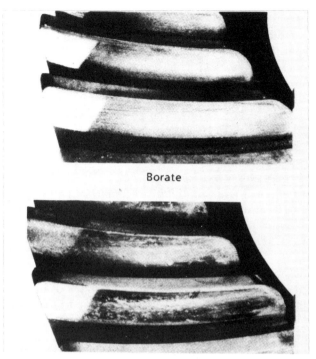

Fig. 7.8 State of tooth surface of truck bevel gear after long-term test with borate-containing gear oil (116)

Table 7.7 Effect of graphite fluoride on FZG damage force level (117)

		A1	*A2*
		Mineral oil SAE 90	Oil A1 + 0.5 percent CF_x
Standard FZG test weighing + cleaning	Damage force level	6	9–10
15 min per force level	Specific wear (mg/MJ)	0.1	0.04
FZG test without interruption 15 min per force level	Damage force level	7	12
FZG test without interruption 1 h per force level	Damage force level level	8	>12

7.3.4 Pitting

Pitting is a fatigue failure, involving the separation of particles of material from the surface of the tooth flank after a minimum number of contacts. Increased gear oil viscosity has proved to be a successful measure for counteracting pitting. With regard to the additives in the gear oil, it may be concluded that the surface-friction-reducing effect of some compounds delays pitting. The positive effect of solid lubricants, especially MoS_2, has long been discussed, and relevant results from field tests have been produced, but so far it has not proved possible to verify it definitely and systematically.

The performance of MoS_2 has also been examined in a comprehensive study of the effect of lubricant additives on pitting (18) (19). These tests were performed on the FZG test machine with a special gear with a low scoring susceptibility, which permitted long-term tests to be carried out without scoring. Details of the test procedure and some results are given in Section 3.3.

The addition of MoS_2 alone allows improved pitting service life of case-hardened gears if the concentration is increased. If only 0.5 percent is added, the tooth surface suffers early failure due to the appearance of grooves.

The high additive concentration of 1.5 percent gives better results for service life and surface quality than using the commercial additives SP1 and SP3 (sulphur/phosphorus type).

Adding tricresyl phosphate to oils containing 0.5 percent MoS_2 produces contradictory results. The service life at the lower load $k_c = 3.58$ da N mm^{-2} is about the same as in the tests without TCP, but is lower than that achieved with only tricresyl phosphate. However, in tests at the higher load $k_c = 4.42$ da N mm^{-2}, there was no pitting.

The graphite-containing industrial gear oil already described produced a markedly longer operating time to the start of pitting at low Hertzian stresses than conventionally doped oils (112). These results can be seen in Fig. 7.9.

Pitting tests have also been carried out with graphite fluoride (117). Figure 7.10 shows the good performance of these solid lubricants with regard to pitting life. Better results compared with MoS_2 and graphite have been achieved on a model apparatus. The borate-containing gear oil already described has also been found to have superior pitting performance to other gear oils.

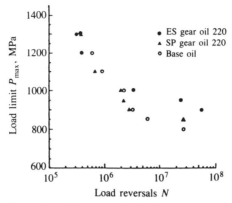

Fig. 7.9   Effect of graphite on pitting load capacity of gears (112)

7.3.5   Reduction of friction

There is a variety of results for the effect of solid lubricants on energy losses in gears. Reducing mechanical friction improves efficiency, and the result is lower energy consumption in industrial gears and lower fuel consumption in the case of vehicle gears.

Figure 7.11 shows that with a graphite-containing industrial gear oil (ES Gear Oil 320), the efficiency of a worm gear can be increased by 2 to 3 percent compared with a conventionally doped gear oil (112). An energy saving of between 2.7 and 9.0 percent has been obtained in actual

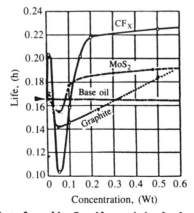

Fig. 7.10   Effect of graphite fluoride on pitting load capacity (117)

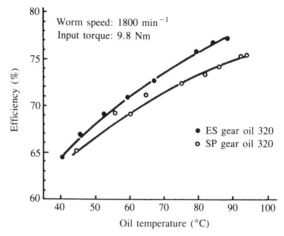

Fig. 7.11   Effect of graphite-containing gear oil on the efficiency of a worm gear (112)

service in crane gears (see Table 7.8). This oil contains 0.5 percent graphite together with a combination of sulphur/phosphorus-based and organometallic EP additives. It is interesting that, in these tests, MoS_2 was inferior to graphite.

Temperature reduction is a measure of reduced friction losses. Figure 7.12 shows this effect in a comparison between an SAE 85W/90 gear oil and an SAE 75 oil with MoS_2 at various ambient temperatures (119). Unfortunately, there is no indication of the MoS_2 concentration. It should be noted that fuel savings resulting from this are relevant only during hot running after a cold start.

The relevant principles are reported elsewhere (121). As can be seen from Fig. 7.13, during this hot running period, average fuel savings of between 1.7 and 5.2 percent can be achieved with the gear oil containing

Table 7.8   Reduction of energy consumption by a crane when a graphite-containing gear oil was used (112)

Oil	SP gear oil	ES gear oil 320
Date	11.8.1982*	31.8.1982
Measured consumption of electrical energy (kW)	220.8	208.0

* Replacement of SP with ES oil.

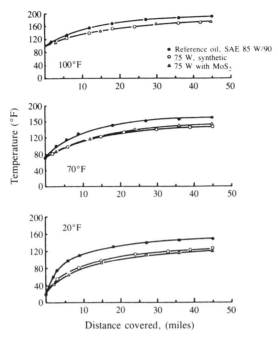

Fig. 7.12   Oil temperature in a differential gear with gear oil containing MoS₂ (119)

Table 7.9   Test oils and test conditions for determining the effect of solid lubricants on fuel consumption (121)

Oil	RGO	FMO–1	FMO–2	FMO–3	FMO–4
Type of friction modifier	–	Contains phosphorus	Borate	Sulphurized ester	MoS₂
Viscosity (mm² s⁻¹)					
at 40°C	84.36	83.50	84.14	84.94	83.10
at 100°C	9.88	9.82	9.87	9.99	9.87
Viscosity index	96	96	96	96	97

Conditions: – 10 mode test (initial temperature 23°C);
– steady-state operation at 60 lm/h and 100 lm/h;
– RGO: SAE 80 gear oil;
– FMO: RGO + 1 percent friction modifier;
– engine oil: SAE 30;
– rear axle gear oil: SAE 90.

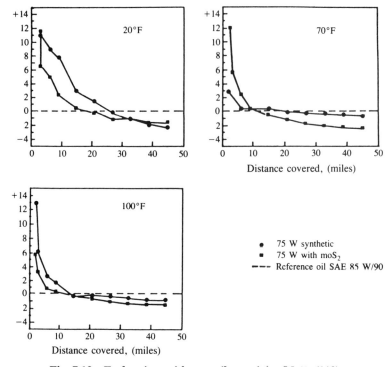

Fig. 7.13   Fuel savings with gear oil containing MoS₂ (119)

solid lubricant; the savings become smaller as the ambient temperature
rises.

Another study has shown that the addition of so-called friction modifiers to gear oils can effect a general increase in fuel economy. Table 7.9

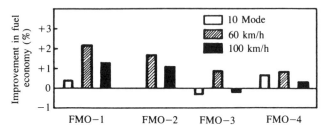

Fig. 7.14   Fuel savings with gear oil containing MoS₂ (FMO–4) (121)

contains details of the test and of the friction modifier types, which also include the two solid lubricants borate and MoS_2. Figure 7.14 shows that savings were obtained with the gear oil containing MoS_2 in all three test programmes. On the other hand, the borate solid lubricant was found to be superior to MoS_2 with regard to fuel economy at the higher road speeds.

8 Lubricant Supply

8.1 INTRODUCTION

Effective lubrication by means of the formation of viscosity-dependent pressure films or additive-dependent reaction films depends on the lubrication method and thus possibly on the design of the oil circuit on the one hand, and on the state of the gear lubricant and thus also on the quantity of oil and the oil change intervals on the other. Naturally, attention must also be paid to achieving a good contact pattern, so commissioning and running-in are very important.

8.2 LUBRICATION METHODS

8.2.1 General principles

When selecting lubrication methods, it has to be borne in mind that, in addition to the gear pair, other friction points have to be supplied with lubricant. Besides once-only lubrication, intermittent and continuous lubrication must be differentiated. The most important types are splash and circulation lubrication, the peripheral velocity being the criterion for differentiating between the two. With circulation lubrication, attention must be paid to suitable design of the oil supply unit (**122**).

8.2.2 Once-only lubrication

Gears, open drives, in which temperature rise, load on the lubricant film and wear stress on the tooth flanks are certain to be low due to the method of operation and/or the operating conditions, can most easily be lubricated by a single application of an adhesive lubricant before commissioning, with repeated applications at long intervals if necessary (Fig. 8.1). This type of gear includes:

- short-life gears;
- gears which operate rarely and for short periods;
- very slow running gears;
- gears subjected to load only when stationary.

With once-only lubrication, the peripheral speed should not exceed 0.8 m/s. Gears with a very short life are an exception; once-only lubricated gears with peripheral speeds of 10 m/s and more have been made

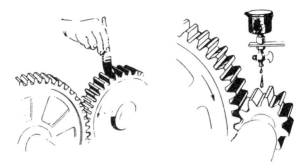

Fig. 8.1 Lubrication of open gears by hand and with a drip feed

in this category. Increased wear and tooth flank damage towards the end of the service life are accepted.

Lubricants for once-only lubrication are (a) solid lubricant powders (rubbed into the tooth flanks), and (b) bonded lubricants, greases and pastes, especially those based on solid lubricants.

8.2.3 Intermittent and semi-continuous lubrication

If higher demands are made on gears which are regarded as suitable for once-only lubrication, it is advisable to use intermittent lubrication. Higher demands are:

- longer service life in the case of short-life gears;
- longer individual operating periods in the case of rarely used gears;
- higher peripheral speeds in the case of slow running gears;
- the occurrence of part loads during operation in the case of gears fully loaded only while stationary.

Measures suitable for performing intermittent lubrication include daily application of grease, regular renewal of a coating of bonded lubricant, and regular supply of grease or oil with the aid of metering devices at intervals of from around 1 min to approximately 1 h. The peripheral speed should not exceed 1 m/s as a rule (Fig. 8.2).

Semi-continuous lubrication should be used if, on the one hand, lubricant supplied once only or at intervals does not protect the tooth surfaces adequately against wear due to long-term operation, high load or high peripheral speed with a corresponding number of load reversals, or, on the other hand, if it is not necessary for the lubricant to remove dissipated heat. Semi-continuous lubrication can be performed most easily with liquid lubricants by means of drip application. Another possi-

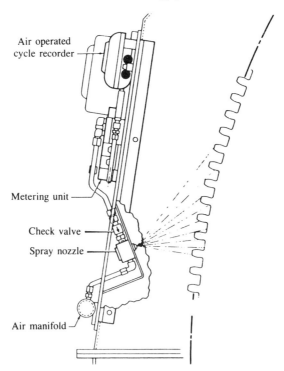

Air operated cycle recorder

Metering unit

Check valve

Spray nozzle

Air manifold

Fig. 8.2 Intermittent spray lubrication of open gears

bility is spray application of grease at intervals. This can be used up to a peripheral speed of about 3 m/s.

There is no clear dividing line between intermittent and semi-continuous lubrication. The former merges into the latter by the shortening of lubrication intervals. In the case of semi-continuous lubrication and at the boundary of intermittent lubrication, arrangements must be made to collect the used lubricant and to dispose of it or use it in another way.

8.2.4 Continuous lubrication

For true power-transmission gears with enclosed housings, continuous lubrication is virtually the only solution, and it can be performed as splash lubrication, pump-assisted splash lubrication, or pressure-fed (spray) lubrication.

8.2.4.1 Splash lubrication

The provision of a lubricant reservoir in the gear housing, into which the gears dip, is the simplest method of ensuring that gears receive a continuous, uniform supply of lubricant. The lubricant can be consistent (mainly grease) or liquid (mainly mineral oil).

8.2.4.1.1 Grease splash lubrication

If grease is used as the lubricant, it must be borne in mind that it cannot remove much of the heat which is given off. To prevent the rotating gears cutting free and running dry, it is advisable mostly to fill the housing with grease; the casing should not be completely packed full, because of the roller bearings. The grease should have a certain amount of space for displacement. Fluid greases, which were developed specifically for gear lubrication, are particularly suitable.

The poor heat-removal properties of grease restrict its use to small-to-medium power transmission levels. As heat production and dissipation depend greatly on operating conditions (e.g., intermittent operation) and the structural design of the casing, it is not possible to indicate a set power limit for grease lubrication. If the power transmitted is not small, tests to check gear heat production are essential. The upper limit for peripheral velocity is 4 m/s.

8.2.4.1.2 Oil splash lubrication

Oil splash lubrication is the most common method of lubrication for toothed gear sets. It combines the advantages of good economy and simple implementation with those of reliable, continuous lubrication and a cooling effect which is satisfactory over a wide range. The latter can be effectively assisted by suitable housing design and additional measures, e.g., a cooling fan or water cooling coils in the housing (Figs. 8.3 and 8.4).

Effective oil splash lubrication should be so designed that at least one gearwheel of each range dips into the oil reservoir, if possible. If this is not possible, gears which do not dip into the oil must be lubricated by the oil which is splashed up. The same applies to roller bearings. In the case of gears which extend a long way vertically, it has proved to be useful to install intermediate levels, which form secondary oil troughs. These auxiliary oil troughs are usually supplied with oil splashed up from below, without the need for complex arrangements. The upwards supply of oil and run-off from the upper oil troughs provide temperature equalization in the gear set and optimize cooling.

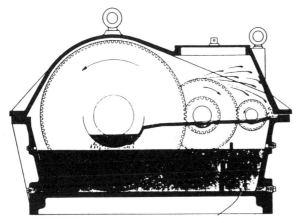

Fig. 8.3   Splash lubrication of enclosed gears

Oil splash lubrication has so far been used up to velocities of ca. × 15 m/s. In special cases, splash lubrication has also been used at higher peripheral velocities. In the upper speed range, oil deflectors have proved effective aids to lubrication and thermal balance in the gear. Recent investigations have shown that splash lubrication can work even with a peripheral speed of 60 m/s.

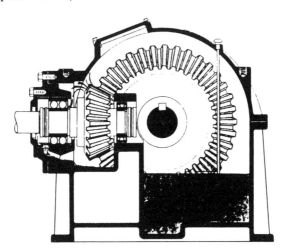

Fig. 8.4   Splash lubrication of enclosed gears with provision for bearing lubrication

In view of splash losses, the immersion depth of the gears is generally reduced as the peripheral speed increases. The following rule of thumb may be used:

immersion depth: 3–5 m up to 5 m/s

1–3 m up to 15 m/s

As adequate wetting of the tooth surfaces becomes more difficult to ensure as the peripheral speed rises and the oil level becomes lower during operation, the immersion depth of the gears must be increased at boundary peripheral speeds without considering the splash losses. Gears with fully immersed pinions (e.g., in bevel gears) are not rare. Great differences in gearwheel diameter in the individual steps may also make it necessary to deviate from the rule of thumb given above in certain circumstances.

The quantity of oil required for a filling is established as a function of tooth power loss as follows

$$Q = 3.0 \div 10.P_z \quad (l)$$

with

$$P_{z1} = P_{z1} \frac{0.1}{\cos \beta} + \frac{0.03}{v + 1} \text{ (kW)} \quad (v \text{ in m/s})$$

In the case of vehicle gearboxes, considerably smaller quantities of oil are used. Smaller quantities of oil mean shorter intervals between oil changes. As the interior of the gearbox is linked to the outside air and its humidity due to the ventilation necessary in connection with heat build-up, the water content of the oil increases during operation. Therefore, even where a gearbox contains a large quantity of oil, the oil should be changed at least once per year.

When splash lubrication is used, care should be taken to ensure that all friction points are supplied with lubricant. Special design measures are necessary to ensure that all stages of a multi-step transmission are lubricated (Fig. 8.5) to maintain a continuous oil supply to the bearings (Figs. 8.4, 8.6, and 8.7).

8.2.4.2   Pump-assisted splash lubrication
If measures such as cooling fins on the casing, cooling fans, or cooling coils in the casing are not sufficient to remove the waste heat from a

Oil collector

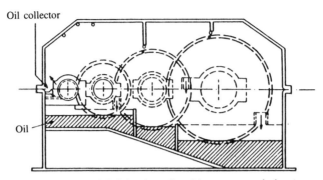

Fig. 8.5   Splash lubrication of multi-step transmissions

splash-lubricated transmission, the cooling effect can be increased considerably by using pump-assisted splash lubrication. In the simplest form, an oil pump driven directly by the gears removes an appropriate quantity of oil from the oil reservoir and passes it through a cooler and back into the gear casing.

Electrically driven pumps (as the main or back-up pumps) are also used. Thermostatically controlled coolers and oil filters in the circuit make it possible to extend oil change intervals.

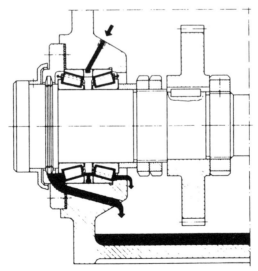

Fig. 8.6   Oil supply to bearings – 1

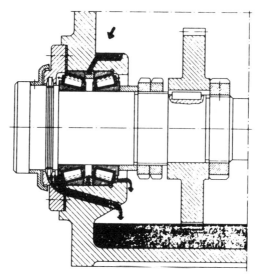

Fig. 8.7 Oil supply to bearings – 2

The immersion depth is determined as for simple splash lubrication. The quantity of oil should be somewhat greater

$$Q = 4.0 \div 12.P_z \quad (l)$$

8.2.4.3 Pressure-feed lubrication

In the velocity ranges in which splash lubrication is inadequate, and for most plain bearing transmissions, it is necessary to use pressure-feed lubrication with oil sprays at the tooth contact points. This method of lubrication may be used up to the maximum peripheral velocities which occur (used up to ca. 250 m/s). The tooth surfaces are supplied with oil by spray heads or port nozzles (Fig. 8.8). It is advisable to increase the oil pressure as peripheral speed increases. In practice, oil pressures of between 1 bar and 3.5 bar are mainly used. In special cases pressures of up to 10 bar are used. It is not necessary, however, to make the velocity of the oil jet equal to the peripheral velocity of the gears.

The oil can be sprayed before or after the tooth contact point. The former aids lubrication, while the latter aids cooling. The quantity to be sprayed is determined as follows

$$Q_e = \frac{P_z}{\vartheta_a - \vartheta_e} \quad (l/\text{min})$$

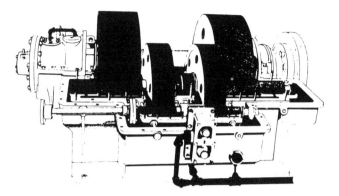

Fig. 8.8   Pressure-feed lubrication with feed to the gears and bearings

Recent, as-yet unpublished studies by the Institute for Engineering Design and Transmissions (Institut für Maschinenkonstruktion and Getriebebau) of the University of Stuttgart show, however, that considerably smaller quantities of oil are adequate to lubricate the tooth contact points. This quantity of oil only takes account of the gears. The oil requirement of the bearings must be determined separately. The total quantity of oil to be pumped can be found with

$$Q_{ges} = 40 \frac{P_{vges}}{\vartheta_a - \vartheta_e} \quad (l/min)$$

It is advisable to maintain the following temperature limits

$$\vartheta_e \leqslant 60°C; \quad \vartheta_a \leqslant 90°C; \quad \vartheta_a - \vartheta_e \leqslant 30°C$$

8.2.4.4   Splash lubrication at higher peripheral speeds
For reasons of design simplicity it is normal to try to lubricate meshing gears with splash lubrication, and to use spray lubrication only at high speeds, when the danger of the oil being thrown off means that adequate lubrication of the meshing tooth surfaces cannot be guaranteed. As can be seen in Fig. 8.9, peripheral speeds of 15–20 m/s² have been used as the limit for splash lubrication in the past (**123**).

Recent investigations have shown, however, that a reliable supply of lubricant to the tooth contact area is possible at peripheral speeds of up to 60 m/s (**123**). Depending on the location of the gear pair and the

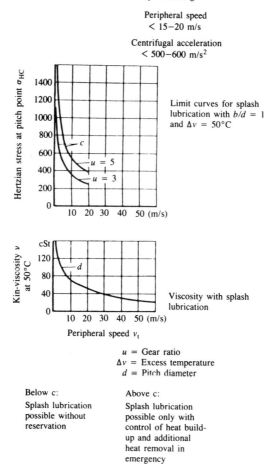

Fig. 8.9 **Limits of splash lubrication – state of the art** (123)

driving gearwheel, however, care must be taken to ensure that the oil-supply gear is immersed to a sufficient depth, which may reach values of up to 25 times the module. Figure 8.10 gives some recommendations on this point.

It must, however, be borne in mind that with deep immersion, temperatures may rise considerably due to the higher splash losses. The result is premature ageing due to the formation of oxidation products.

Recommended immersion depths (with m < 5 mm, minimum immersion depth must be $e = 12$ mm)

Arrangement	Drive	Hertzian stress (N/mm^2)	Peripheral speed range (m/s)	Immersion depth (mm)
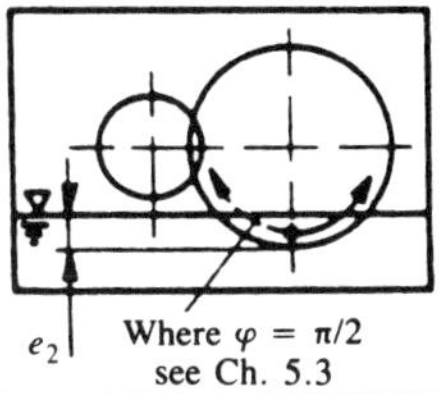 e_2 — Where $\varphi = \pi/2$ see Ch. 5.3	Gear driving	$\sigma_{HC} \leq 800$	$20 \leq v_t \leq 35$	$e_2 = 2.5$ m
			$35 < v_t \leq 60$	$e_2 = 5$ m
		$\sigma_{HC} = 1434$	$20 \leq v_t \leq 35$	$e_2 = 6$ m
			$35 < v_t = 60$	$e_2 = 12$ m
	Gear driving	$\sigma_{HC} \leq 800$	$20 \leq v_t \leq 60$	$e_2 = 2.5$ m
		$\sigma_{HC} = 1434$	$20 \leq v_t \leq 60$	$e_2 = 6$ m
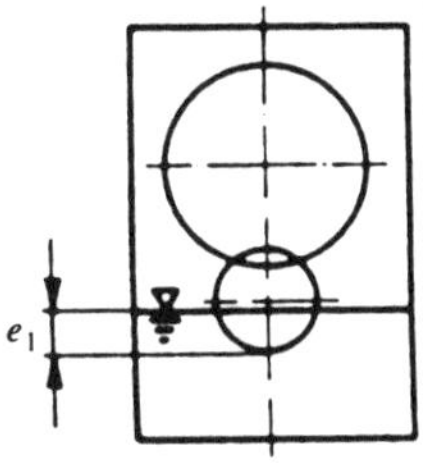 e_1	Pinion driving	$\sigma_{HC} \leq 800$	$20 \leq v_t \leq 60$	$e_1 = 2.5$ m
		$\sigma_{HC} = 1434$	$20 \leq v_t \leq 35$	$e_1 = 6$ m
			$35 < v_t \leq 60$	$e_1 = 15$ m
	Gear driving	$\sigma_{HC} \leq 800$	$20 \leq v_t \leq 35$	$e_1 = 6$ m
			$35 < v_t \leq 60$	$e_1 = 10$ m
		$\sigma_{HC} = 1434$	$20 \leq v_t \leq 35$	$e_1 = 10$ m
			$35 < v_t \leq 60$	$e_1 = 18$ m
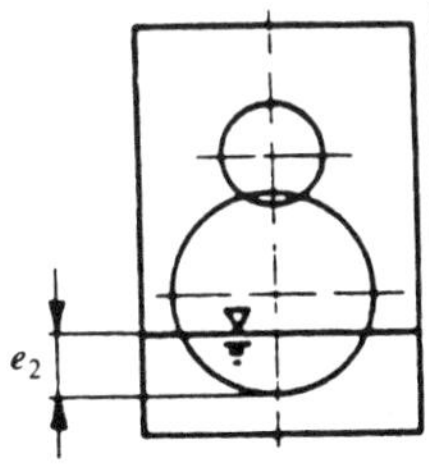 e_2	Pinion driving	$\sigma_{HC} \leq 800$	$20 \leq v_t \leq 35$	$e_2 = 6$ m
			$35 < v_t \leq 60$	$e_2 = 18$ m
		$\sigma_{HC} = 1434$	$20 \leq v_t \leq 35$	$e_2 = 18$ m
			$35 < v_t \leq 60$	$e_2 = 25$ m
	Gear driving	$\sigma_{HC} \leq 800$	$20 \leq v_t \leq 60$	$e_2 = 6$ m
		$\sigma_{HC} = 1434$	$20 \leq v_t \leq 60$	$e_2 = 21$ m

$a = 180$, $u = z_2/z_1 = 1.5$

$m = 4.5$

$b = 30$ mm with Hertzian stress $\sigma_{HC} = 800$ N/mm^2

$b = 20$ mm with Hertzian stress $\sigma_{HC} = 1434$ N/mm^2

Fig. 8.10 Recommended immersion depths for higher peripheral velocities (123)

8.3 DESIGN OF OIL SUPPLY SYSTEMS

In its simplest form, the oil supply system for a transmission with pressure feed consists of an oil reservoir, an electrically or mechanically driven oil pump, an oil filter, and an oil cooler. The capacity of the oil reservoir must be such that there is sufficient resting time for the circulated oil for air separation in the oil reservoir and that the oil has a suitable service life. In general the capacity is set as $Q = 5Q_{eges}$ (l). This means that the oil circulation time $T = Q/Q_{eges}$ is 5 min. If there is a shortage of space, e.g., in mobile systems, Q should be made smaller and an air separator should be incorporated.

Large oil-supply systems should be fitted with additional assemblies: an electric back-up pump (possibly to be used as a starter pump where the main pump is driven mechanically by the transmission), a pressure limitation valve, a suction valve, a by-pass filter (for changing a filter during operation), automatic control for the oil cooler, and possibly a heater to achieve operational preparedness quickly at low ambient temperatures (Fig. 8.11).

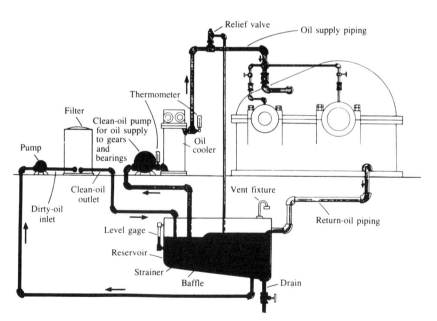

Fig. 8.11 Pressure-feed lubrication – central supply system

The safety of the whole system is ensured by: oil level measurement in the reservoir, pressure measurement at the oil supply unit and at the transmission, differential pressure measurement at the oil filter, oil flow monitoring between the oil supply unit and the transmission, temperature measurement before and after the oil cooler, immediately before the transmission, in the oil return and in the oil reservoir. All measure-

Problems

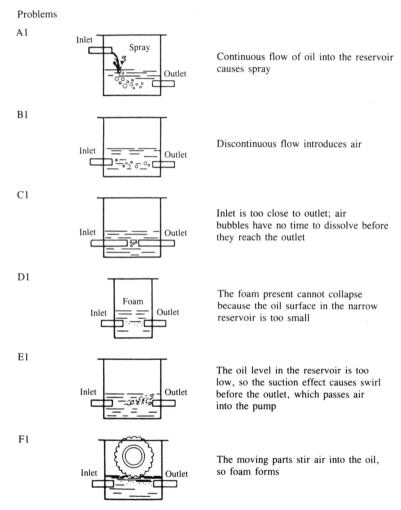

A1

Continuous flow of oil into the reservoir causes spray

B1

Discontinuous flow introduces air

C1

Inlet is too close to outlet; air bubbles have no time to dissolve before they reach the outlet

D1

The foam present cannot collapse because the oil surface in the narrow reservoir is too small

E1

The oil level in the reservoir is too low, so the suction effect causes swirl before the outlet, which passes air into the pump

F1

The moving parts stir air into the oil, so foam forms

Fig. 8.12(a)   Problems in the design of oil inlets and outlets

Solutions

A2

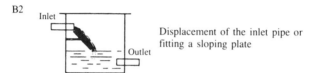

Displacement of inlet pipe or fitting a sloping plate, which permits 'gentle' return of the oil. A sieve or wire mesh can also slow the flow and prevent spray

B2

Displacement of the inlet pipe or fitting a sloping plate

C2

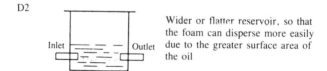

Displacement of the inlet and outlet pipes or fitting a baffle between the inlet and outlet pipes

D2

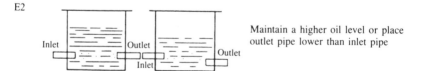

Wider or flatter reservoir, so that the foam can disperse more easily due to the greater surface area of the oil

E2

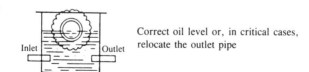

Maintain a higher oil level or place outlet pipe lower than inlet pipe

F2

Correct oil level or, in critical cases, relocate the outlet pipe

Fig. 8.12(b)   Solutions in the design of oil inlets and outlets

ment and monitoring equipment should be fitted with contact sensors which actuate an alarm if deviations from the set point values with permissible variation are detected, and switch on back-up pumps and filters, etc. (where fitted), and switch off the plant or prevent start-up if limits are not complied with.

When designing oil inlets and outlets for the oil reservoir, measures must be taken to ensure that air cannot swirl through the oil and that foam formation is not excessive. Figure 8.12(a) shows some problems, and Fig. 8.12(b) suggests suitable solutions.

8.4 OIL CHANGE INTERVALS

The frequency with which oil must be changed depends on the mechanical and thermal load as well as on the quantity of oil being circulated and the oil maintenance, that is, the presence or absence of oil filters, separation devices for impurities etc. Criteria for the necessity of an oil change are the solid impurity content and the reaching of a certain saponification number, which is a measure of the degree of ageing of the oil. The latter applies exclusively to undoped gear oils, that is, to oils which contain no extreme-pressure additives. The limits are 0.2 percent solid contaminants and a saponification number of 3 mg KOH/g oil. In each case, this requires periodic analysis of the oil in the transmission; this is a complex measure which is only worthwhile in the case of large-capacity systems, mainly pressure-feed systems. Usually the first analysis is performed after about 5000 h operation, and then repeated every 1500 h operation. In the case of splash lubrication, the oil change intervals are usually set purely empirically on the basis of practical experience. With industrial transmissions it is between 250 and 2500 h operation after the completion of running-in, but service lives of 5000 to 7500 h operation are possible even with undoped oils. When doped oils are used, these values can be exceeded. In the case of vehicle transmissions, oil change intervals are generally linked to the distance covered. Oil change intervals of 20 000 km are quite normal.

8.5 COMMISSIONING AND RUNNING-IN

Before commissioning, the transmission casing and all areas of the oil circuit should be cleaned. This can be done by flushing with a low-viscosity flushing oil. Care must be taken to ensure that the gear set is only subject to part load, to avoid tooth surface damage.

After new gears are commissioned, local overloads must be expected due to production-related gear-cutting errors. Generally the desired surface finish on which the calculation of load capacity is based is achieved only during running-in. This applies especially to large gear-wheels.

To prevent damage due to local overload on the tooth flanks, the load must be applied carefully during the running-in process, that is, until the roughness on the tooth flanks has been worn off or smoothed down. Shock loads should be avoided. As the contact pattern slowly becomes larger, the rated load can gradually be applied. This running-in process may take quite a long time. It can, therefore, be beneficial to use so-called running-in oils to improve the contact pattern more quickly. They cause increased, but intentional, local wear, which makes the tooth surface smoother much more quickly than using gear oils intended for continuous operation. They are special oils which easily react chemically and chemico-physically with the flank surfaces and thus allow removal of metal, without any possibility of scoring or any other type of flank damage. When the desired contact pattern has been achieved, the running-in oils should be replaced with continuous-service gear oils to prevent further chemically induced wear during subsequent gear operation. The running-in process can be aided by copper-coating or phosphatizing, with or without the use of a special running-in gear oil.

9 Gear and Transmission Failure

9.1 INTRODUCTION

Losses due to damage to machine elements are classed as tribologically induced losses. This means that, in the final analysis, they are caused by friction and wear. In a recently completed study (124), the extent of these losses is estimated as follows

Germany (former FRG)	c. DM 14–17 billion pa
USA	c. DM 100 billion pa
UK	c. DM 8.5 billion pa

Rigorous application of tribological knowledge can produce the following savings

Germany (former FRG)	c. DM 2 billion pa
USA	c. DM 10 billion pa
UK	< DM 1 billion pa

A prerequisite for achieving these savings is a reduction in the frequency of failure of lubricated machine components. This requires failure analysis, however.

Increasing mechanization and automation thus makes ever-growing demands on the reliable operation of all machine parts. This applies to gears and transmissions.

The following high costs result from interruption of operation due to gear failure:

- the cost of repair or replacement of gears;
- consequential costs, e.g., loss of production due to downtime.

Some of the failures can be traced back to tribological or tribotechnical circumstances.

9.2 FAILURE STATISTICS

Figures 9.1 and 9.2 (125)(126) shows the results of failure evaluation of turbine gears. While, in the case of turbine spur gears, the gear set accounted for only 31 percent of the failures, in the case of turbine planetary gears it was the major element, accounting for 39 percent. Figure

442   *Lubrication of Gearing*

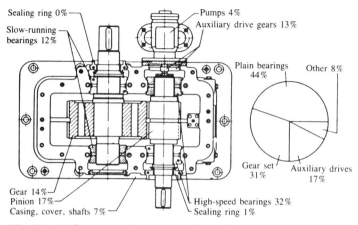

Fig. 9.1   Failure rates of components in turbine spur gears (124)(125)

9.3 shows the reasons for these failures compared with a gear manufacturer's assessment. It can be seen that about 50 percent of all transmission damage can be detected by improved monitoring, that is, by inspection of gears and bearings, before serious damage occurs.

As gear failures accounted for 40–60 percent of transmission failures, they were investigated separately. They are divided into 'causes of damage' and 'failure modes' (Table 9.1).

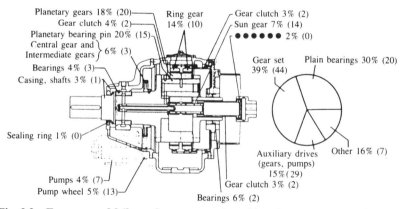

Fig. 9.2   Frequency of failure of components in turbine planetary gears with rotating sun wheel. The corresponding values for planetary gears of all designs by various manufacturers are given in parentheses (124)(125)

Fig. 9.3 Causes of failure in turbine gears according to statistics of Allianz-Versicherungs-AG compared with assessment of a gear manufacturer (values in parentheses (125))

Table 9.1 Survey of causes of failure and failure modes for gears (125)(126)

Cause of failure (gears)	No. of failures Distribution (%)	Failure modes (gears)	No. of failures Distribution (%)
Product failure	41	Sudden fracture	56
Planning and design error	11	Fatigue fracture	17
		Changes in contact pattern	16
Material failure	8		
Overhaul error	4	Incipient cracks	7
Assembly error	10	Deformations	4
Production error	8		
Operating error	40		
Maintenance error	21		
Control error	19		
Outside influences	19		
of which			
foreign bodies	6		
input and output machines	10		
Overload due to mains switching	3		

Table 9.2 **Distribution of causes of failure and failure locations in stationary transmissions** (125)(126)

Cause of failure	Frequency (%)	Location	Frequency (%)
Product failure	53.7	Gears	58.2
Production error	17.4	Bearings,	12.5
Planning, design,	15.4	radial	11.1
and analysis error		axial	1.4
Material failure	10.9	Housing	9.7
Assembly error	7	Shafts	6.4
Overhaul error	3	Running gear	5.7
Operating error	39.3	Other	7.5
Control error	21.4		
Maintenance error	7.9		
Outside influences	7		
Foreign bodies	2.5		
Mains	1.5		
Other external	3		

An evaluation of the damage to the insured stock of transmissions by a machinery insurer gives the distribution of causes of failure and failure locations in Table 9.2 (**125**)(**126**).

The dominant significance of product faults at 53.7 percent is obvious. Furthermore, it was found that the gears in transmissions are in fact the machine components which are most susceptible to failure. Table 9.3

Table 9.3 **Most important conditions for failure mechanism and failure mode in stationary transmissions** (125)(126)

Failure mechanisms	Failure frequency (%)	Failure modes	Failure frequency (%)
Mechanical overload	41.0	Sudden fracture	47.8
Seizing	12.8	Fatigue, creep	12.5
Erratic operation	10.3	fracture	
Lack of lubricant	10.3	Change in contact	9.5
Loosening	9.4	pattern	
Deformation	5.1	Scoring	8.8
Thermal overload	3.9	Mechanical	6.5
Wear	3.0	surface damage	
Other	4.2	Deformation	3.7
		Abrasion	3.7
		Incipient cracks	2.0

contains a survey of failure mechanisms and failure modes. The predominance of mechanical overload at 41 percent and seizing at 11.8 percent is reflected in the high fracture rate of 60.4 percent in total.

9.3 SYSTEMS FOR CLASSIFYING GEAR FAILURE

There are various classification systems for the various types of failure. It would seem most appropriate to class them according to the failure mode, and, after its identification, to analyse the course of failure. Frequency, however, failures are classed and arranged according to the cause of failure or the failure mechanism. In standardizing types of failure, the two classification systems may be mixed.

9.3.1 Classification according to failure mode
A system for classing failures according to failure mode contains the following main types of failure:

- wear evidence;
- deformations;
- fatigue evidence;
- cracks;
- fractures;
- corrosion;
- other (erosion, cavitation, current passage etc.).

9.3.2 Classification according to failure process or cause
A system for classing failures according to failure process or cause contains the following main types of failure.

- Failures due to manufacture:
  - design errors;
  - material failure;
  - production errors;
  - assembly errors.
- Failures due to operation:
  - lubrication failure;
  - overload;
  - outside influences.

9.3.3 Standardization
To facilitate failure analysis, DIN 3979 was devised for cross-referencing failure mode, cause of failure, and remedies (Table 9.4).

Table 9.4 Tooth failures in toothed gears to DIN 3979

Tooth flank damage
Wear failure:
– normal wear
– abrasive wear
– wear due to meshing interference
– scratches
– grooves
– scoring
Fatigue damage:
– pitting
  – initial pitting
  – progressive pitting
– spalling
Deformation:
– indentations
– rippling
– rolling and peening
– hot flow
Cracks in teeth:
– quenching cracks
– material cracks
– grinding cracks
– incipient fatigue fractures
Corrosive damage:
  chemical corrosion, rust
– abrasive corrosion
– scaling
Annealing:
– erosion, cavitation
– current passage

Tooth fractures
Sudden fracture
Fatigue or endurance fracture

9.4   FAILURE ANALYSIS PROCEDURES

9.4.1   Aims

In order to keep operational interruptions and unexpected costs due to premature failure of machine components to a minimum, the following aims may be set.

Basic avoidance of damage:
design measures
preventive maintenance
Prevention of repetition of damage:
failure analysis
The necessary measures can be described as follows.

Design measures:
Design with lubrication and maintenance in mind, taking account of tribological requirements.

Preventive maintenance:
Checks and monitoring to detect potential failure points, remove causes of failure mechanisms, prompt replacement of parts.

Failure analysis:
Establish failure mode, reconstruct failure process, find out cause of failure.

The following comments relate primarily to failure analysis.

9.4.2 Procedure

The procedure for investigating a failure contains the following three stages:

- identification of failure mode;
- reconstruction of failure process;
- identification of causes of failure.

It must be borne in mind that several groups of causes (cause) may be responsible for each failure mode, and that several individual causes (reason) may be responsible for each group of causes. Figure 9.4 clarifies this.

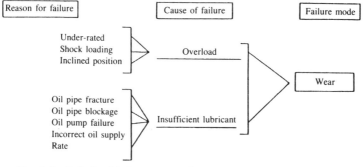

Fig. 9.4 Relationship between mode, cause, and reason for failure

9.5   THE EFFECT OF THE LUBRICANT ON GEAR FAILURE

There are several ways in which the lubricant can exert an influence on the occurrence or prevention of damage to gears and transmissions, which may be classed in the following groups in a simplified manner (**127**):

(a)   wrong lubricant;
(b)   wrong type of lubrication;
(c)   excess lubricant;
(d)   insufficient lubricant;
(e)   wrong method of lubricant application;
(f)   contaminated lubricant.

(a)   Incorrect lubricant
By this we mean the selection of a lubricant which is incorrect in type or kind, in quality or viscosity, for a given application, which can then cause failure in certain circumstances. An incorrect lubricant can cause tooth flank damage, such as wear, scoring and pitting.

(b)   Incorrect type of lubrication
Machine components can also be damaged if the limits for splash, pressure-feed (spray), and atomization lubrication are not observed. Often there is a failure to consider that gears may not receive sufficient oil at higher peripheral speeds. The result can be flank damage caused by insufficient oil.

(c)   Excess lubricant
It is a fact that if the oil level in the gear casing is too high, increased splash losses can upset the thermal balance of the machine and eventually contribute to damage to machine components. Overheating damage can be the result.

(d)   Insufficient lubricant
Undoubtedly, an insufficient lubricant is more damaging than an excess of lubricant. An adequate quantity of lubricant must reach a friction point in a certain unit of time in order to form a hydrodynamic or elasto-hydrodynamic contact film, and to remove heat. If this does not occur, the result can be tooth flank damage as well as tooth deformation due to overheating.

(e) *Incorrect supply of lubricant*

This group includes failures caused by incorrect design of the 'internal lubrication system', that is, incorrectly designed and arranged oil supply passages, grooves, and pockets. This type of damage is found repeatedly in plain bearings.

(f) *Contaminated lubricants*

Solid and liquid impurities can cause gear and transmission damage. Solid contaminants include abraded metal particles, dirt particles which have come in from outside, and also certain oil-insoluble reaction products of the ageing process of the gear oil. Liquid contaminants include products of oil ageing, often acidic, and water of condensation, as well as liquids which have come in from outside, such as water, acids, alkalis etc.

As the failures listed under (b), (c), and (e) are due to design inadequacies, and the failures described under (c), (d), and (f) are primarily the result of inadequate maintenance, they cannot be blamed on the lubricant. Thus only the factors mentioned under point (a) are causes of failure which are directly due to the lubricant.

9.6 RELATIONSHIP BETWEEN CAUSE OF FAILURE AND FAILURE MODE

The most important types of gear failure and their causes are summarized in Fig. 9.5. The dominant importance which the lubricant can have as a cause of failure is obvious. Incorrect viscosity or unsuitable oil quality, contamination, and insufficient oil can cause the typical wear and scoring, as well as other surface damage on the teeth, such as corrosion. If we disregard, for the moment, unsuitable operating conditions and manufacturing errors, which play an important part in the case of roller bearings and plain bearings, we can state, with certain reservations, that the greatest demands are made on the quality of gear lubricants compared with the situation relating to bearings.

As there are almost always areas of mixed lubrication on the tooth surface due to geometric factors, operating conditions, and the kinematics of tooth engagement, the EP (extreme pressure) performance of a lubricant is usually very important. Comparing the failure modes and failure causes for plain bearings, roller bearings, and gears in this schematic way naturally involves the danger of over-simplification. Furthermore, this representation cannot claim to be comprehensive, even if the most important aspects are certainly demonstrated. The aim of this

Cause of failure \\ Failure mode	Tooth fractures		Wear						Loss of material			Cracks				Deformation				Corrosion		Other damage				
	Sudden fracture	Fatigue fracture	Normal wear	Abrasive wear	Wear due to meshing interference	Scratches	Grooves	Scoring	Pitting	Spalling	Flaking	Grinding cracks	Hardening cracks	Material cracks	Incipient fatigue cracks	Indentations	Rippling	Hot flow	Cold flow	Chemical corrosion	Friction corrosion	Cavitation	Erosion	Current passage	Scaling	Annealing
Material failures — Slag inclusions	●								●				●	●												
Forging seams	●												●	●												
Non-metallic inclusions														●												
Unsuitable material pairs								●	●											●	●					
Design faults — Under-rating	●								●	●					●				●							
Incorrect tooth geometry				●				●																		
Meshing interference				●																						
Incorrect tooth clearance								●																		●
Production faults — Forging faults	●																									
Heat production too great during machining								●				●														
Inappropriate heat treatment								●				●	●						●							●
Assembly faults — Poor surface quality			●					●	●																	
Out of true			●					●	●	●					●											
Insufficient insulation																								●		
Operating conditions — Frequent load changes	●								●																	
Shock, vibration loading	●	●							●	●						●		●		●	●					
Overload	●							●	●	●	●					●		●	●							●
Incorrect running-in								●	●																	
Speed too high/too low								●	●	●								●								
Lubrication faults — Insufficient lubricant								●										●	●	●						●
Incorrect viscosity								●	●	●	●							●				●				●
Poor quality								●	●	●	●							●								
Contaminant, solid/liquid				●				●	●											●			●	●	●	
Oil supply																									●	

Fig. 9.5 Gear failures and their causes

survey is to make a contribution to facilitating differentiated treatment of failure of machine components, because only if the failure modes can be linked fairly definitely to the relevant causes is it possible to take measures to prevent a repetition. In addition, only then is it possible to identify the actual role of the lubricant in the occurrence of the failure.

In the case of sets of gears, gearwheels which have been manufactured without defects must be assembled correctly to ensure good contact patterns, which are a prerequisite for long-term fault-free operation. Figure 9.6 shows possible contact pattern defects, which can be related to the causes given in Fig. 9.7 (125)(126).

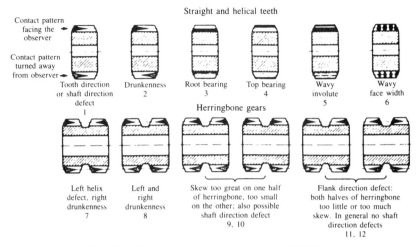

Fig. 9.6   Typical contact pattern defects (125)(126)

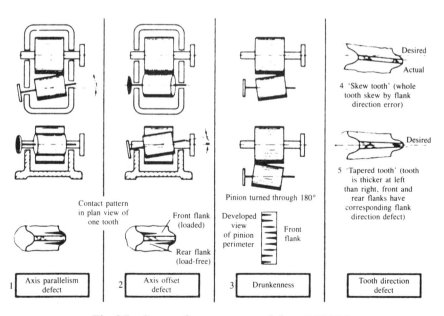

Fig. 9.7   Causes of contact pattern defects (125)(126)

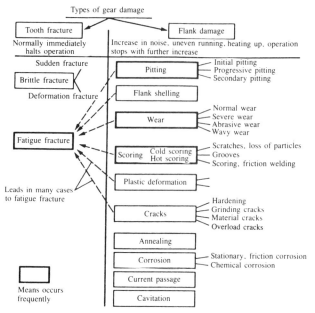

Fig. 9.8 Summary of types of damage occurring on gears (125)(126)

Figure 9.8 gives a frequently used summary of the types of damage found on gears (125)(126). It can be seen that most types of damage finally cause fatigue fractures.

9.7 EXAMPLES OF GEAR FAILURE

The above systematic observations on the classification of the various types of failure are explained below with the aid of characteristic examples of gear failure (126)–(129).

9.7.1 Classification by failure mode

9.7.1.1 Systematics

Table 9.5 shows the systematics of gear failures, arranged by failure mode. The causes of failure are assigned to these failure modes in accordance with Table 9.6.

9.7.1.2 Examples

Examples of gear damage, classified according to failure mode, are shown in Figs 9.9–9.48 (see pages 456–490).

Table 9.5 Gear damage – classification by failure mode

Tooth fracture
Sudden fracture
Fatigue fracture

Tooth flank damage
Wear
  – normal wear
  – abrasive wear
  – wear due to meshing interference
  – scratches
  – grooves
  – scoring
Loss of material
  – pitting
  – spalling
  – flaking
Cracks
  – grinding cracks
  – hardening cracks
  – material cracks
Deformation
  – indentations
  – rippling
  – hot flow
  – cold flow
Corrosion
  – chemical corrosion
  – friction corrosion
Other damage
  – cavitation
  – erosion
  – scaling
  – annealing
  – current passage

9.7.2 Classification according to failure process or cause

9.7.2.1 Systematics
Table 9.7 shows the systematics of gear damage, arranged according to failure process or cause.

9.7.2.2 Examples
Examples of gear damage, arranged according to failure process or cause, are shown in Figs. 9.36–9.48 (see pages 478–490).

Table 9.6   Causes of gear damage

Material defects
- inclusions
  - slag
  - non-metallic substances
- forging seams
- unsuitable materials pairs

Design faults
- under-rating
- incorrect tooth geometry
- meshing interference
- incorrect tooth clearance

Production defects
- forging defects
- thermal overload during machining
- unsuitable heat treatment
- poor surface quality

Assembly defects
- out of true
- inadequate insulation

Operating conditions
- frequent load changes
- shock and vibration loading
- overloading
- inappropriate running-in
- speed too high, too low

Lubrication defects
- incorrect type of lubrication
- type of oil supply
- insufficient lubricant
- excess lubricant
- incorrect viscosity
- incorrect type of lubricant
- contaminants
  - solid
  - liquid

Table 9.7 **Gear failure – classification according to cause and mechanism**

Failure due to manufacture
Material defects
– slag inclusions
– forging seams
Production defects
– hardening cracks
– grinding cracks
– production grooves
– hardening defects
Assembly defects
– insufficient straightening, out of true
– insufficient insulation
Design defects
– underrating
– incorrect tooth geometry
– meshing interference
– incorrect tooth clearance

Failure due to operation
Lubrication defects
– incorrect type of lubrication
– incorrect type of lubricant supply
– incorrect lubricant
  – type
  – quality
  – viscosity
– contaminated lubricant
  – solid
  – liquid
– insufficient lubricant
– excess lubricant
Overloads
– rated load too high
– too many load changes
– shock and vibration loads
Outside influences
– current passage

Fig. 9.9   Gears/tooth fractures
Gear type: straight spur gear
Failure type
  general: fractures
  specific: sudden fractures (brittle fracture)
Features,   over the whole cross-section, rough and fissured frac-
  appearance: ture surface (brittle fracture) of sometimes bead
                   formed over whole fracture surface (deformation
                   fracture)
Causes: severe, often brief overload, e.g., due to seizing of
                   gears

Fig. 9.10   Gears/tooth fractures

Gear type: case-hardened bevel gear pinion
Failure type
  general: fractures
  specific: fatigue fractures (almost 100 percent fatigue fracture)
Features,   fracture surface with two zones – fatigue fracture
  appearance: surface and sudden fracture surface. Fatigue fracture
                   surface appears fine-grained, and is finer grained
                   with slower fracture.
Causes: repeated load reversals; the fracture propagates from
                   a crack, bore etc., until the remaining cross-section is
                   inadequate.

Fig. 9.11 Gears/wear

Gear type:	Case-hardened helical spur gear with shaved teeth
Damage type	
general:	wear
specific:	normal wear
Features,	
appearance:	asperities worn off and surface structure disappeared
Causes:	mixed friction on the tooth flank, particularly during running-in

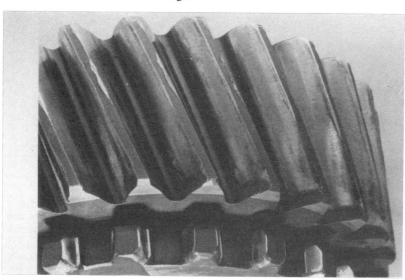

Fig. 9.12 Gears/wear

Gear type:	helical spur gear
Damage type	
general:	wear
specific:	abrasive wear
Features,	uniform abrasion of flank surface, so that original
appearance:	structure is largely destroyed. Mat appearance
Causes:	solid foreign bodies, such as sand, abraded metal etc. in the lubricating oil

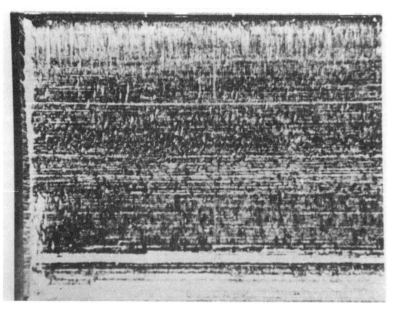

Fig. 9.13 Gears/wear

Gear type: quenched and tempered spur gear
Damage type
  general: wear
  specific: wear due to meshing interference
Features,   shaving marks running vertically on the tooth, later,
  appearance: rounding of the tooth edge
Causes: defects in tooth geometry or centre distance less than
                  specified

Fig. 9.14 Gears/wear

Gear type: case-hardened straight spur gear

Damage type

  general: wear

  specific: severe scratches

Features, individual streak-like depressions at irregular inter-

  appearance: vals in the direction of sliding

Causes: solid contaminants, such as dust, abraded material,
  rust, scale, mould sand etc. in the lubricating oil

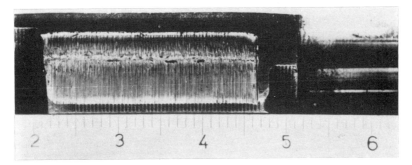

Fig. 9.15 Gears/wear

Gear type:	case-hardened straight spur gear
Damage type	
general:	wear
specific:	grooves
Features, appearance:	smooth, streaky depressions, which, unlike scratches, extend from the tip to the root and have a peak-to-valley height of 3–5 μm
Causes:	high load on the tooth flank. Foreign bodies in the oil increase the stress

Fig. 9.16 Gears/wear

Gear type:	helical spur gear
Damage type	
general:	wear
specific:	scoring
Features, appearance:	streaky roughness of varying width and depth extending in the vertical direction on the tooth, and more or less covering the tooth width
Causes:	combined effect of high specific load and high speed; reinforced by unsuitable lubricant or insufficient oil

Fig. 9.17 Gears/shelling

Gear type: helical spur gear

Damage type

  general: shelling

  specific: pores/grey spots

Features, mat, spotted areas of fine pits

  appearance:

Causes: oil viscosity too low or unsuitable additives in lubricant

Fig. 9.18   Gears/shelling

Gear type: soft-nitrided straight spur gear

Damage type

  general: shelling

  specific: pitting

Features,

  appearance: incipient destruction of the tooth flank by shelling

Causes: fatigue of the material due to overloading

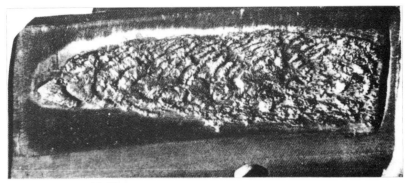

Fig. 9.19   Gears/shelling

Gear type: case-hardened, ground helical spur gear
Damage type
  general: shelling
  specific: pitting
Features,   total destruction of the tooth flanks due to extensive
  appearance: shelling propagating from pits
Causes: overloading due to one-sided contact

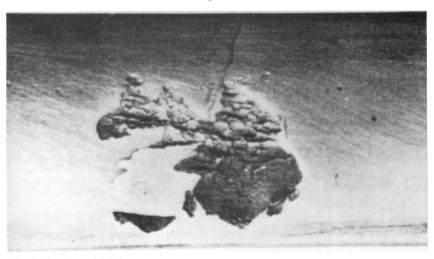

Fig. 9.20   Gears/shelling

Gear type: straight spur gear, case-hardened and ground
Damage type
  general: shelling
  specific: spalling
Features,
  appearance: local surface shelling from the flank
Causes: brittle behaviour of the material

465

Fig. 9.21   Gears/shelling

Gear type:	case-hardened, shaved straight spur gear
Damage type	
general:	shelling
specific:	spalling
Features,	surface shelling, not very deep
appearance:	
Causes:	overloading of material surface during machining (cracks form under the surface)

Fig. 9.22   Gears/shelling

Gear type:	case-hardened, straight spur gear, generated by planing
Damage type	
general:	shelling
specific:	flaking
Features, appearance:	local tooth fractures
Causes:	sudden overloading due to gear-changing error

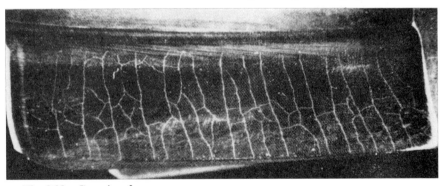

Fig. 9.23   Gears/cracks

Gear type:	case-hardened straight spur gear
Damage type	
general:	cracks
specific:	grinding cracks
Features, appearance:	fine network of linked cracks on the flank surface
Causes:	incorrect grinding (overheating)

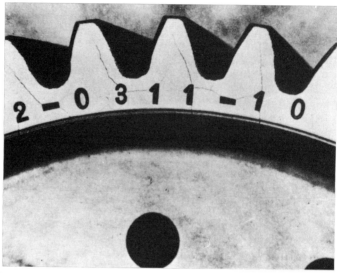

Fig. 9.24 Gears/cracks
Gear type:	case-hardened gear
Damage type	
general:	cracks
specific:	hardening cracks
Features,	linear cracks, often extending over large areas of the
appearance:	gear
Causes:	incorrect hardening. Transformations between surface zones and core zones, displaced by time, during air cooling

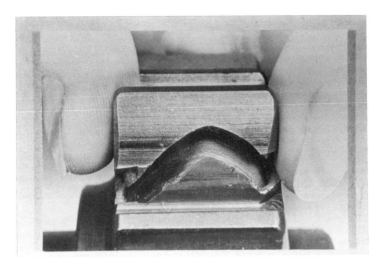

Fig. 9.25 Gears/cracks

Gear type: straight spur gear
Damage type
 general: cracks
 specific: material cracks
Features,   isolated crack across the whole cross-section of the
 appearance: tooth or extending to a considerable depth
Causes: inclusions of foreign bodies in the material or forging
      seams

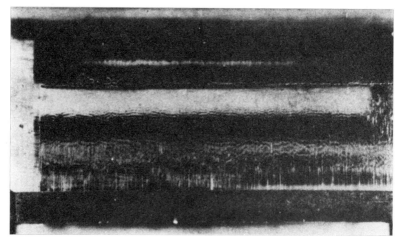

Fig. 9.26   Gears/deformations

Gear type: case-hardened spur gear

Damage type

  general: deformations

  specific: rippling

Features,   wavy surface variation vertical to the sliding direc-

  appearance: tion

Causes: inadequate lubrication at high specific loads and/or vibration

Fig. 9.27   Gears/deformations

Gear type: case-hardened spur gear

Damage type

  general: deformations

  specific: cold flow

Features,   flattening due to plastic deformation and squeezing

  appearance: of the material with formation of ridges, shelling and rolling out of material

Causes: very high constant loads, coupled with inadequate surface hardening

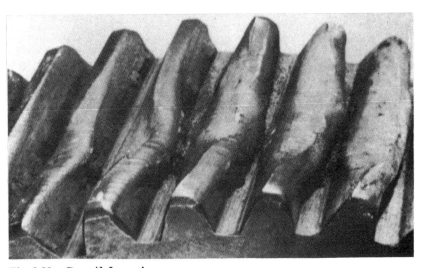

Fig. 9.28 Gears/deformations

Gear type: case-hardened helical spur gear
Damage type
  general: deformations
  specific: hot flow
Features,   whole teeth or parts of teeth become soft and assume
  appearance: a doughy state, become deformed and lose their orig-
    inal shape
Causes: severe overheating due to lack of oil, e.g., if lubrica-
    tion stops

Fig. 9.29　Gears/corrosion
Gear type:	internal change gear
Damage type	
general:	corrosion
specific:	chemical corrosion
Features,	pitted, uniform depressions over the whole surface of
appearance:	the flanks
Causes:	action of chemically active substances on the flank,
	e.g., acidification of the lubricating oil

Fig. 9.30 Gears/corrosion

Gear type:	toothed clutch
Damage type	
general:	corrosion
specific:	friction corrosion
Features,	reddish brown areas and uniform powdery abraded
appearance:	areas on the flank
Causes:	vibratory motion under load in the presence of oxygen without a separating lubricant contact film

Fig. 9.31 Gears/other damage

Gear type: gear pinion
Damage type
  general: other damage
  specific: cavitation
Features,   uniformly distributed, locally delimited erosion or
  appearance: sand-blasted appearance of flank
Causes: vibratory load in high-speed gears, coupled with
                  water and gases in the lubricating oil

Fig. 9.32   Gears/other damage

Gear type:	spur gear
Damage type	
general:	other damage
specific:	erosion
Features,	locally delimited erosion
appearance:	
Causes:	high-pressure impingement of the oil jet on the flank in high-speed gears, coupled with foreign bodies in the oil

Fig. 9.33   Gears/other damage

Gear type:	case-hardened spur gear
Damage type	
general:	other damage
specific:	scaling
Features,	raised marks with a metallic sheen
appearance:	
Causes:	oxidation processes on the tooth flank during heat treatment

Fig. 9.34 Gears/other damage

Gear type: helical spur gear

Damage type

  general: other damage

  specific: annealing

Features,   severe discoloration of the flank with groove-like

  appearance: depressions in the direction of the sliding motion

Causes: overheating caused by severe friction due to overloading, too high a speed, too little flank clearance,
incorrect lubrication

Fig. 9.35 Gears/other damage

Gear type:	planetary gear
Damage type	
general:	other damage
specific:	current passage
Features,	numerous small craters, sometimes also larger burn
appearance:	marks, annealing marks distributed over the flanks
Causes:	electrical potentials between the gears, caused by potentials of shafts in electrical machines, lead to sparking

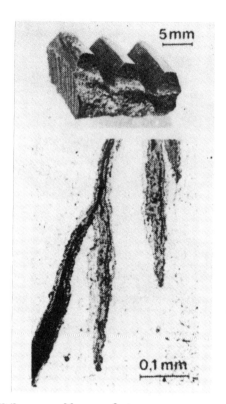

Fig. 9.36   Gears/failure caused by manufacture

Gear type: ring gear of a planetary gear set
Damage type
   general: material defect
   specific: slag inclusions
Features,   forging cracks, filled with scale. Clearly visible oxide
   appearance: seam along the material boundary
Causes: non-metallic inclusions in the material

Fig. 9.37 Gears/failure caused by manufacture
Gear type: planetary gear
Damage type
  general: production defect
  specific: production grooves
Features,   gear fracture
  appearance:
Causes: fatigue fracture propagated from production grooves
at weight-reducing hole

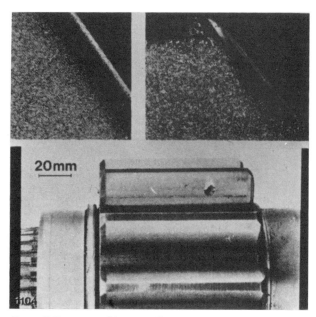

Fig. 9.38   Gears/failure caused by manufacture
Gear type: case-hardened pinion
Damage type
    general: production defect
    specific: hardening defect
Features,
    appearance: linear shelling in the region of the pitch circle
Causes: inadequate hardening

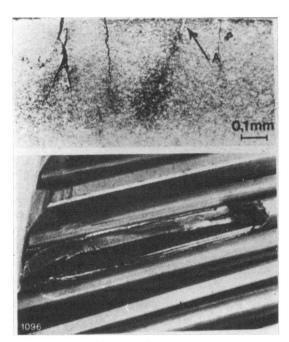

Fig. 9.39   Gears/failure caused by manufacture

Gear type:	bevel gear
Damage type	
general:	production defect
specific:	grinding cracks
Features,	cracks extending from the surface downwards, which
appearance:	formed the starting point for shelling
Causes:	insufficient cooling before nitriding or probably grinding wheel advance too great

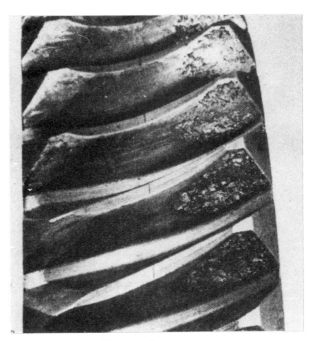

Fig. 9.40   Gears/failure caused by manufacture

Gear type: bronze worm gear

Damage type

   general: assembly defect

   specific: inaccurate alignment

Features,   premature fatigue (pitting) and excessive wear,

   appearance: locally delimited

Causes: contact pattern defect due to production and
                     assembly errors lead to one-sided overload

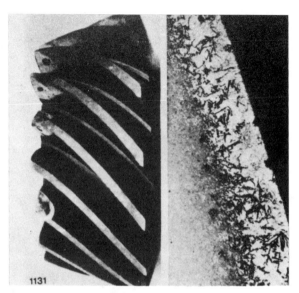

Fig. 9.41   Gears/failure caused by manufacture
Gear type: case-hardened bevel gear
Damage type
   general: design defect, production defect
   specific: meshing interference
Features,   premature fatigue (pitting) and excessive wear,
   appearance: locally delimited
Causes: contact pattern defect due to design, production and
                    assembly errors lead to one-sided overload

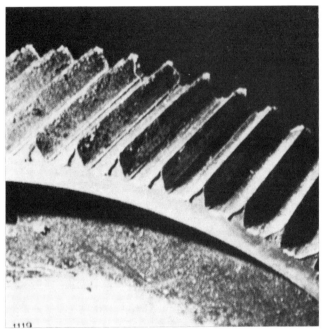

Fig. 9.42   Gears/failure due to manufacture
Gear type: straight spur gear
Damage type
  general: overload
  specific: rated loads too high and too many load reversals
Features,   extreme wear on all tooth flanks
  appearance:
Causes: shear reversal strength exceeded because of excess
                    flank pressure

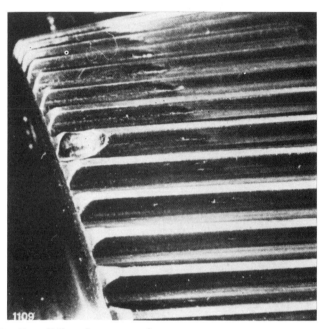

Fig. 9.43   **Gears/failure due to operation**
  Gear type: pinion
  Damage type
    general: overload
    specific: shock loading
  Features,   fracture (corner fracture), pitting
    appearance:
  Causes: shock and vibration loading

(a)

(b)

Fig. 9.44   Gears/failure due to operation/manufacture
Gear type: helical gear pair
Damage type
   general: overload
   specific: underrating/too high a continuous load
Features,   tooth deformation
   appearance:
Causes: continual overloading over a long period of time

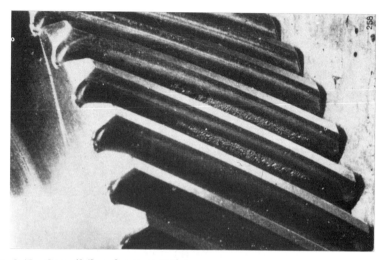

Fig. 9.45   Gears/failure due to operation
Gear type: quenched and tempered helical spur gear
Damage type
  general: overload
  specific: rated load too high
Features,   pitting
  appearance:
Causes: material fatigue due to excessive pressure and sliding

Fig. 9.46   Gears/failure due to operation
Gear type: straight spur gear
Damage type
  general: overload
  specific: shock loading
Features,   spalling of flank with ridging due to plastic deforma
  appearance: tion
Causes: additional dynamic forces in operation

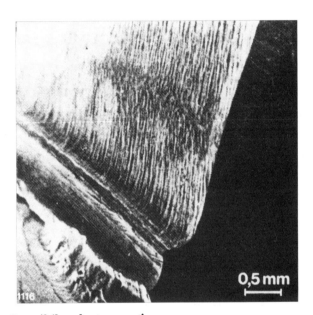

Fig. 9.47 Gears/failure due to operation
Gear type: quenched and tempered planet gear
Damage type
  general: errors in lubrication
  specific: incorrect lubricant/contaminated lubricant
Features,   extreme wear of tooth flank with scratches and
  appearance: grooves
Causes: lubricant with inadequate load capacity (additives,
                 viscosity) and/or contaminated lubricant

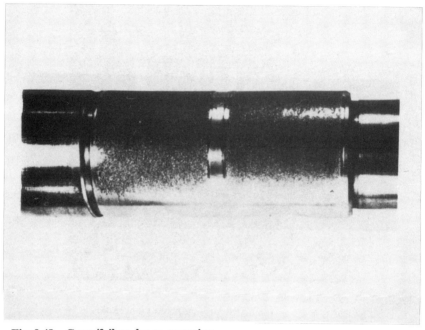

Fig. 9.48   Gears/failure due to operation
Gear type: planetary gear bearing pin
Damage type
  general: outside influences
  specific: current passage
Features,
  appearance: uniformly distributed craters over whole surface
Causes: current passage due to inadequate insulation

REFERENCES

(1) Bartz, W. J. (1984) Gear lubrication – current state of the art and future develop-
     ment, Parts 1 and 2 (in German), *Tribologie und Schmierungstechnik*, **31**, 126–131,
     198–203.
(2) Bartz, W. J. (1974) Gear lubrication, *Applications of motive power engineering*, (in
     German), Krausskopf-Verlag, Mainz, Germany, Vol. 3, pp. 391–455.
(3) Bartz, W. J. (1970) Principles of gear lubrication (in German), *Die Maschine*, 4,
     70–74; **5**, 35–39.
(4) Bartz, W. J. (1971) Gear oil as a functional element (in German), *Mineralltechnik*,
     1971, **16**, 1–34.
(5) Niemann, G. and Rettig, H. (1965) Gear lubrication problems (in German), *Der
     Maschinenschaden*, **38**, 37–50.
(6) Dudley, D. W. and Winter, H. (1961) *Gears* (in German), Springer-Verlag, Berlin/
     Heidelberg/New York.
(7) Borsoff, V. N. (1959) On the mechanism of gear lubrication, *J. Bas. Engng*, **81**,
     79–93.
(8) Industrial gear lubrication (in German), Technical publication, Deutsche BP AG.
(9) Bartel, A. (1962) Gear lubrication (in German), VDI-Verlag, Düsseldorf, Germany.
(10) Kruppke, E. (1965) Quality ratings for hypoid oils, measured in hypoid truck rear
     axles (in German), *Schmiertechnik*, **12**, 21–27, 84–94, 160–162.
(11) Powell, D. L. and Hoyi, J. R. (1966) Automotive hypoid gear loading and sliding
     relationship, *Gear lubrication*, The Institute of Petroleum, pp. 111–117.
(12) Niemann, G. and Lechner, G. (1967) Temperature increase of gears during oper-
     ation (in German), *Schmiertechnik*, **14**, 13–20.
(13) Theyse, F. H. (1967) Blok's flash temperature hypothesis and its practical applica-
     tion to gears (in German), *Schmiertechnik*, **14**, 22–29.
(14) Niemann, G. and Seitzinger, K. (1971) Temperature rise of case-hardened gears as
     an indicator of their scoring load capacity (in German), *VDI-Zeitschrift*, **113**,
     97–105.
(15) Niemann, G. (1965) *Machine components* (in German), Springer-Verlag, Berlin/
     Heidelberg/New York, Vol. 3.
(16) Winter, H. and Michaelis, K. (1983) Friction coefficients and losses of lubricants (in
     German), *Minerall-Technik*, **30**, 1–22.
(17) Ohlendorf, H. (1959) *Losses and temperature rise in spur gears* (in German), Disser-
     tation, Munich Technical University, Germany.
(18) Bartz, W. J. (1971) Studies of the effect of modern lubricants on pitting with rolling
     and sliding stress (in German), *Forschungsheft*, No. 2, Forschungsvereinigung
     Betriebstechnik (FVA), Frankfurt, Germany.
(19) Bartz, W. J., Krüger, V., and Käser, W. (1976) Studies of the effect of modern lubri-
     cants on pitting with rolling and sliding stress (in German), *Forschungsheft*, No. 39,
     Forschungsvereinigung Betriebstechnik (FVA), Frankfurt.
(20) Dawson, P. H. (1962) Effect of metallic contact on the pitting of lubricated rolling
     surfaces, *J. mech. Engng. Sci.*, **4**, 16–21.
(21) Niemann, G. and Rettig, H. (1961) Characteristics of lubricants for gears (in
     German), *Erdöl und Kohle*, **19**, 809–817.
(22) Bartz, W. J. (1980) The effect of the lubricant on pitting in various gears (in
     German), *Maschinenmarkt*, **86**, 224–227, 537–538.
(23) DIN 3990 Calculation of the load capacity of spur and bevel gears (in German),
     Sheets 1 and 2.

(24) Lechner, G. (1986) Gear lubrication (in German), *Maschinenmarkt*, **74**, 24–31.

(25) Lechner, G. (1962) Determining the load capacity of oils as a basis for calculating the scoring resistance of gears (in German), *Minerall-Technik*, **7**, 1–19.

(26) Dowson, D. and Higginson, G. R. (1966) *Elastohydrodynamic lubrication*, Pergamon Press, Oxford, UK.

(27) Dyson, A., Naylor, H., and Wilson, A. R. (1965/66) The measurement of oil-film thickness in elastohydrodynamic contact, *Proc. Instn mech. Engrs.*, **180**, Part 3B, 119–134.

(28) Bartz, W. J. (1971) The significance of elastohydrodynamics for the design of gear pairs (in German), *Konstruktion*, **23**, 257–262.

(29) Bartz, W. J. (1973) The significance of elastohydrodynamics in gear lubrication (in German), *VDI-Berichte 195* (Gear meeting), VDI-Verlag, Düsseldorf, pp. 87–102.

(30) Martin, H. M. (1916) Lubrication of gear teeth, *Engineering*, **102**, 119–121.

(31) Lechner, G. (1973) Calculation of the scoring load capacity of spur and bevel gears (in German), Tech.-Sci. Publication No. 8. of ZF Friedrichshafen AG.

(32) DIN 51 354 Mechanical testing of lubricants in the FZG gear test machine (in German), 1977, Parts 1 and 2.

(33) Michaelis, K. (1987) *The integral temperature for evaluating the scoring load capacity of spur gears* (in German), Dissertation, University of Munich, Germany.

(34) Michaelis, K. (1980) Calculation of the scoring load capacity of spur gears and bevel gears (in German), *Maschinenmarkt*, **86**, 660–662.

(35) Michaelis, K. Analysis and design of gear pairs, with reference to the EP behaviour (in German), Gear lubrication course, Technische Akademie Esslingen, Ostfildern, Germany.

(36) Bartz, W. J., Dammeyer, D., and Walch, J. (1979) *Mineral oil school* (in German), Seventh edition, Perlach Verlag, Augsburg, Germany.

(37) Reinhardt, G. P. and Seidel, G. H. (1981) Fuels, *Manual of fuels and lubricants for vehicles*, Part 1 (in German), pp. 27–99.

(38) Bartz, W. J. (1983) Chemical composition and production of lubricants, *Manual of fuels and lubricants for vehicles*, Part 2 (in German), First edition, Expert-Verlag, Ehningen bei Böblingen, pp. 45–127.

(39) Bellingen, S. (1969) Principles of engine lubrication (in German), course on 'The lubrication of heat engines', Technische Akademie Wuppertal, Germany, pp. 1–7.

(40) Kristen, H. (1982) The manufacture, structure, chemical, and physical characteristics of mineral-oil-based hydraulic fluids, *Hydraulic fluids*, First edition, Vincentz Verlag, Hanover, pp. 153–170.

(41) BP Benzin und Petroleum AG (1978) *The book of crude oil* (in German), Fourth edition, Reuter and Klöckner Verlagsbuchhandlung, Hamburg, Germany.

(42) Bartz, W. J. (1981) Reasons for using synthetic lubricants and operational fluids, 'The use of synthetic lubricants and operational fluids in industry and vehicles', course at Technische Akademie Esslingen, Ostfildern, Germany.

(43) Schmid, W. A. Base fluids in synthetic lubricants – comparative consideration of their characteristics, ibid.

(44) Szydywar, J. Synthetic hydrocarbons, e.g., polyalphaolefins, ibid.

(45) Kussi, S. Polyether in lubrication engineering, ibid.

(46) Prey, G. Ester fluids for lubrication, ibid.

(47) Szydywar, J. Aviation turbine lubrication, ibid.

(48) Lonsky, P. Silicon lubricants and operational fluids, ibid.

(49) Wunsch, F. The use of polyphenyl ethers and polyfluoroalkyl ethers in lubrication technology, ibid.

(50) Mader, W. (1979) *The application of lubricating greases* (in German), Vincentz Verlag, Hanover, Germany.

(51) Boner, C. J. (1976) *Modern lubricating greases*, Scientific Publications, Broseley.

(52) Schmidt, G. (1983) Lubricating greases, *Manual of fuels and lubricants for vehicles*, Part 2 (in German), First edition, Expert-Verlag, Ehningen bei Böblingen, Germany, pp. 347–399.

(53) Schmidt, G. (1981) The chemistry and manufacture of lubricating greases, 'The application of lubricating greases in industrial practice', course at Technische Akademie Esslingen, Ostfildern, Germany.

(54) Augustiniak, H. The characteristics of lubricating grease and their analysis, ibid.

(55) Endom, L. Flow characteristics of lubricating greases and their importance in practical applications, ibid.

(56) Augustiniak, H. Metallic-soap lubricating greases, ibid.

(57) Schmidt, G. Characteristics and applications of gel, bentonite and polyurea greases, ibid.

(58) Wunsch, F. Synthetic lubricating greases and their importance, ibid.

(59) DIN 51 509 Selection of lubricants for toothed gears. Plastic lubricants (in German), Part 2 (draft).

(60) Bartz, W. J. *et al.* (1983) Chemical summary and production of lubricants, *Manual of fuels and lubricants for vehicles*, Part 2 (in German), Expert-Verlag, Ehningen bei Böblingen, Germany, pp. 45–127.

(61) Gegner, E. (1984) Basic observations on the use of additives, *Additives for lubricants* (in German), Vincentz-Verlag, Hanover, Germany, pp. 11–18.

(62) Kristen, U. Oxidation inhibitors, copper deactivators and rust inhibitors, ibid., pp. 19–46.

(63) Eckert, R. J. A. Pour point improvers, foam inhibitors and adhesion improvers, ibid., pp. 219–231.

(64) Bartz, W. J. Viscosity and flow behaviour of lubricating oils and the effect of VI improvers, ibid., pp. 161–184.

(65) Schdel, U. Polymethacrylates and polyisobutylenes, ibid., pp. 185–195.

(66) Eckert, R. J. A. Hydrogenated diene copolymers as VI improvers, ibid., pp. 197–209.

(67) Bartz, W. J. Polyolefins, ibid., pp. 211–217.

(68) Lindstaedt, B. Ash-forming extreme-pressure and anti-wear additives, ibid., pp. 47–74.

(69) Kristen, U. Ash-free extreme-pressure and anti-wear additives, ibid., pp. 75–92.

(70) Bartz, W. J. Reducing friction with lubricants and additives, ibid., pp. 107–120.

(71) Raddatz, J. Detergent/dispersant additives – manufacture, function and applications, ibid., pp. 139–160.

(72) Schaper, W. Additives for metal machining fluids, ibid., pp. 233–245.

(73) Diehl, K.-H. Chemical preservatives for coolants, ibid., pp. 247–281.

(74) Hubmann, A. Base oils – matching them to additives, and their importance for the applications characteristics of lubricants, ibid., pp. 283–318.

(75) Bartz, W. J. (1969) Lubrication with molybdenum disulphide (in German), reprint from *Technische Rundschau*, Nos 44, 45, and 49, Halwag-Verlag, Bern (and references cited therein).

(76) Bartz, W. J. (1983) Principles of tribology with special reference to solid lubricants, 'Reducing friction and wear and solid lubricants, course (in German) at Technische Akademie Esslingen, Ostfildern, Germany.

(77) Wunsch, F. (1974/75) Solid lubricants – theory and practice (in German), *Ingenieur-Digest*, **12/14**.

(78) Killeffer, D. H. and Linz, A. (1952) Molybdenum compounds, their chemistry and technology, *Interscience*, New York.

(79) Knappwost, A. (1965) Mechanical/chemical surface reactions in lubrication with solid lubricants (in German), Meeting Report 1, Symposium on solid lubricants, Munich, Germany, pp. 40–51.

(80) Stock, A. J. (1966) Evaluation of solid lubricant dispersions on a four-ball tester, *Lubric. Engng*, **22**, 146–152.

(81) Bartz, W. J. (1972) Some investigations on the influence of particle size on the lubricating effectiveness of molybdenum disulfide, *ASLE Trans*, **15**, 207–215.

(82) Johnson, R. L., Swikert, M. A., and Bisson, E. E. Friction and wear of hot-pressed bearing materials containing molybdenum disulphide, NACA TN 2027, 1950.

(83) Gnsheimer, J. (1960) The mechanism of lubrication with molybdenum disulphide in the presence of liquids (in German), *Schmiertechnik*, **7**, 270–286.

(84) Gnsheimer, J. (1967) Influence of certain vapors and liquids on the frictional properties of molybdenum disulphide, *ASLE Trans*, **10**, 390–399.

(85) Bartz, W. J., Schultze, G. R., and Göttner, G. H. (1968) The lubrication efficiency of MoS$_2$ in engine and gear oils (in German), *Erdl-Erdgas-Zeitschrift*, **84**, 16–25.

(86) Bartz, W. J. (1973) Studies of the lubrication efficiency of molybdenum disulphide (in German), *Mineralltechnik*, **18**, 1–33.

(87) DIN Taschenbuch 192, Lubricants – standards for characteristics, requirements, testing (in German), 1983, Beuth-verlag, Berlin/Cologne, Germany.

(88) Bouchert, D. (1982) *The design of radial plain bearings lubricated with grease* (in German), Dissertation, Technical University of Clausthal, Germany.

(89) SAE Handbook, Vol. 3, Section 23, 1984, Society of Automotive Engineers, Warrendale, PA.

(90) Lubricant Service Designation for Automotive Manual Transmissions and Axles, API Publication No. 1560, 1981, American Petroleum Institute.

(91) Laboratory Performance Tests for Automotive Gear Lubricants, ASTM STP 152, intended for API–GL–4 and GL–5 Services.

(92) MB Fuel and Lubricant Regulations (in German), Daimler-Benz AG, Stuttgart, Germany.

(93) Asseff, P. A. (1979) *Modern Automotive Lubricants*, Lubrizol Publication.

(94) Schiemann, L. F. and Schwind, J. J. *Fundamentals of Automotive Gear Lubrication*, Lubrizol Publication.

(95) Hubmann, A. (1982) Transmission lubricants – structure, characteristics, testing, and standardisation (in German), *Tribologie und Schmierungstechnik*, **29**, 186–195; 255 260, 1983, **30**, 20–26.

(96) Hamann, D. (1982) *Lubricants and related products* (in German), Verlag Chemie, Basel, Germany.

(97) Möller, V. J. (1985) Changes in lubricating oil during service and used oil analysis of transmission oils, 'Transmission lubrication', course (in German) at Technische Akademie Esslingen, Ostfildern, Germany.

(98) DIN 51 509 Selecting lubricants for toothed gears (in German), Part 1, 1976.

(99) Michaelis, K. (1985) Gear lubrication for special operating conditions 'Transmission lubrication', course (in German), at Technische Akademie Esslingen, Ostfildern.

(100) Plewe, H.-J. (1980) *Studies on the abrasive wear of lubricated low-speed gears* (in German), Dissertation, Technical University of Munich, Germany.

(101) Winter, H. and Michaelis, K. (1981) Studies on the thermal balance of transmissions (in German), *Antriebstechnik*, **20**, 70–74.

(102) Papay, A. G. and Jayne, G. J. (1980) New developments in gear oil additive technology, *Proc. Lubrication of toothing gears and vehicle gears*, Technische Akademie Esslingen, Ostfildern, Germany, pp. 145–165.

(103) Greene, A. B. and Risdon, T. J. The effect of molybdenum-containing oil-soluble friction modifiers on engine fuel economy and gear oil efficiency, SAE Paper 81187.

(104) Bartz, W. J. and Oppelt, J. (1979) The effect of air inclusions on the scoring load capacity and wear of gear pairs (in German), Final report, *Forschungsheft*, No. 59, Forschungsvereinigung Antriebstechnik.

(105) Laukotka, E. M. (1985) The lubrication of worm gears with suitable lubricants, 'Transmission lubrication', course (in German) at Technische Akademie Esslingen, Ostfildern.

(106) Schnnenbeck, G. (1985) Test methods for studying the effect of lubricants on the occurrence of grey spots (in German), *Antriebstechnik*, **24**, 31–36.

(107) Gemeinholzer, G. (1985) New high-performance sintered friction quality HS43 for plate clutches in EP gear oil and synthetic lubricants (in German), *Antriebstechnik*, **24**, 18–28.

(108) DIN 51 509 The selection of lubricants for toothed gears, plastic lubricants (in German), Part 2 (draft).

(109) Blanke, H.-J. (1979) Critical evaluation of lubricant spray equipment for lubricating crown gear drives with consistent sprayed adhesive lubricants, with particular reference to failure susceptibility and the consequent impairment of reliability, 'Increasing the reliability of open crown gear drives lubricated with spray-on adhesive lubricants, course (in German) at Technische Akademie Esslingen, Ostfildern.

(110) Michaelis, K. FZG tests for consistent lubricants, ibid.

(111) Fenneberger, K. Basic types of graphite and their suitability as lubricants, ASLE Preprint 75 AM–5C–2.

(112) Matsuo, K. and Maeda, Y. (1984) Application of graphite to energy-saving industrial gear oil, Third International Conference on Solid Lubrication, Denver, CO.

(113) Chmelik, G. F. (1966) MoS_2-containing greases and gear lubricants – today and looking ahead, NLGI Paper 23–26.

(114) Grünberg, V. (1972) Use of solid lubricants in the lubrication of rolling bearings, *Molysulfide Newsletter*, **15**, 1–2.

(115) Adams, J. H. and Godfrey, D. (1980) Borate additive for gear lubricants – performance and E.P. mechanism, *Proc. Lubrication of toothed gears and vehicle gears*, Technische Akademie Esslingen, Ostfildern, pp. 167–178.

(116) Callis, G. E. and Suh, G. Y. Durability testing of low viscosity borate gear lubricants, SAE Paper 831731.

(117) Ayel, J. F., Gervason, P. C., and Eudeline, J. P. (1972) Improving load-carrying capacity and antiwear properties of gear transmission oils by adding graphite fluoride, *Molysulfide Newsletter*, **15**, 1–2.

(118) Dumdum, J. M., Aldorf, H. E., and Barnum, E. C. (1983) Lubricant-grade cerium fluoride – a new solid lubricant additive for grease, pastes, and suspensions, NLGI Paper.

(119) Naman, T. M. Automotive fuel economy – potential improvement through selected engine and gear lubricants, SAE Paper 800438.

(120) Bartz, W. J. (1979) Potential energy saving by tribological means (in German), Working group literature evaluation lubrication no. 895/996, Report no. 13/14-79.

(121) Ikemoto, Y. Performance advantages of multigrade manual transmission oil, SAE Paper 811207.

(122) Langenbeck, K. (1983) Practical gear lubrication, 'Transmission lubrication', course (in German) at Technische Akademie Esslingen, Ostfildern.

(123) Walter, P. (1982) Studies of splash lubrication of spur gears at peripheral speeds of up to 60 m/s (in German), Reports of the Institut fur Maschinenkonstruktion und Getriebebau, University of Stuttgart, Report No. 60.

(124) Bartz, W. J. (1988) Energy saving by tribological means, *Manual of tribology and lubrication* (in German), Expert Verlag, Ehningen bei Böblingen, Germany, Vol. 2.

(125) Ehrlenspiel, K. (1972) Service experience with spur gears and planetary gears – measures for failure prevention (in German), *Der Maschinenschaden*, **45**, 139–144.

(126) Ehrlenspiel, K. (1976) Stationary gears, *Allianz manual of failure prevention* (in

German), Second edition, Allianz Versicherungs AG, Munich and Berlin, Germany, pp. 719–745.

(127) Bartz, W. J. (1979) The influence of lubricants in the failure of machine components, *Failures of lubricated machine components* (in German), Expert Verlag, Ehningen bei Böblingen, pp. 13–55.

(128) DIN 3979 Tooth failure in toothed gears (in German), 1976, Beuth-verlag, Berlin/Cologne.

(129) Definitions of possible gear failures, *Designation and causes as per ZF Standard 201* (in German), (1975) Third edition, Zahnradfabrik Friedrichshafen AG, Germany.

INDEX